Pediatric Injury Biomechanics

Jeff R. Crandall • Barry S. Myers • David F. Meaney
Salena Zellers Schmidtke
Editors

Pediatric Injury Biomechanics

Archive & Textbook

Foreword by Stephen A. Ridella

Editors
Jeff R. Crandall, PhD
Department of Biomedical Engineering
University of Virginia
Charlottesville, VA, USA

David F. Meaney, PhD
Department of Bioengineering
University of Pennsylvania
Philadelphia, PA, USA

Barry S. Myers, MD, PhD, MBA
Department of Biomedical Engineering
Duke University
Durham, NC, USA

Salena Zellers Schmidtke, MS, BME
The Southern Consortium for
  Injury Biomechanics
Alexandria, VA, USA

ISBN 978-1-4614-4153-3   ISBN 978-1-4614-4154-0 (eBook)
DOI 10.1007/978-1-4614-4154-0
Springer New York Heidelberg Dordrecht London

Library of Congress Control Number: 2012942797

© Springer Science+Business Media New York 2013
This work is subject to copyright. All rights are reserved by the Publisher, whether the whole or part of the material is concerned, specifically the rights of translation, reprinting, reuse of illustrations, recitation, broadcasting, reproduction on microfilms or in any other physical way, and transmission or information storage and retrieval, electronic adaptation, computer software, or by similar or dissimilar methodology now known or hereafter developed. Exempted from this legal reservation are brief excerpts in connection with reviews or scholarly analysis or material supplied specifically for the purpose of being entered and executed on a computer system, for exclusive use by the purchaser of the work. Duplication of this publication or parts thereof is permitted only under the provisions of the Copyright Law of the Publisher's location, in its current version, and permission for use must always be obtained from Springer. Permissions for use may be obtained through RightsLink at the Copyright Clearance Center. Violations are liable to prosecution under the respective Copyright Law.
The use of general descriptive names, registered names, trademarks, service marks, etc. in this publication does not imply, even in the absence of a specific statement, that such names are exempt from the relevant protective laws and regulations and therefore free for general use.
While the advice and information in this book are believed to be true and accurate at the date of publication, neither the authors nor the editors nor the publisher can accept any legal responsibility for any errors or omissions that may be made. The publisher makes no warranty, express or implied, with respect to the material contained herein.

Printed on acid-free paper

Springer is part of Springer Science+Business Media (www.springer.com)

Foreword

Since the beginning of the new millennium, most developed countries have experienced remarkable reductions in injuries and fatalities resulting from motor vehicle crashes. In presentations made during the 22nd International Technical Conference on the Enhanced Safety of Vehicles, June 13–16, 2011, Washington, DC, representatives of governments from throughout the world reported 25–50% reductions in motor vehicle crash-related fatalities since the year 2000. Numerous factors are responsible for these reductions. They include but are not limited to cultural and legislative reforms targeting driving, regulatory pressures, improved vehicle technology, and global economic conditions. Despite these hard-fought, impressive improvements in vehicle safety, motor vehicle crashes continue to be a major public health concern, particularly for children. In the most recent available data, the Centers for Disease Control and Prevention posited that motor vehicle crashes are the leading cause of unintentional injury deaths for children 5–14 years of age and the second leading cause of death of unintentional injury deaths for children less than 4 years of age. Nonfatal injury risk is low for children compared to adults, but the fact remains that over 100,000 children under 14 years of age are treated for motor vehicle crash-related injuries each year.

Statistics of this magnitude reflect not only the scope and impact of pediatric injury in the motor vehicle crash environment but also provided much of the impetus for this project. The automotive safety community treats the issue of child occupant protection very seriously. The Southern Consortium for Injury Biomechanics recognized the opportunity to bring together a unique group of expert injury biomechanists to create a living laboratory of injury data, anthropometry, and biomechanical test data with the capacity to provide state-of-the-art pediatric knowledge. From head to toe, each chapter provides the most current information on pediatric response to deceleration and impact events. Such data that will promote and facilitate development of better child crash test dummies and more accurate child computer models. Better and more "life-like" biofidelic test tools will permit vehicle and child seat manufacturers to design and construct even safer environments for children, and ultimately, reduce the incidence of death and injury for children involved in motor vehicle crashes.

The National Highway Traffic Safety Administration was pleased to help sponsor this important effort. We appreciate the contributions of the chapter authors and acknowledge the leadership of the Southern Consortium for Biomechanics at the University of Alabama at Birmingham for conceptualizing and then bringing this important idea to a successful completion.

Washington, DC, USA   Stephen A. Ridella

Preface

There's no question "Necessity is the Mother of Invention." And so it was in the late 1990s when to me, as Director and Principal Investigator of the University of Alabama at Birmingham's Injury Control Research Center (UAB-ICRC), *Dame Necessity* presented herself early one Monday morning. The ICRC's biomechanist suddenly and unexpectedly accepted a position elsewhere. My initial thoughts (following a fleeting assessment of the pros and cons of flinging myself from the top of our parking deck) involved trying to figure out how to turn this mini-catastrophe into something positive. Then, almost magically, a goal I long believed in became worth trying. I recalled a familiar political admonition: *Never allow a crisis to go to waste when it is an opportunity to do some things you have never done before.*

Failing to convince our biomechanist to change his mind, I decided on the spot to seize the opportunity to attempt to create something I had never before been in a position to try. I would establish a *voluntary* consortium of the brightest and best scientists from a few prestigious institutions whom I would (somehow) convince to work together collegially instead of competitively on the most pressing and important injury biomechanics problems of the day.

The caliber of person I needed to help me pull off this plan was not at the time at UAB. I must look elsewhere to find the leading scholar with the credentials and experience to attract the kind of extraordinary scientists who would need to agree to become part of this consortium if it were to have even one chance in a hundred to succeed. The enterprise would require someone willing to risk, as I was, innovative thinking, and unorthodox practices. Based on years of experience, I understood if talented, high-profile scientists from other institutions were to gamble on my idea, I myself must be willing to occupy the place where predictable disfavor, even wrath, would be aimed if our venture should fail. Why wouldn't my university be skittish? The infrastructure of the model I had in mind would call on my university to share indirect dollars, the lifeblood of virtually all academic research settings. Still, given the lack of depth in injury biomechanics on our School of Engineering's bench during that era, I had little choice. I would ask forgiveness rather than seek permission. I believed a consortium would succeed and UAB's administrative hierarchy would eventually forgive us, even amid the predictable cacophony of murmured displeasure from accountants and grant managers who cared not a wit about the quality of the science, but only about the bottom line. I was not dissuaded. I know that success has many fathers. When we succeeded many would emerge to bask in the shared glory of having made it all possible. I also knew those tentative fathers were totally insulated against failure. If the experiment did not work, it would be entirely my fault.

The immediate objective was to identify and engage a highly regarded engineer with impeccable credentials to help establish what I envisioned becoming the *Gold Standard* motor vehicle crash injury biomechanics research *initiative* in the nation. I stress "initiative" because the model in my mind was not based on bricks and mortar. Rather, it was Gestalt-driven with the whole becoming greater than the sum of its parts. Experts working on research we mutually determined necessary would remain exactly where they were, in some of the finest university-based injury biomechanics laboratories throughout the country.

Soon I placed the first and one of the most important of all the phone calls that would be made on behalf of an idea that would come to be known as "The Southern Consortium for Injury Biomechanics." I contacted Dr. Barry S. Myers of Duke University, a highly regarded biomechanist. Dr. Myers had heard of me but that was about all. Still, he listened to what I had to offer: possible funding for a small scale injury biomechanics proposal to be included in our ICRC's forthcoming competitive renewal application. His acceptance of that funding opportunity paved the way for a collaboration that was quite unlike anything ever before attempted in the CDC-sponsored injury control research center program.

We developed a blueprint linking the UAB-ICRC with Dr. Myers' Biomechanics Engineering Group at Duke University. That was only Phase 1 of three distinct phases to the new collaborative relationship. Phases 1 and 2 were confined to the collaboration between the UAB-ICRC and Barry S. Myers' group at Duke. Phase 3 extended beyond the UAB-Duke nidus and, within a short period of time, was populated by a handful of equally well-known and well-regarded injury biomechanists from several other highly esteemed research institutions.

Our shared vision was bold: significantly reduce traffic fatalities and injuries by bringing together top scientists to conduct collaborative rather than competitive safety research. For all practical purposes, the approach we took had little reason to succeed. Traditionally, most scientists work in their own laboratories on specific aspects of large problems. Even with communication through published manuscripts and professional meetings, it can take years for important knowledge products to result in safety innovations. SCIB participants chose to cooperate to identify and define problems, select and engage the most qualified scientists to work on each aspect of each problem, and remain in touch in a constant, deliberate manner to provide regular feedback, input, ideas, and solutions to problems. Freed from the fear that revealing individual discoveries could threaten future funding, SCIB promoted a stimulating, creative environment.

With shared commitment to the concept, we witnessed the emergence of loosely knit, discipline-driven project teams composed of prominent scientists who previously had been, at best, friendly competitors. In retrospect, it appears a primary reason the collaborations began to thrive was that collectively scientists had the resources and tools needed to address some of the most challenging injury biomechanics problems of the day. This unique multidisciplinary mix resulted in the teams enjoying state-of-the-art research capacity that facilitated rigorous investigation of a group of problems from a multiple scientific perspectives. This capacity yielded more complete understanding of the problems. Moreover it yielded applicable solutions. Unlike independent workers, the SCIB research teams had the ability to apply a spectrum of research tools and techniques to a broad range of vexing problems impacting both human and mechanical systems. The teams were fluid, so any given project could be staffed by experienced scientists whose skill sets were ideally matched to specific project requirements at particular times. This diverse supply of "big hitter" personnel is not typically found at a single institution. A reasonable analogy might be a professional sports all-star team. In retrospect, the comparison seems entirely apropos.

Between 2000 and 2009 SCIB research filled in critical knowledge gaps in the understanding of how the human body reacts to the rapid, destructive impact of a high-velocity crash. The SCIB's research portfolio focused on preventing injuries to the head, neck, and extremities in adults and in children. SCIB members appreciated that children are different and could not continue to be studied simply as "small adults." This served as the underlying inspiration for pioneering work on a digital pediatric crash dummy known as the Digital Child. It also lead to improving computer modeling of pediatric crash injuries through development of a comprehensive online pediatric biomechanics data archive that evaluates and documents the current state of the most relevant research as well as policies affecting the automobile industry.

Admittedly, my vision of this initiative eventually becoming the *Gold Standard* motor vehicle crash injury biomechanics research activity in the country proved overly ambitious. We did not attain it. We tried hard, did some exciting and outstanding work, and were quite productive. The taxpayers whose hard-earned dollars supported this effort received an enormous return on their investment including a lengthy list of significant successful projects resulting in important contributions to injury biomechanics, the simultaneous training of a new generation of injury scientists who were able to participate in these unique collaborative research projects and who were trained to carry on this work and, finally, this Archive and Textbook which is the overall effort's crowning achievement.

As it is formally referenced, *Pediatric Injury Biomechanics: Motor Vehicle Crash Injury Research* is designed to be updated continuously as research advances and new data emerge. It is intended to become an indispensible tool for research scientists, vehicle designers, and rule, policy, and decision makers working on to make the motor vehicle environment as safe as possible for children.

From the outside looking in at the end results of this effort, the extent of the challenges overcome during this decade-long odyssey may be hard to see. To those of us who lived it (and survived it) memories of some of the more harrowing moments have been attenuated by time and the sheer joy of being able, at long last, to hold in our hands one of the first printed copies of a genuine labor of love. The Textbook's incubator, the Southern Consortium for Injury Biomechanics, grew from a mere idea in late 1999 and early 2000 into an engine of leadership in injury biomechanics research for the remainder of the decade.

In retrospect, there is no question the unconventional research alliance between 2 and eventually as many as 11 prestigious medical research universities represented *the* very best in innovative, strategic thinking and creative management. That SCIB came into existence at all reflects an unselfish willingness among this group of extraordinary scientists to work voluntarily in a manner that recognized the potential cost-effectiveness of collaboration while exploiting the enormous benefits for the field of injury biomechanics.

In the years since the SCIB was established in 2000, extramural funding for injury biomechanics research has become near nonexistent. Denied the financial resources to continue groundbreaking work together, SCIB leaders made an important decision: to identify a single area in which there was consensus regarding greatest need and greatest potential benefit. A thoughtful, deliberative process generated the concept, design, and name of SCIBs final effort, the *Pediatric Injury Biomechanics: Motor Vehicle Crash Injury Research* textbook.

A number of factors influenced the selection of the final project. For example, one of the National Highway Traffic Safety Administration's (NHTSA) highest priorities is child injury prevention in the automotive crash environment. Automobile and child restraint manufacturers around the world are working diligently to improve child safety in cars. Unfortunately, researchers report the lack of available, reliable data needed to predict pediatric occupant kinematics and mechanisms of injury, to determine appropriate injury criteria, to develop age appropriate crash test dummies, and to evaluate effectively the potential for injury to children in the automotive crash environment. Also the study of when, how, and why pediatric injuries occur is one of the most rapidly emerging areas of crash injury research. Little published data exist that describe the structural, material, and mechanical properties of children as they age. Consolidation and analysis of this information is intended to provide essential information needed by researchers working in the field of pediatric injury including those involved in rulemaking activities, injury criteria development, child dummy development, and child injury interventions development, all of which will markedly improve the motor vehicle safety effectiveness for the younger population. Consolidation of this information also will help avoid duplication of previous research and identify critical gaps in the data and areas for future research.

This Archive and Textbook is particularly responsive to the needs of the automotive safety community because it consolidates all publicly available pediatric injury biomechanics research in one place, making it easily accessible. The book also helps identify exactly what research needs to be done to address the remaining issues related to child safety in the automotive crash environment.

The data archive component should prove to be an invaluable tool that will assist in identification of gaps in research as well as provide insight regarding future research direction decisions. It is intended to be a dynamic, *living* document that will be updated with current research and advancements in pediatric injury biomechanics on a regular, on-going basis. It is our hope the Archive and Textbook quickly becomes the "go-to" Best Reference for the epidemiology of motor vehicle-related childhood injury data, pediatric anthropometry, pediatric biomechanical properties, tissue tolerance, and computational models.

The textbook is divided into three sections. The first is entitled *Introduction and Scope of the Problem*. It contains two chapters, leading with information on pediatric anthropometry. The second chapter is devoted to the epidemiology of motor vehicle crashes in which children were injured or killed. The book's second section is entitled *Experimental Data* and consists of four chapters. Together, they provide a critical review, analysis, and summary of experimental testing organized by body regions. These four chapters also address age-based material, physical properties, and structural properties organized by body regions. By title, Chap. 3 is "Experimental Injury Biomechanics of the Pediatric Extremities and Pelvis." Chapter 4 is "Experimental Injury Biomechanics of the Pediatric Head and Brain." Chapter 5 is "Experimental Injury Biomechanics of the Pediatric Neck," and Chap. 6 is "Experimental Injury Biomechanics of the Pediatric Thorax and Abdomen." The third section is entitled *Computational Models* and consists of a single, comprehensive chapter entitled "Pediatric Computational Models." It provides information describing models that are currently available and explains how they are being used. In addition it devotes considerable attention to innovative computational modeling techniques that can be used to improve efficiency and data analysis.

In closing I wish to offer my most profound thanks to Mr. Stephen A. Ridella and Dr. Erik G. Takhounts who are both affiliated with the National Highway Traffic Safety Administration (NHTSA) and who are respected scientists in their own right. Steve and Erik reviewed, edited, corrected, and ultimately gave their seal of approval to the textbook. The importance of their assistance and the contributions they made cannot be underestimated. And, of course, my sincere thanks to an impressive assemblage of topic-specific expert contributors who were invited to be part of this effort because of their professional experience and impeccable scientific reputations. The excellence of their science is reflected in the chapters of this book.

Finally, a few words about the extraordinary group of gifted scientists who are ultimately responsible for this landmark contribution to pediatric injury biomechanics coming to fruition. My cherished friends and colleagues, Dr. Jeff R. Crandall (UVA), Dr. Barry S. Myers (Duke University), Dr. David Meany (U. Penn), Dr. Bharat K. Soni, and Dr. David Littlefield (UAB) have done a remarkable job of writing topic-specific chapters and editing the work of other chapter contributors.

The final coeditor, Salena Zellers Schmidtke, a UAB School of Engineering Masters Degree recipient who is the founder, President, and CEO of Bioinjury

LLC, of Alexandria, Virginia, has been the project's *CPU*. This amazing woman understood, embraced, and nurtured the project from the moment she learned of the idea and was offered the opportunity. Notwithstanding the enormous contributions of her coeditors and chapter contributors, there is consensus that this book would not have happened had it not been for her tenacity, her commitment, and her dedication to the field of injury biomechanics… and all that that implies. Being able to think of these people not only as colleagues but as friends is a distinct honor and privilege that I acknowledge with humble appreciation.

Birmingham, AL, USA   Philip R. (Russ) Fine

Acknowledgements

The painful error of inadvertent omission is a risk associated with listing, by name, those persons who deserve being acknowledged for their labor on behalf of a worthy undertaking. In this particular instance we refer to important individual roles in an effort spanning several years that have culminated in publication of this Pediatric Injury Biomechanics Archive & Textbook. But, it is a risk we must take. Thus, to any person whose name may have been unintentionally omitted, we sincerely apologize and ask, in advance, for your understanding and forgiveness.

At the onset we wish to thank and acknowledge United States Senator Richard Shelby, whose fundamental appreciation for the importance of injury biomechanics research, particularly that which is motor vehicle related, made the SCIB possible at all. Because of Senator Shelby's commitment to and efforts on behalf of highway and traffic safety, the University of Alabama at Birmingham was able to lead an effort resulting in the establishment of the Southern Consortium for Injury Biomechanics; engage a small group of highly regarded injury biomechanists, clinicians, and other associated scientists from some of the country's most prestigious institutions to work collaboratively on some of the most pressing injury biomechanics issues; conduct much-needed but previously unfunded or underfunded research; and, culminate more than a decade of enormous productivity by publication of this book. Thank you, Senator Shelby, for having gone "the whole nine yards," and then some.

We also acknowledge and thank the National Highway Traffic Safety Administration (NHTSA) and the Federal Highway Administration (FHWA) for their invaluable participation in the effort, both as prudent stewards of the taxpayers' dollars that underwrote the Southern Consortium's work, and for their technical guidance, input, and assistance with the publication of this textbook.

Also, we wish to thank and acknowledge UAB SCIB leadership and central office personnel. An expression of gratitude is due Dr. Jay Goldman, Dean Emeritus, UAB School of Engineering, Mr. Jeff Foster, SCIB Associate Director for Administration and Finance, Dr. Andrea Underhill, Ms. Gail Hardin, Ms. Joy Fleisher, Dr. Despina Stavrinos, and Ms. Crystal Franklin. We would be remiss if we failed to acknowledge and thank UAB School of Medicine leadership which includes the Office of the Vice President of Health Affairs and Dean, UAB School of Medicine; UAB Department of Medicine leaders such as Dr. Bill Koopman and Dr. Bob Kimberly and Dr. Lou Bridges. Dr. Koopman's vision, advocacy, and initial support coupled with Dr. Bridges' limitless encouragement and generous additional financial support helped sustain the SCIB's UAB component during some difficult and uncertain periods.

We also wish to acknowledge and thank Dr. Albert King, Distinguished Professor and Chair, Department of Biomedical Engineering, Dr. King-Hay Yang, Professor and Director of the Bioengineering Center, and Ms. Christina D. Wagner, all of whom are affiliated with Wayne State University's prestigious and highly regarded College of Engineering.

Finally we wish to thank and acknowledge our colleague and friend Dr. Russ Fine. It was he who envisioned the entity that came to be known as the Southern Consortium for Injury Biomechanics. It was he who was willing to take the chance to try something that had not been tried before. Because of Russ Fine's ability to see out of the box and take a chance on this new approach with UAB-ICRC's Southern Consortium for Injury Biomechanics (SCIB), the community of US biomechanics experts who can provide industry and government with guidance on design of safety equipment has increased tremendously. We cannot thank Russ enough.

Charlottesville, VA, USA	Jeff R. Crandall
Durham, NC, USA	Barry S. Myers
Philadelphia, PA, USA	David F. Meaney
Alexandria, VA, USA	Salena Zellers Schmidtke

Contents

1. **Pediatric Anthropometry** 1
   Kathleen D. Klinich and Matthew P. Reed

2. **Epidemiology of Child Motor Vehicle Crash Injuries and Fatalities** 33
   Kristy B. Arbogast and Dennis R. Durbin

3. **Experimental Injury Biomechanics of the Pediatric Extremities and Pelvis** 87
   Johan Ivarsson, Masayoshi Okamoto, and Yukou Takahashi

4. **Experimental Injury Biomechanics of the Pediatric Head and Brain** 157
   Susan Margulies and Brittany Coats

5. **Experimental Injury Biomechanics of the Pediatric Neck** 191
   Roger W. Nightingale and Jason F. Luck

6. **Experimental Injury Biomechanics of the Pediatric Thorax and Abdomen** 221
   Richard Kent, Johan Ivarsson, and Matthew R. Maltese

7. **Pediatric Computational Models** 287
   Bharat K. Soni, Jong-Eun Kim, Yasushi Ito,
   Christina D. Wagner, and King-Hay Yang

Index 335

Contributors

Kristy B. Arbogast, PhD Division of Emergency Medicine, Center for Injury Research and Prevention, The Children's Hospital of Philadelphia, Philadelphia, PA, USA

Brittany Coats, PhD Department of Mechanical Engineering, University of Utah, Salt Lake City, UT, USA

Jeff R. Crandall, PhD Nancy and Neal Wade Professor of Engineering and Applied Science, University of Virginia, Charlottesville, VA, USA

Dennis R. Durbin, MD, MSCE Division of Emergency Medicine, Center for Injury Research and Prevention, The Children's Hospital of Philadelphia, Philadelphia, PA, USA

Philip R. Fine, PhD, MSPH The Southern Consortium for Injury Biomechanics, UAB School of Medicine, The University of Alabama at Birmingham, Birmingham, AL, USA

Jay Goldman, DSc, PE The Southern Consortium for Injury Biomechanics, UAB School of Engineering, University of Alabama at Birmingham, Birmingham, AL, USA

Yasushi Ito, PhD Department of Mechanical Engineering, University of Alabama at Birmingham, Birmingham, AL, USA

Johan Ivarsson, PhD Biomechanics Practice, Exponent, Inc., Phoenix, AZ, USA

Richard Kent, PhD Center for Applied Biomechanics, University of Virginia, Charlottesville, VA, USA

Jong-Eun Kim, PhD Department of Mechanical Engineering, University of Alabama at Birmingham, Birmingham, AL, USA

Kathleen D. Klinich, PhD Biosciences Group, University of Michigan Transportation Research Institute, Ann Arbor, MI, USA

Jason F. Luck Department of Biomedical Engineering, Pratt School of Engineering, Duke University, Durham, NC, USA

Matthew R. Maltese, MS Children's Hospital of Philadelphia, Philadelphia, PA, USA

Susan Margulies, PhD Department of Bioengineering, University of Pennsylvania, Philadelphia, PA, USA

David F. Meaney, PhD Department of Bioengineering, University of Pennsylvania, Philadelphia, PA, USA

Barry S. Myers, MD, PhD, MBA Department of Biomedical Engineering, Duke University, Durham, NC, USA

Roger W. Nightingale, PhD Department of Biomedical Engineering, Pratt School of Engineering, Duke University, Durham, NC, USA

Masayoshi Okamoto Honda R&D Co. Ltd, Tochigi, Japan

Matthew P. Reed, PhD Biosciences Group, University of Michigan Transportation Research Institute, Ann Arbor, MI, USA

Stephen A. Ridella Director, Office of Applied Vehicle Safety Research, National Highway Traffic Safety Administration, United States Department of Transportation, Washington, DC, USA

Salena Zellers Schmidtke, MS, BME The Southern Consortium for Injury Biomechanics BioInjury, LLC Alexandria, VA, USA

Bharat K. Soni, PhD Department of Mechanical Engineering, University of Alabama at Birmingham, Birmingham, AL, USA

Yukou Takahashi Honda R&D Co. Ltd, Tochigi, Japan

Erik G. Takhounts, PhD Human Injury Research Division, National Highway Traffic Safety Administration, U.S. Department of Transportation, Washington, DC, USA

Christina D. Wagner, PhD Computational Biomechanics, Biomedical Engineering, Wayne State University, Detroit, MI, USA

King-Hay Yang, PhD Computational Biomechanics, Biomedical Engineering, Wayne State University, Detroit, MI, USA

Chapter 1
Pediatric Anthropometry

Kathleen D. Klinich and Matthew P. Reed

Introduction

Anthropometry is the measurement of human size, shape, and physical capabilities. Most pediatric anthropometry data are gathered to describe child growth patterns, but data on body size, mass distribution, range of motion, and posture are used to develop crash test dummies and computational models of child occupants. Pediatric anthropometry data are also used to determine child restraint dimensions, so they will accommodate the applicable population of child occupants.

This chapter summarizes anthropometric data that have application to pediatric impact biomechanics, beginning with studies that have measured body segment dimensions of different child populations. Several studies have gone beyond documentation of size to describe the three-dimensional shape of different body segments. Other studies have measured mass distribution of human body segments and developed methods for estimating mass, center of gravity, and inertial properties for children of different ages based on external dimensions. Range-of-motion characteristics of the spine and shoulder are also reported. Posture data on how children sit in vehicles can be used to position crash test dummies in a realistic manner.

The chapter concludes with a discussion of how these data have been applied to design of crash dummies along with recommendations for future research needs in pediatric anthropometry.

K.D. Klinich, Ph.D. (✉) • M.P. Reed, Ph.D.
Biosciences Group, University of Michigan Transportation Research Institute,
2901 Baxter Rd, Ann Arbor, MI 48109, USA
e-mail: kklinich@umich.edu

Background

The development of engineering anthropometry dates from the 1950s, when the first systematic application of human measurement to the design of products and workspaces was undertaken. Studies of child body size in the United States in the 1960s led to the first compilations of data for application to child occupant safety assessments in the early 1970s. The most comprehensive data on child anthropometry currently available were gathered during the mid-1970s by researchers at the University of Michigan Transportation Research Institute, led by Richard (Gerry) Snyder. These landmark studies, funded by the US Consumer Product Safety Commission, form the basis for most contemporary assessments of child body dimensions in the United States.

Anthropometry texts (e.g., Roebuck 1995) differentiate between *structural* anthropometry, intended to characterize the body dimensions independent of any particular posture or task, and *functional* anthropometry that explicitly considers measurements of the body in the context of task performance. In reality, even the traditional structural measurement definitions include subject tasks, such as sitting maximally erect on a rigid planar surface. Functional measures, such as seated reach, are strongly related to structural measures that describe body segment length. For safety applications, both structural and functional measures are important.

This chapter reviews some of the important aspects of anthropometric procedures and ways in which data from children differ systematically from adult data. A range of studies relating to growth patterns, body dimensions, body segment masses, and applications of these data to crash safety are summarized.

Anthropometric Methods

Measurement methods used for child anthropometry studies have been derived from well-established procedures used to conduct adult studies. Roebuck et al. (1975) provides a detailed review of measurement methods and definitions. Snyder et al. (1977) documents methods to adapt procedures for measuring small children who are not able to follow verbal instructions.

Most anthropometric data currently available for children were gathered using basic tools, including calipers and tape measures. Following documented procedures, the measurer located body landmarks defining the measurement location, adjusted a caliper, tape measure, or other tool to span the desired anatomical region, and read a value from the instrument. Snyder et al. (1977) incorporated computer-recording equipment, but the measurement methodology was based on standard techniques. An important innovation was the use of pressure-sensing anthropometers to standardize the deformation of soft tissues during recording.

The reliability of anthropometric measurement is strongly dependent on the consistency of the postures the subjects assume and the accuracy and precision with which measurers locate body landmarks. Roebuck et al. (1975) addresses landmark identification in detail.

Advances in computer technology have allowed three-dimensional measurements to be added to the conventional lengths, breadths, and circumferences. Reed et al. (2006) used a three-dimensional coordinate measurement machine to quantify body landmark locations for children in automotive seated postures. Whole-body scanning, which has been applied to large-scale adult anthropometric studies (Robinette et al. 2002) provides an opportunity to obtain much more detailed data on child size and shape, but large-scale pediatric body-scan studies have not yet been conducted.

Because internal skeletal geometry is important for crash safety applications, recent studies have used medical imaging data to create three-dimensional models of pediatric and external anthropometry Reed et al. (2009), Loyd et al. (2010).

Measurement techniques of segmental inertial properties typically use external measurements of subjects coupled with mathematical models that approximate the human body as a group of geometric shapes. The mass, center of gravity, and inertial characteristics are estimated using assumptions of constant density, or variable density estimates derived from medical scans. Some researchers have performed experimental tests to estimate inertial properties of limbs. Medical imaging data could be used to make more precise estimates of segmental inertial properties, but only a few researchers (Ganley and Powers 2004; Loyd et al. 2010) have applied these techniques to pediatric subjects.

Quantifying the voluntary range of motion of body segments, such as the spine, poses challenges because it is difficult to limit the subject's motion only to the segment of interest. Measurement techniques typically use either specialized angle measurement devices affixed to the subject, or video analysis of subject landmarks. For pediatric subjects, measurement errors and between-subject variability are often the same order of magnitude as the changes with age or gender that are of interest. Comparing results of different studies is often complicated by results varying with measurement technique.

Pediatric Growth Patterns

The size of different body regions as a proportion of total body height changes from birth to adulthood (Burdi et al. 1969). Figure 1.1 illustrates these changes. The most notable region is the head, which makes up approximately one-fourth of the stature of a newborn. These fractions shift such that the head is one-seventh the total stature at adulthood, which is reached near age 20. The head size of a 5-year-old child is approximately 90% of the size of the adult head. Significant differences in size with gender typically do not occur until age 10.

Sitting height as a function of total height (stature) also varies with age (Burdi et al. 1969). At birth, sitting height is approximately 70% of total height. By age 3, sitting height averages 57% of total height, reaching approximately 52% of total

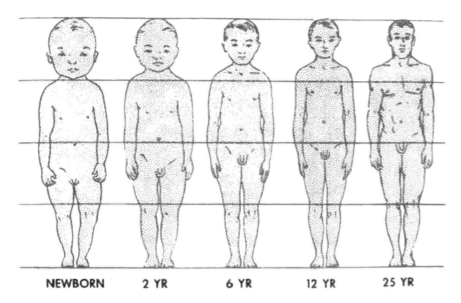

Fig. 1.1 Changes in body proportion from birth to adulthood relative to 25, 50, and 75 % references to total stature (Burdi et al. 1969)

body height at adulthood. With increasing relative lower extremity length, the body's center of gravity shifts downward relative to the torso as children age.

The amount of subcutaneous tissue, or body fat, varies considerably among children (Burdi et al. 1969). During the first 9 months, this tissue accumulates rapidly, but by age 5, the subcutaneous tissue layer is about half the thickness of that of the 9-month-old infant.

Size

The UMTRI anthropometry study (Snyder et al. 1977) is the most comprehensive database on child anthropometry available. A total of 87 anthropometric and functional body measurements were taken on a sample of 4,127 children representing the US population from age 2 weeks through 18 years. The 1977 publication extended results from an earlier study (Snyder et al. 1975) that measured anthropometry of 4,000 different children up to age 12 but did not include the functional measurements.

Data were collected across the United States by two teams using an automated anthropometric data acquisition system. To complete the measurements in a limited amount of time, a set of 22 core measurements (Group I) were collected on every subject. The remaining measurements were divided into Group II: body shape, Group III: linkage and center-of-gravity, and Group IV: head, face and hand measurements. Table 1.1 lists the specific measurements within each group. Data for one

1 Pediatric Anthropometry 5

Table 1.1 Measurements collected in UMTRI child anthropometry study (adapted from Snyder et al. 1977)

	Core (I)	Body shape (II)	Linkage and CG (III)	Head, face, hand (IV)
Circumference	Head Chest Waist Hip Upper Arm Forearm Upper thigh Calf	Neck Natural waist Wrist Ankle Calf		Ball of fist
Breadth	Foot Hand Shoulder Max Hip (seated) Head	Chest Waist Ankle Calf Upper thigh Wrist Neck Max thigh (seated)	Bispinous Biocromial Hip (at trochanter)	Bizygomatic Max frontal Mouth Bitragion Maximum fist
Length	Foot Hand Shoulder–elbow Elbow–hand Buttock–knee	Head	Clavicle–acromion Acromion–radiale Radiale–stylion	Nose Thumb crotch–middle finger Middle finger–thumb grip Thumb Index finger Middle finger
Depth		Upper thigh Forearm Upper arm		Ear–sellion Maximum fist
Clearance		Thigh		Minimum hand
Reach			Vertical arm Frontal arm Lateral arm	
Height	Stature Knee Sitting	Chest Waist Hip Eye	Step Suprasternale Iliocristale Iliospinale Trochanteric Gluteal furrow Sphyrion Tibiale Supine Supine Sitting	Lower face Face Head
Diameter				Middle finger Index finger Thumb
Other	Weight		Standing CG Seated CG	Tragion to back of head Tragion to top of head

additional measurement group was collected for each subject. A slightly different set of measurements was collected for children under age 2. The measurements in each of the noncore groups can be correlated to the core measurements.

The Snyder et al. report contains summary data for 16 age groups, along with scatter plots for each age and by gender. The report provides detailed definitions of each measurement, along with a photo and illustration for each. Selected measurements are presented as a function of weight or stature. Although the original data from the 1975 study have been lost, individual data records from the 1977 study for 3,900 children ages 2 months to 20 years are available and are being used for contemporary analyses (Huang and Reed 2006). The aggregate data presented in the report are available online. The online dataset also includes illustrations of each measurement (http://ovrt.nist.gov/projects/anthrokids/).

Weber and Schneider (1985) compiled selected measures of child anthropometry gathered in three prior UMTRI pediatric anthropometry studies (Snyder et al. 1975, 1977; Schneider et al. 1985). The measures chosen were those most relevant for designing child restraints and ATDs. Data are presented for ages ranging from newborn through age 10 years. Results include number of subjects, mean, maximum, minimum, and standard deviation. In addition to weight, the measures included in this study are indicated in Table 1.2.

Dreyfuss et al. (1993) used the UMTRI anthropometry data to construct graphical representations of children aged 2 months through 18 years. Along with the figures showing key body dimensions at each age, they describe milestones in motor development, expression of emotions, social development, language, and cognitive development at each age.

Table 1.2 Measures compiled by Weber and Schneider (1985)

	Breadth	Depth	Circumference	Length	Height
Standing					X
Sitting					X
Head	X	X	X		
Neck		X	X		
Shoulder (seated)	X	X			X
Shoulder–elbow				X	
Elbow–hand				X	
Chest	X	X	X		
Waist	X		X		
Seated abdomen		X			
Hip (seated)	X				
Thigh (seated)	X	X			
Thigh (standing)	X				
Crotch					X
Rump–sole				X	
Rump–knee				X	
Knee–sole				X	

Beusenberg et al. (1993) reported on a proprietary child anthropometry database developed by TNO (Twisk 1994; Note: The rights to this report were transferred to FTSS and are not publicly available). This database includes measures from the various UMTRI anthropometry studies, as well as results of studies from Germany, Great Britain, and the Netherlands. They derived regression equations for each anthropometric variable as a function of age. They also calculated segmental moment of inertia properties and mass distributions using the methods of Jensen and Nassas (1988).

In the mid-1990s, the UK Department of Trade and Industry compiled an extensive summary of child anthropometric data from a range of sources (Norris and Wilson 1995). They relied extensively on Snyder et al. (1977) for overall body dimensions, but also brought in many other sources. The extensive bibliography in this work covers most of the important child anthropometric studies that appeared prior to 1994.

The US National Center for Health Statistics within the Centers for Disease Control and Prevention (CDC) publishes growth charts for children from birth through age 20 (Kuczmarski et al. 2002). The CDC growth charts were originally developed in 1977 and most recently updated in 2000. The updated charts now include height, weight, and body mass index (BMI) by age. Head circumference is also charted for children through age 36 months. Compared to the originals, the updated charts use a more nationally representative sample for the infant data, and use improved statistical techniques in developing the percentile curves.

The primary use of these charts is by health care providers to ensure that their patients are achieving appropriate levels of growth. In addition, clinicians use the curves to diagnose whether their patients are overweight by determining if the patient exceeds the 85th percentile BMI for their age. The curves would be expected to change over time to reflect growing levels of childhood obesity. However, because the primary use of these charts is by clinicians, a decision was made to exclude weight data from the most recently collected dataset (1988–1994) for children over age 6, to make the growth charts reflect a population with a healthier range of weights rather than the true distribution of height and weight in the current pediatric population. Since development of the charts excluded obese children, they would not be useful in the pediatric impact biomechanics application of evaluating whether a child restraint designed for a certain height and weight value would accommodate a target proportion of children sizes at a particular age (i.e., 90% of 4-year olds). The most recent charts can be downloaded, in addition to coefficients used to generate the percentiles (http://www.cdc.gov/GROWTHCHARTS/).

Figure 1.2 compares weight for stature from the 1977 UMTRI study to the 2000 NCHS (often called CDC) growth curves. The 50th percentile curves for boys and girls from each study are nearly identical. The 5th and 95th percentile curves from the UMTRI study are slightly wider than the NCHS study, most likely because of smaller sample size in the UMTRI study.

The best available data on stature and body weight for US children are gathered in the on-going National Health and Nutrition Examination Survey (NHANES). Child body dimensions are recorded as part of a nationwide study of nutritional

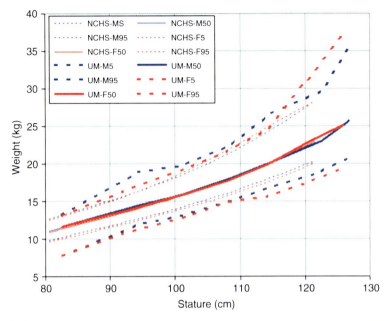

Fig. 1.2 Comparison of 5th, 50th, and 95th percentile weight vs. stature for boys (M5, M50, M95) and girls (F5, F50, F95) from UMTRI (1977) and NCHS growth curves (2000)

status and related health variables (http://www.cdc.gov/nchs/nhanes.htm). Analysis of data from NHANES has shown a marked increase in the prevalence of obesity among children (Hedley et al. 2004). Using data from 1999 to 2002, Hedley et al. estimated that 16% of children ages 6–19 exceeded the 95th-percentile BMI-for-age on the corresponding CDC growth curve, indicating an approximately threefold increase in children at this level of BMI-for-age relative to the data on which the CDC curves are based. For children ages 2–5, 10.3% exceeded the 95th percentile in the earlier data. These analyses indicate an important trend toward increased overweight and obesity in US children, which may have implications for child restraint design. Ogden et al. (2010), using data up to 2008, found that the prevalence of obesity among US children ages 2–19 did not change significantly between 1999 and 2008, leveling off at about 17%. Significant differences across race/ethnicity groups were observed.

The WHO growth curves (de Onis et al. 2007) are primarily based on the original 1977 NCHS data, with adjustments to remove outliers and be consistent with other growth curves. In addition, BMI curves by age were constructed and checked to ensure that the upper age levels of children correspond with obesity recommendations for adults. The goal of these curves was to provide references for a healthy population, one not distorted by obesity or malnutrition (http://www.who.int/childgrowth/en/).

The WHO also developed standards for head circumference-for-age, arm circumference-for-age, triceps skinfold-for-age and subscapular skinfold-for-age

1 Pediatric Anthropometry

Table 1.3 Comparison of stature and weight for 3- and 6-year-old children from several international studies

Country	Serre et al. (2006) France	Smith and Norris (2001) Great Britain	Snyder et al. (1977) United States	Dreyfuss and Tilley (2002)
3YO stature (cm)	97	101	93	93
6YO stature (cm)	118	119	113	114
3YO weight (kg)	15.7		13.5	14
6YO weight (kg)	22.8		20.1	20

(WHO Multicentre Growth Reference Study Group 2007). A selective strategy for developing these curves was adopted to avoid underdiagnosis of obesity and overdiagnosis of malnutrition. The most recent curves were based on 8,440 healthy breastfed children from six diverse countries (Brazil, Ghana, India, Norway, Oman and USA). The study was designed to provide growth targets for healthy children living in conditions that would allow them to achieve their growth potential. Thus they do not represent the true range of child sizes present in the world because they exclude both malnourished and obese children (http://www.who.int/childgrowth/standards/en/).

Serre et al. (2006) collected 41 anthropometric measures on 69 3-year-old and 80 6-year-old French children. Mean, max, minimum, and standard deviations are reported for each age group and measure. Table 1.3 compares the mean stature and weight values from this study to others in the literature.

Shape

Body

In an initial effort to develop anthropometry specifications for 3- and 6-year-old crash dummies, Young et al. (1975) developed physical three-dimensional models of 3- and 6-year-old children in an upright seated posture. They identified 98 measures needed to specify the shape and size of a child. Values for 30 of these measures were taken from a variety of sources in the literature, as the initial UMTRI anthropometry study was not yet completed. For the remaining two-thirds of the measures, values were based on measures from two children close to the target size (aged 3.5 and 5.5 years) and best estimates of the research team. The three-dimensional models were segmented into body regions to provide an estimate of the mass and moments of inertia for each body segment. Figure 1.3 illustrates the complete 3-year-old model, while Fig. 1.4 shows the 6-year-old model after segmentation.

Reed et al. (2001) developed specifications for a 6-year-old occupant classification anthropometric test device (OCATD), shown in Fig. 1.5. This device was developed to provide a realistic mass distribution, external contour, internal skeleton, and seating pressure to allow testing of sensors for advanced airbag systems.

Fig. 1.3 3YO model (Young et al. 1975)

Fig. 1.4 Segmented 6YO model (Young et al. 1975)

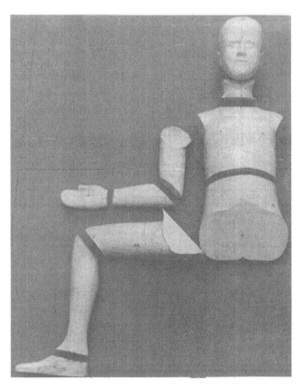

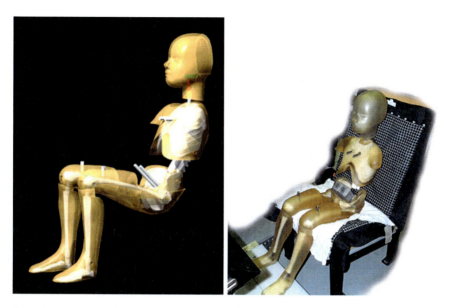

Fig. 1.5 Electronic and physical representations of OCATD representing a 6-year-old child

Reference dimensions for a typical 6-year-old child were developed from existing child anthropometry databases, and a child who closely matched these dimensions was recruited for the study. To specify the subject geometry, surface scans of the child were taken while he sat in a custom molded measurement seat, as were three-dimensional coordinates of key skeletal landmarks. The geometry was scaled to the target reference dimensions in a few locations where necessary. Both physical and CAD versions of the 6YO OCATD were produced that have exterior shape, internal skeleton, and body segment dimensions that represent a typical 6-year-old child. The surface model and landmarks are available at: http://mreed.umtri.umich.edu/mreed/downloads.html#ocatd.

Head

The shape of the head changes from birth to adulthood. The facial volume is traditionally defined as the mandible plus the 14 facial bones and the adjacent soft tissue (excluding the brain). In the infant, the face makes up a smaller proportion of the total head volume. For newborns, the face-to-cranium volume ratio is 1:8, while it is 1:2.5 in the adult. The chin becomes more prominent over time. Figures 1.6 and 1.7 show illustrations of a newborn and adult skull, while Fig. 1.8 shows a qualitative diagram of the proportional changes with age.

A detailed study of the anthropometry and shape of children's heads, necks, and shoulders was conducted at UMTRI by Schneider et al. (1985). The primary goal of this study was to prevent head and neck entrapment in products used by young

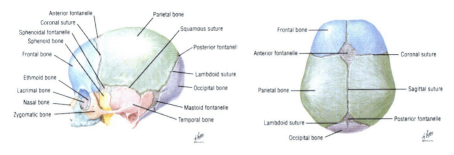

Fig. 1.6 Illustrations of newborn skull (Netter 1997)

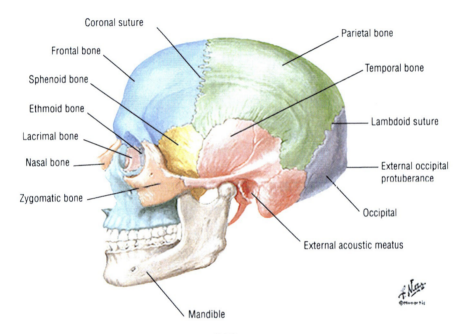

Fig. 1.7 Illustrations of adult skull (Netter 1997)

Fig. 1.8 Changes in proportion of face and head from newborn to adult (Burdi et al. 1969)

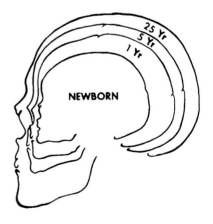

children such as cribs. One of the results of the study was the requirement for cribs that maximum slat spacing be 2.375″.

The study took extensive measurements on 300 children ranging from age 0 to 4 years that represented the racial and socioeconomic distribution of children within the population. The children were categorized into ten age groups, with 6-month age groups used for the older children and 3- or 4-month age groups used with children under 1 year of age. Each group had equal numbers of female and male subjects. Thirty-four dimensions were taken on all subjects. In addition, stereophotogrammetry was used with 100 subjects spanning four different age groups (4–6, 13–18, 25–30, and 43–48 months) to document head contours and locations of key anatomical landmarks. As shown in Fig. 1.9, 36 landmarks and six arcs were collected to describe the shape of the head, neck, and shoulders.

Results include the mean, standard deviation, minimum, maximum and 5th, 50th, and 95th percentile values of each measurement by age group, as well as scatter plots of each measure as a function of age. For the contours and landmarks, mean values for each group are presented, as well as a contour representing a small and large subject within each age group.

Bradtmiller (1996) reported a study to measure the size and shapes of children's heads to guide safety helmet design. The study used a three-dimensional laser scanner to capture the shapes of the subject's heads, as well as six traditional anthropometric measures for each child. A total of 1,035 children aged 2–19 from six different geographic regions were measured in the study, providing a representative racial distribution. The scanned data were reduced to 1,830 vectors relative to a central head origin, spaced at 3° intervals. These vectors were averaged to construct four different headform shapes based on head breadth and length to guide safety helmet design. Table 1.4 shows the head sizes defined by each headform. These four sizes are expected to accommodate the range of child's head sizes up to those who can fit in adult-sized helmets.

Loyd et al. (2010) analyzed CT scans of 185 children to estimate external head contours for children aged 1 month to 10 years. They described each contour using radial coordinates of the head surface, and developed average contours for each age group by averaging the vector length at each pair of angles. The dataset of external contours is available from: http://biomechanics.pratt.duke.edu/data.

Neck

Levick et al. (2001) use three-dimensional surface scanning, MRI and two-dimensional measures to quantify pediatric neck internal structures. Data were gathered from 117 children aged 4 weeks to 14 years. Weight and stature were collected for all subjects, as were five dimensions to characterize neck anthropometry in neutral posture. For four of the subjects from the age range 3 to 14 years, three-dimensional laser surface scans of the external surfaces of the head and neck were collected. The study also analyzed three available MRI scans of infants aged 4, 14, and 16 months. They constructed representations of heads and necks of select subject

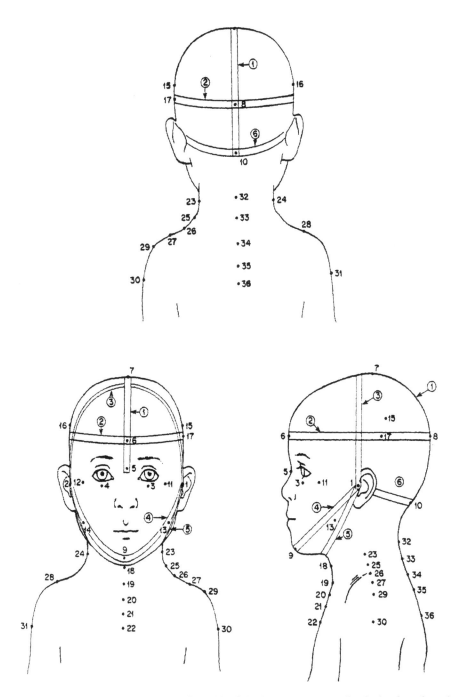

Fig. 1.9 Landmarks and contours collected by Schneider et al. to describe the head, neck, and shoulder shape for young children

1 Pediatric Anthropometry

Table 1.4 Headform sizes (adapted from Bradtmiller 1996)

Size	Head length (mm)	Head breadth (mm)
Small	160–175	135–145
Medium-narrow	175–190	135–145
Medium-wide	175–190	145–155
Large	190–205	145–155

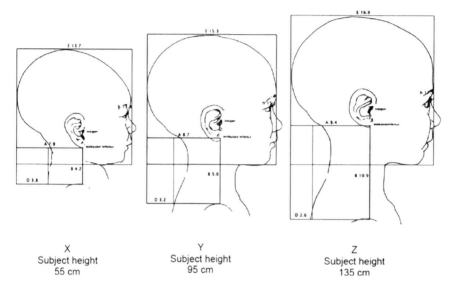

Fig. 1.10 Changes in head–neck proportion with stature (Levick et al. 2001)

sizes as shown in Fig. 1.10. The authors noted that the simpler, less invasive techniques can be used on all children but provide the least amount of information. While scanning provides much more detail, they are inaccurate when used with children too young to remain still. MRI allows visualization of internal structures, but usually requires sedation of pediatric subjects.

Pelvis

Chamouard et al. (1996) measured the pelvis and thigh dimensions of 54 children between 2.5 and 12 years old. They developed regressions for the height of the iliac crest, the distance between the back and anterior superior iliac spine (ASIS), pelvis width, and thigh thickness as a function of weight. They also measured the stiffness of the thigh and abdomen flesh of child volunteers under quasi-static lap belt loading.

Fig. 1.11 Target pelvis for a typical child matching the reference anthropometry of the Hybrid III 6YO ATD based on principal component analysis of CT data (*blue*) in comparison to the current pelvis of the Hybrid III 6YO ATD (Reed et al. 2009)

UMTRI researchers used statistical analysis of CT scans to develop size and shape specifications for typical pelves of children between 6 and 10 years of age (Reed et al. 2009). A uniform polygonal mesh was fit to pelvis CT data, including key pelvis landmarks, from 81 children. A principal component analysis of the mesh vertices was conducted to reduce the dimension of the dataset. The first 12 principal components account for 95% of the variance in the data. The first component accounts for size differences, while the second and third components deal with the breadth and angle of the iliac wings. Design targets for the 6YO and 10YO Hybrid-III ATDs were obtained by regression analysis using the bispinous breadth as the predictor of both size and shape. Figure 1.11 shows the target pelvis bone shape in comparison to the current 6YO ATD pelvis.

Thorax

In a recent study at UMTRI (Reed et al. 2010), three-dimensional geometry for the ribcage was extracted from CT data for 52 children from 5 to 12 years of age. A statistical analysis of the rib data was used to generate a ribcage size and shape typical of 6YO children for use in dummy design (Fig. 1.12).

Shoulder

Using the same methods previously applied to the pelvis and ribcage, a three-dimensional statistical model of clavicle shape was created (Reed et al. 2010). Figure 1.13 shows the clavicle model for the 6YO in the original as-scanned (supine) location and adjusted to the appropriate position for a seated child. More research is needed to indentify the clavicle orientation in different seated postures. The child clavicle shape differs substantially from the shape of the skeletal representation in the current Hybrid-III crash dummies.

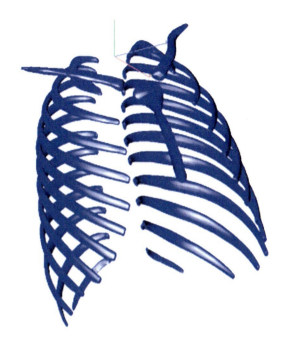

Fig. 1.12 Ribcage model for children the size of the 6YO Hybrid-III ATD based on statistical analysis of CT data (Reed et al. 2010)

Fig. 1.13 Clavicle model for 6YO child in supine (*magenta*) and seated (*blue*) postures relative to the geometry of the current Hybrid-III 6YO shoulder (*red*) (Reed et al. 2010)

Mass Distribution

The 1977 UMTRI anthropometry study (Snyder et al. 1977) includes measures of overall body center of gravity location for children up to age 10. The center of gravity locations are provided for sitting and standing postures.

Prange et al. (2004) performed mechanical tests on the heads of three neonate specimens. As part of the study, they report for each subject the head mass, key head dimensions, and locations of condylar landmarks and the head center of gravity.

They also report the mass moment of inertia about the y-axis; the mean value for three subjects is 4,945 g cm^2.

Jensen (1981a, b, 1986a, b; Jensen and Nassas 1988) performed a longitudinal study of the segmental masses and moments of inertia for 12 boys aged 4–20 years, with most subjects measured annually over 9 years. Data collection involved photographing each subject in a supine picture and digitizing the outline. The subjects were represented with 15 segments composed of transverse ellipses of known density. For children less than age 10, the rate of change of moment of inertia was approximately 30% of the rate of change for older children. BMI provided the best correlation with moment of inertia. Variation of segment mass proportion with age showed a decrease in head proportion and increase in the thigh, shank, foot, and upper arm proportions. Analysis includes variations with body types (endomorph, mesomorph, and ectomorph). Plots of mass parameters diverged after age 10.

Schneider and Zernicke (1992) estimated the mass, center of mass, and transverse moment of inertia of the upper and lower limb segments of infants from 0.04 to 1.5 years of age. They used a 17-segment mathematical model of the human body based on 44 infants for upper extremities and 70 infants for lower extremities. Segmental masses and moments of inertia increase with age, while center of mass location (as a percentage of segment length) is fairly constant. Age, body mass, segment length, and segment circumference were considered potential predictors of the mass parameters. Regression equations were developed to predict mass distribution measures as a function of body mass, segment lengths, and segmental circumferences.

Li and Dangerfield (1993) used the water displacement method and mathematical models to estimate the center of gravity, radius of gyration, principal moment of inertia, and volume of the upper and lower extremity segments for 280 boys and girls aged 8 through 16. Limb segments were modeled as truncated cones. For center of gravity (CG), there was no difference with gender for single segments (hand, forearm, upper arm), but CG were higher for girls when comparing multisegments (hand and forearm, entire upper limb). For moment of inertia, values for segments involving the hand (hand, hand and forearm, entire upper limb) were larger for boys. Radius of gyration did not differ with gender for single segments, but was larger for boys when considering multisegments. Hand volume was larger for boys at each age. For the calf and thigh, CG locations as a percentage of length were similar for all gender and age groups, at 59 and 54%, respectively. Moment of inertia of the calf was larger for boys after age 13. Values for the thigh, collected only up to age 13, were generally larger for girls. Radius of gyration for the calf and thigh were approximately 27 and 31% of length for all subject groups.

Sun and Jensen (1994) used photographs of 27 infants aged 2–9 months to estimate segment mass and principle moments of inertial in infants. For each subject, front and side photographs were taken monthly for 6 months. The photos were digitized and used to create ten-segment models of each subject. Relationships

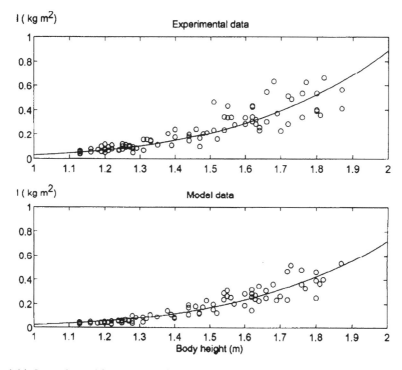

Fig. 1.14 Lower leg and foot moment of inertia vs. body weight for experimental and model data (Lebiedowska and Polisiakiewicz 1997)

between mass, radius of gyration, segment length, and moment of inertia for ten body segments are presented as a function of age. Only five of the 40 measures had nonlinear relationships with age. The nonlinear measures were foot mass, lower trunk A-P moment of inertia, and neck, lower trunk, and foot transverse moments of inertia.

Lebiedowska and Polisiakiewicz (1997) measured the moment of inertia of the lower leg (including the foot) of 90 children from aged 5–18 using the free oscillation technique, with and without external mass loading. Anthropometric measures were also collected. Modeling the lower leg as a two-cylinder representation with constant density predicted moments of inertia that were close to the experimental data. Figure 1.14 shows a fifth power fit of moment of inertia as a function of body height for both the experimental and model data.

Wells et al. (2002) analyzed the body-segment length, diameter, circumference, and skinfold thickness of 168 pediatric subjects aged 1–18 months. Infants were modeled using 17 geometric shapes with dimensions corresponding to the collected measurements. These representations were used to estimate the center of mass for

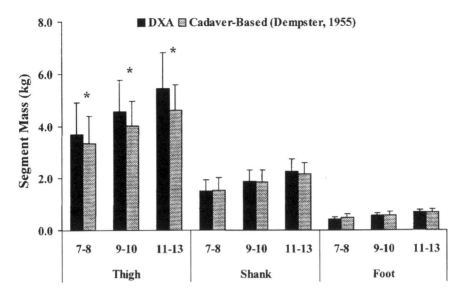

Fig. 1.15 Comparison of segment mass estimates using DXA and cadaver-based estimates for children aged 7–13. *Asterisk* significant differences ($p<0.05$) (Ganley and Powers 2004)

each body segment, which they assumed to be the center of volume. For this range of subject ages, all measures increased linearly with age.

Ganley and Powers (2004) compared two different techniques to estimate the mass, center of mass, and moment of inertia for the thigh, shank, and foot of 50 children aged 7–13. The lower extremities of subjects were scanned using dual energy X-ray absorptiometry, which identifies proportions of bone, muscle, and fat for each 3.9 cm thick segment. The mass and CG of each segment were calculated using known densities and compared to cadaver-based estimates from the literature. Figure 1.15 shows the differences in segment masses using the two methods. Although there were differences in inertial measures using the two different techniques, the authors report that the effect on calculating net joint moments for gait analysis would be negligible.

Loyd et al. (2010) estimated head mass, moment of inertia, and center of gravity location for children using a 14 pediatric cadaver specimens. Equations for estimating anatomical landmark locations and head center of gravity are provided.

Although many researchers have studied mass distribution in pediatric populations, it is difficult to directly compare results from different studies because of varying age groups, body segments studied, and measures calculated. Table 1.5 summarizes the pediatric mass distribution studies described in the preceding paragraphs to assist in identifying the differences in scope among them.

Table 1.5 Summary of pediatric mass distribution studies

Study	Age range	Body region	Calculated measures
Prange et al. (2004)	<1 months	Head	Mass, CG, inertia
Schneider and Zernicke (1992)	0–18 months	Upper and lower limb segments	Mass, CG inertia
Sun and Jensen (1994)	2–9 months	10 body segments	Mass, radius of gyration, inertia
Wells et al. (2002)	1–18 months	17 body segments	CG
Snyder et al. (1977)	2–10 years	Whole body, standing and seated	CG
Jensen (1981a, b, 1986a, b; Jensen and Nassas 1988	4–20 years, boys	15 body segments	Mass, inertia
Li and Dangerfield (1993)	8–16 years	Upper and lower limb segments	CG, radius of gyration, volume, inertia
Lebiedowska and Polisiakiewicz (1997)	5–18 years	Lower leg	Inertia
Ganley and Powers (2004)	7–13 years	Lower limb segments	Mass, CG, inertia
Loyd et al. (2010)	0–16 years	Head	Mass, CG, inertia

Spine Range of Motion

Ohman and Beckung (2008) measured the rotational and lateral bending cervical spine range of motion of 38 infants aged 2–10 months. The mean value of rotation to one side in all infants was 100° (STD 6.2°). For lateral flexion, mean value is 70° (STD 2.4°). For infants measured at more than one session, there were no significant trends between range of motion and age. Overall, infants at 2 months of age averaged 5° less rotation than the older infants.

Lewandowski and Szulc (2003) measured the cervical spine range of motion on 300 Polish children ranging in age from 3 to 7 years. Measurements were taken with a tensiometric electrogoniometer. Mean values and standard deviations by age and gender for each motion are shown in Table 1.6. They did not find statistically significant differences with gender. The authors do not present information on whether differences with age are statistically significant.

Lynch-Caris et al. (2006) measured the cervical spine range of motion in 75 children aged 8–10 years of age using the Cervical Range of Motion (CROM) device. Results from their study, for boys and girls combined, are shown in Table 1.7. While there are some statistically significant differences between age groups for some measures, there are no consistent trends with age.

Arbogast et al. (2007) performed a study of cervical spine range of motion on 67 children within three age groups: 3–5, 6–8, and 9–12 years. The maximum flexion/extension, lateral bending, and horizontal rotation were recorded using a range of motion measurement device as well as video. Results from the measures with the

Table 1.6 Cervical spine range of motion mean values (standard deviations) by age and gender

Age (years)	3		4		5		6		7	
	Girls	Boys	Girls	Boys	Girls	Boys	Girls	Boys	Girls	Boys
Flexion	60.9 (8.0)	68.1 (8.0)	56.8 (6.3)	58.4 (6.3)	55.7 (5.1)	60.4 (4.8)	58.9 (9.7)	61.9 (9.8)	61.4 (7.8)	60.1 (7.9)
Extension	68.8 (6.4)	69.6 (6.4)	59.9 (5.8)	67.1 (5.2)	58.1 (5.4)	65.4 (4.6)	60.4 (7.5)	66.8 (7.3)	57.8 (6.9)	63.9 (6.9)
Right-side bending	35.5 (9.3)	35.6 (9.3)	46.5 (7.1)	41.5 (6.5)	48.6 (6.4)	38.0 (7.6)	44.6 (8.2)	38.5 (8.3)	49.7 (6.2)	44.4 (6.6)
Left-side bending	35.8 (9.0)	36.2 (9.0)	46.5 (6.7)	41.3 (6.5)	48.3 (6.3)	38.3 (7.1)	44.6 (7.5)	38.9 (7.9)	50.1 (5.8)	44.8 (6.3)
Right rotation	70.4 (5.3)	74.1 (5.3)	74.5 (5.8)	76.6 (8.1)	77.2 (6.3)	78.4 (6.5)	73.1 (6.3)	73.9 (4.1)	75.2 (7.0)	75.1 (5.9)
Left rotation	70.3 (6.0)	74.3 (6.0)	74.7 (5.6)	76.4 (8.1)	76.6 (7.5)	78.8 (6.2)	73.1 (6.5)	74.3 (3.9)	74.9 (6.9)	75.5 (5.8)

Table 1.7 Cervical spine range of motion mean values (standard deviations) by age (Lynch-Caris et al. 2006)

Age (years)	8	9	10
Flexion	69 (9.7)	72 (9.3)	68 (11.1)
Extension	94 (12.1)	84 (15.6)	89 (11.7)
Lateral bending	65 (7.6)	58 (7.2)	60 (5.8)
Rotation	80 (6.4)	77 (7.1)	77 (8.1)

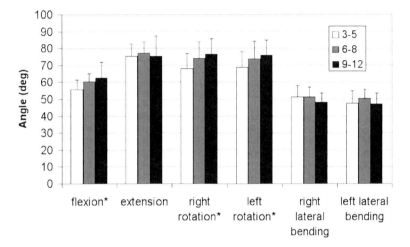

Fig. 1.16 Summary of mean values obtained with ROM instrument by age group; error bars indicate standard deviation (based on Arbogast et al. 2007)

ROM instrument are shown in Fig. 1.16. Measures indicated with an asterisk were significantly different with age. None of the results differed with gender, which differs from cervical range of motion trends with adults (Youdas et al. 1992). Results were not fully consistent between the two different measurement methods, as the measures derived from video tended to be greater than those measured with the device, except for lateral bending. Using the ROM measures, flexion and left and right rotation increased by a mean value of 7° from the youngest to oldest age groups. The video data did not show a significant difference with age. Their findings are similar to other pediatric neck range-of-motion studies, given the range of values for each measure.

Kasai et al. (1996) conducted a radiological study of pediatric spines to identify changes in lordosis and range of motion. They evaluated the lateral radiographs of 180 boys and 180 girls ranging in age from 1–18 years, with ten children in each age/gender group. The cervical lordosis angle (measured from C3 to C7) decreased until age 9 then rapidly increased, without any gender differences. For each pair of adjacent vertebrae, tilting angles defined by the distal endplates were calculated in flexion, neutral and extension postures. Forward tilting angle was defined as the

Fig. 1.17 Illustration of back range of motion device (BROM II, Sammons Preston)

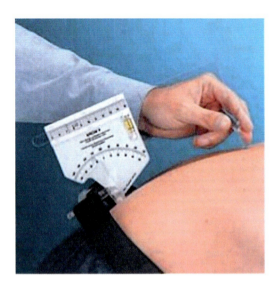

difference between flexion and neutral postures, while backward tilting angle was defined as the difference between neutral and extension postures. There were no substantial changes in tilting angle with age at any vertebral height. However, the amount of sliding (relative fore-and-aft motion of adjacent vertebral bodies) in the flexion, neutral, and extension postures increased significantly with age.

Kondratek et al. (2007) measured lumbar range of motion on 225 children from aged 5 through 9. The Back Range of Motion (BROM II) device, shown in Fig. 1.17, was used to measure flexion, lateral bending, and rotation of the lumbar spine. Table 1.8 shows the mean values and standard deviation of each measure by age. Boys did not demonstrate any measures that consistently varied over the entire age range, although there were a few instances of statistically significant differences between two age groups. Girls tended toward reduced range of motion with increasing age, although not all measures were statistically different. They did not find any relationships between lumbar range of motion measures and either sitting or standing height. The authors question the validity of the flexion data because of large standard deviations, low values from power analysis, and low intrarater reliability.

Posture

Klinich et al. (1994) examined belt fit of 155 children ranging in age from 6 to 12 years while they were seated in three different booster seats in three different vehicles, as well as when they were seated on the vehicle seat alone. Belt fit and posture were subjectively rated by viewing videos of the volunteers in each test condition. Booster seats showed improved belt fit for most of the children in most

Table 1.8 Lumbar range of motion mean values (standard deviations) by age and gender

Age (years)	5		7		9		11	
	Girls	Boys	Girls	Boys	Girls	Boys	Girls	Boys
Flexion (degrees)	23.7 (6.8)	21.6 (7.7)	25.2 (8.1)	23.9 (6.0)	21.8 (6.6)	24.3 (7.5)	19.8 (5.8)	21.4 (6.6)
Extension (degrees)	17.2 (5.6)	16.6 (6.6)	15.6 (6.3)	14.1 (5.6)	14.9 (5.5)	13.4 (7.4)	15.0 (6.1)	16.4 (5.8)
Right-side bending (degrees)	25.2 (8.1)	23.9 (8.0)	20.3 (6.7)	19.6 (3.8)	17.1 (6.9)	18.3 (4.9)	15.6 (5.4)	20.4 (8.3)
Left-side bending (degrees)	26.2 (9.0)	22.8 (8.2)	19.2 (7.6)	19.2 (4.9)	16.7 (6.0)	18.0 (5.7)	14.9 (4.6)	20.9 (7.1)
Right rotation (degrees)	24.4 (10.3)	19.2 (7.5)	19.5 (7.5)	20.2 (10.5)	16.1 (6.3)	19.2 (9.3)	13.0 (5.7)	13.8 (4.0)
Left rotation (degrees)	23.8 (9.9)	18.1 (9.7)	16.7 (7.6)	21.5 (10.4)	17.3 (8.0)	19.9 (9.6)	14.0 (6.0)	14.4 (6.0)

of the test conditions compared to using the three-point belt while seated on the vehicle seat. When children were seated on the vehicle seat, one factor contributing to poor belt fit was the tendency for the children to slouch forward, rather than sitting upright with the buttocks contacting the vehicle seatback. Children appeared to choose this posture because of discomfort when the backs of the calves were resting on the front edge of the vehicle seat cushion. The shorter cushion length provided by a booster allowed the children to be more comfortable while sitting in a more upright posture.

Reed et al. (2006) performed a laboratory study of seatbelt fit, seated posture, and body dimensions on 62 children with masses ranging from 18 to 45 kg. Measurements were taken with the children seated in three vehicle seats with and without each of three belt-positioning boosters. In addition to standard anthropometric measurements, three-dimensional body landmark locations were recorded with a coordinate digitizer in sitter-selected and standardized postures. The database quantifies the vehicle-seated postures of children and demonstrates how belt-positioning boosters can improve belt fit. In addition, analysis of the data allowed quantification of hip and head locations and pelvis and head angles for both sitter-selected and standardized postures. Results indicate that in the sitter-selected posture, children sit with their hips further forward and with more reclined pelvis orientations than in the standardized posture. The hip locations of the 6YO and 10YO ATDs, when seated using the standardized posture, are positioned further rearward than the sitter-selected postures of children by 20 mm in boosters and 40 mm on the vehicle seat. The head positions of the dummies are 20 mm higher on boosters and 40 mm higher on vehicle seats than the head positions of children of comparable size. The 10YO head position is also forward of the head locations of children. Raising the child up on a booster seat improves their posture as well as increases the angle of the lap belt to a more favorable orientation. Pelvis angles with both dummies are within the range seen with children. The papers report on development of a seating procedure that places crash dummy heads and hips in more realistic positions.

Application to ATD Design

Chamouard et al. (1996) compared pelvis and thigh dimensions of different pediatric ATDs to dimensions measured on real children as well as pediatric X-rays. ATDs that were assessed include the 3, 6, and 10-year-old TNO P dummies, 3- and 6-year-old part 572 dummies, and 3- and 6-year-old Hybrid III dummies. The peak height of the iliac crest, measured at the side, was acceptable for several dummies, but they did not assess the height of the ASIS, which is more critical for belt assessment. They note that the shape of the front of the pelvis is not realistic on any of the ATDs that were evaluated. In addition, the pelvis and thigh flesh stiffness of all of the dummies evaluated was substantially greater than that measured on child volunteers.

Irwin and Mertz (1997) report on methods used to specify the mass and dimensions of the CRABI and Hybrid III pediatric ATDs. The specifications intend to represent average children aged 6 months, 12 months, 18 months, 3 years, and 6 years. Dimensional specifications for the 3YO are primarily based on the Schneider et al. (1985) and Snyder et al. (1977) child anthropometry studies, with the segment masses taken from the studies by Young et al. (1976) and Wolanin et al. (1982). The 6YO dimensions are primarily based on the studies by Young et al. (1976) and Wolanin et al. (1982). For the smaller dummy sizes, dimensions are primarily based on the UMTRI anthropometry data compiled by Weber and Schneider (1985). The infant body segments were modeled as ellipsoids, and mass distribution estimated from the volume of each segment. Thus the target dimensions of the Hybrid III family of pediatric ATDs are not entirely based on the UMTRI anthropometry study, with the 3- and 6-year-old specifications at least partly based on older data that are not as representative as the UMTRI data.

Van Ratinger et al. (1997) describe anthropometric targets for the 3-year-old Q-series child dummy. The anthropometry for this series of ATDs is based on CANDAT, a database of child anthropometry developed by TNO (Twisk 1994) using several databases of child anthropometry collected in Europe and North America. They use the methods of Jensen and Nassas (1988), Schneider and Zernicke (1992), and Yeadon and Morlock (1989) to estimate segment masses and inertial properties. Values for the Q3 dummy are presented in the paper. Although the Q series of ATDs has been through several iterations to improve impact response, the anthropometry is still designed to meet the original specifications.

Serre et al. (2006) compared key dimensions measured on 3- and 6-year-old children to those measured on the Hybrid III 3- and 6YO ATDs. Most measures were within 1 cm, suggesting that the original anthropometry specifications are still reasonable. Loyd et al. (2010) also compared key head dimensions from their study to those of the CRABI and Hybrid III pediatric dummies. Most dummy head contours were close to those calculated in their study, with the exception of the 12MO and 3-year-old dummies.

Future Research Needs in Child Anthropometry

Child anthropometric data are currently applied in three crash-safety domains, each of which would benefit from more detailed and up-to-date data.

Restraint system design: The design of child restraint systems and boosters is based substantially on estimates of child body dimensions as a function of age. The compilation of Weber and Schneider (1985), which is primarily based on the Snyder et al. (1977) study and other sources, is widely used in the industry to size restraints. However, those data (collected prior to the recent upsurge in pediatric obesity) neglect the substantial secular changes in the upper percentiles of body weight and associated variables such as hip breadth. A reanalysis of the Snyder data based on more recent stature and weight data from NHANES (http://www.cdc.gov/nchs/

products/elec_prods/subject/nhanes3.htm) would provide more accurate dimensions for restraint system design. Further analysis of posture data from older children would also be valuable to provide guidance for improving belt fit for children who are no longer using CRS or boosters (typically ages 8 and up).

Crash dummy development: Crash dummies are intended to represent the size, shape, mass distribution, and posture of occupants who match a particular set of reference characteristics. The child crash dummies are specified on erect sitting height and body weight, among other dimensions. Differences can be demonstrated between the reference values for the Hybrid-III family of dummies (3, 6, and 10YO) and average dimensions for children of the equivalent age in any particular population of children, but these differences are not important as long as the overall proportions of the dummies are reasonable. However, unrealistic body structures are a substantial issue. In addition to the research described above, more work is needed to accurately characterize the skeletal geometry of children similar in size to the ATDs, particularly in the areas of the pelvis, thorax, and shoulder that interact with belt restraint systems.

Computational modeling: Rapid improvements in computer simulation hardware and software provide an opportunity to greatly improve restraint system optimization for children with a wide range of body characteristics. The development of accurate finite-element models of child occupants will require substantial new data on child anatomy.

To achieve this goal, detailed medical imaging data are needed from a large number of children. CT data should be used to create statistical models of skeletal components, including the spine. MRI data are needed to model the soft tissues of the thorax and abdomen. The MRI data should be gathered in seated postures, rather than supine, so that the organ positions and interrelationships are representative of vehicle occupants.

Finally, a large-scale study of whole-body, external body shape is needed. Such a study could be patterned in part on the CAESAR study (Robinette et al. 2002) that recently gathered three-dimensional external body shape data for 2,400 US civilians using a laser scanner. Scanned postures should include standing postures for extracting standardized anthropometric measures and seated postures approximating vehicle occupant postures. Recent advancements in body shape modeling (Reed and Parkinson 2008) provide the opportunity to integrate a statistical model of external body shape with the skeletal and soft tissue models to create an accurately scalable FE model.

References

Arbogast KB, Gholve PA, Friedman JE et al (2007) Normal cervical spine range of motion in children 3–12 years old. Spine 32:309–315

Bradtmiller B (1996) Sizing head forms: design and development. Technologies for occupant protection assessment. SAE Tech Paper # SAE-SP-1174

Beusenberg M, Happee R, Twisk D et al (1993) Status of Injury Biomechanics for the Development of Child Dummies, SAE Technical Paper 933104, doi:10.4271/933104

Burdi AR, Huelke DF, Snyder RG et al (1969) Infants and children in the adult world of automobile safety design: pediatric and anatomical considerations for design of child restraints. J Biomech 2(3):267–280

Chamouard F, Tarriere C, Baudrit P (1996) Protection of children on board vehicles: influence of pelvis design and thigh and abdomen stiffness on the submarining risk for dummies installed on a booster. Presented at the 15th international technical conference on the enhanced safety of vehicles, Washington, DC

de Onis M, Onyango AW, Borghi et al (2007) Development of a WHO growth reference for school-aged children and adolescents. Bull World Health Organ 85:660–667

Dreyfuss H, Tilley AR (1993) The Measure of man and woman: human factors in design. Whitney Library of Design. New York

Dreyfuss H and Tilley AR (2002) The measure of man & woman, edited by John Wiley & Sons, New York

Ganley KJ, Powers CM (2004) Anthropometric parameters in children: a comparison of values obtained from dual energy X-ray absorptiometry and cadaver-based estimates. Gait Posture 19:133–140

Hedley AA, Ogden CL, Johnson CL et al (2004) Overweight and obesity among US children, adolescents, and adults, 1999–2002. JAMA 291:2847–2850

Huang S, Reed MP (2006) Comparison of child body dimensions with rear seat geometry. SAE Transactions: Journal of Passenger Cars – Mechanical Systems, 115:1078–1087

Irwin AL, Mertz HJ (1997) Biomechanical bases for the CRABI and Hybrid III child dummies. SAE Tech Paper #973317

Jensen RK (1986a) The growth of children's moment of inertia. Med Sci Sports Exerc 18:440–445

Jensen RK (1986b) Body segment mass, radius and radius of gyration proportions of children. J Biomech 19:359–368

Jensen RK (1981a) The effect of a 12-month growth period on the body moments of inertia of children. Med Sci Sports Exerc 13:238–242

Jensen RK (1981b) Age and body type comparisons of the mass distributions of children. Growth 45:239–251

Jensen RK, Nassas G (1988) Growth of segment principal moments of inertia between four and twenty years. Med Sci Sports Exerc 20:594–604

Kasai T, Ikata T, Katoh S et al (1996) Growth of the cervical spine with special reference to its lordosis and mobility. Spine 21:2067–2073

Klinich KD, Pritz HB, Beebe MS et al (1994) Survey of older children in automotive restraints. SAE Tech Paper #942222

Kondratek M, Krauss J, Stiller C et al (2007) Normative values for active lumbar range of motion in children. Pediatr Phys Ther 19:236–244

Kuczmarski RJ, Ogden CL, Guo SS et al (2002) 2000 CDC growth charts for the United States: methods and development. National Center for Health Statistics. Vital Health Stat 11(246)

Lebiedowska MK, Polisiakiewicz AJ (1997) Changes in the lower leg moment of inertia due to child's growth. Biomechanics 30:723–728

Levick N, Solaiyappan M, Gentry J, et al (2001) Modalities for constructing 3-D models of head and neck anthropometry, spinal cord and vertebral structures of infants to adolescents – for application to crash test dummy design. SAE Technical Paper 2001-01-0172, doi:10.4271/2001-01-0172

Lewandowski J, Szulc P (2003) The range of motion of the cervical spine in children aged from 3 to 7 years – an electrogoniometric study. Folia Morphol (Warsz) 62(4):459–461

Li Y, Dangerfield PH (1993) Inertial characteristics of children and their application to growth study. Ann Hum Biol 20:433–454

Loyd AM, Nightingale R, Bass CR et al (2010) Pediatric head contours and inertial properties for ATD design. Stapp Car Crash J 54:167–196

Lynch-Caris T, Brelin-Fornari J, Van Pelt C (2006) Cervical range of motion data in children. SAE Tech Paper #2006-01-1140

Norris B, Wilson JR (1995) Childata: the handbook of child measurements and capabilities: data for design safety. Department of Trade and Industry. Consumer Safety Unit, London, UK

Netter, Henry F, Dalley AF (1997) Atlas of Human Anatomy. East Hanover, NJ: Novartis Print.

Ohman AM, Beckung ER (2008) Reference values for range of motion and muscle function of the neck in infants. Pediatr Phys Ther Springer 20(1):53–58

Ogden CL, Lamb MM, Carroll MD et al (2010) Obesity and socioeconomic status in children and adolescents: United States, 2005–2008. NCHS Data Brief (51):1–8

Prange MT, Luck JF, Dibb A et al (2004) Mechanical properties and anthropometry of the human infant head. Stapp Car Crash J 48:279–299

Reed MP, Ebert SM, Rupp JD (2010) Pediatric thoracic and shoulder skeletal geometry. UMTRI Technical Report 2010–40

Reed MP, Sochor MM, Rupp JD et al (2009) Anthropometric specification of child crash dummy pelves through statistical analysis of skeletal geometry. J Biomech 42:1143–1145

Reed MP, Parkinson MB (2008) Modeling variability in torso shape for chair and seat design. Presented at the ASME design engineering technical conferences. DETC2008-49483, New York

Reed MP, Ebert-Hamiton SM, Manary MA et al (2006) Improved positioning procedures for 6YO and 10YO ATD's based on child occupant postures. Stapp Car Crash J 50:337–388

Reed MP, Lehta MM, Schneider LW (2001) Development of anthropometric specifications for the six-year-old OCATD. SAE Tech Paper #2001-01-1057

Robinette K., Blackwell S, Daanen H et al (2002) Civilian American and European surface anthropometry (CAESAR). ASAF Research Lab AFRL-HE-WP-TR-2002-0169

Roebuck JA (1995) Anthropometric methods: designing to fit the human body. Presented at human factors and ergonomics society, California

Roebuck JA, Thomson WG, Kroemer KH et al (1975) Engineering anthropometry methods. Wiley and Sons, New York

Schneider K, Zernicke RF (1992) Mass, center of mass, and moment of inertia estimates for infant limb segments. J Biomech 25:145–148

Schneider LW, Owings CL, Lehman RJ et al (1985) Anthropometry and shape of children's heads, necks, and shoulders for product safety design. University of Michigan Transportation Research Institute, Ann Arbor, MI

Serre T, Lalys L, Brunet C et al (2006) 3 and 6 years old child anthropometry and comparison with crash dummies. SAE Tech Paper #2006-01-2354

Smith S, Norris B (2001) Childata: Assessment of the Validity of Data. http://www.virart.nott.ac.uk/pstg/childchanges.htm.

Snyder RG, Schneider LW, Owings CL et al (1977) Anthopometry of infants, children, and youths to age 18 for product safety design. Highway Safety Research Institute, Ann Arbor, UM-HSRI-77-17

Snyder RG, Spencer ML, Owings CL et al (1975) Physical characteristics of children as related to death and injury for consumer product safety design. SAE Tech Paper #SAE-SP-394

Sun H, Jensen RJ (1994) Body segment growth during infancy. Biomechanics 27:265–275

Twisk D (1994) Anthropometric data of children for the development of dummies. TNO Report 75061275-B

Van Ratinger MR, Schroolen M, Twisk D et al (1997) Biomechanically based design and performance targets for a 3-year-old-child crash dummy for front and side impact. SAE Tech Paper #973316

Youdas JW, Garrett TR, Suman VJ et al (1992) Normal range of motion of the cervical spine: an initial goniometric study. Phys Ther 72(11):770–780

Weber K, Schneider L (1985) Child anthropometry for restraint system design. UMTRI-85-23

Wells JP, Hyler-Both DL, Danley TD et al (2002) Biomechanics of growth and development in the healthy human infant: a pilot study. J Am Osteopath Assoc 102:313–319

WHO Multicentre Growth Reference Study Group (2007) WHO child growth standards: Head circumference-for-age, arm circumference-for-age, triceps skinfold-for-age and subscapular skinfold-for-age: methods and development. Geneva: World Health Organization

Wolanin MJ, Mertz HJ, Nyznyk RS et al (1982) Description and basis of a three-year-old child dummy for evaluation passenger inflatable restraint concepts. Presented at the ninth international technical conference on experimental safety vehicles, Kyoto, Japan, SAE Tech Paper #826040

Yeadon MR, Morlock M (1989) The appropriate use of regression equations for the estimation of segmental inertia parameters. J Biomech 22:683–689

Young JW, McConville JT, Reynolds HM et al (1976) Development and evaluation of masterbody forms for three-year old and six-year-old child dummies. DOT HS-801 811

Chapter 2
Epidemiology of Child Motor Vehicle Crash Injuries and Fatalities

Kristy B. Arbogast and Dennis R. Durbin

Introduction

Although children represent only 10–15% of the overall traffic fatality burden in the United States, motor vehicle crashes (MVCs) remain a leading cause of death and disability for children and young adults, accounting for nearly half of all unintentional injury deaths to children and adolescents (Centers for Disease Control and Prevention National Center for Injury Prevention and Control, Web-based Injury Statistics Query and Reporting System [CDC NCIPC WISQARS] 2010). Moreover, their exposure to motor vehicle risk is significant because they travel by motor vehicles nearly as much as adults. Prevention of the fatalities, injuries and disability associated with MVC must be a priority for ensuring our children's overall health.

Since the mid-1990s, the injury prevention and control community has achieved significant success in reducing this burden. For example, in 2009, 41% fewer children under 14 years of age died as a result of MVCs than in 1996 (National Highway Traffic Safety Administration [NHTSA] 2010a). These reductions have been achieved, in part, through a combination of increased attention to age-appropriate restraint use and rear seating. From 1999 to 2007 there was a nearly threefold increase in child restraint system (CRS) use among 3–8 year old children in crashes, with a specific emphasis on booster seats for those 5–8 years of age (Children's Hospital of Philadelphia [CHOP] 2008). Spurred by the child fatalities associated with passenger air bag deployment in the early 1990s, more children are also

---

K.B. Arbogast, Ph.D. (✉) • D.R. Durbin, M.D, M.S.C.E
Division of Emergency Medicine, Center for Injury Research and Prevention,
The Children's Hospital of Philadelphia, Suite 1150, 3535 Market St., Philadelphia, PA, USA
e-mail: arbogast@email.chop.edu

being seated in the rear seats of vehicles. Specifically, by 2007, ~90% of children 0–7 years of age rode in the rear seat (CHOP 2008). Even with increases in age-appropriate restraint use and rear seating, the protection of children in motor vehicles can be further improved. As many as one-third of 8–12-year-olds were still routinely seated in the front row (CHOP 2008) and as recently as 2009, data indicate that 45% of 4–7-year-olds were not restrained in accordance with best practice recommendations for their age (Pickrell and Ye 2010).

As a first step towards reducing MVC injuries and fatalities, statistical analyses of surveillance and other epidemiologic data can be used to identify specific injury-producing circumstances and provide population-based estimates of the potential effectiveness of safety technologies and other interventions. Such analyses identify and help prioritize problems in motor vehicle safety and define the specific nature and magnitude of these problems by examining risk factors for and outcomes of crashes and injuries. Specifically, they can help put individual cases into a broader population-based context of specific crash injury scenarios. The translation between epidemiology and engineering is important. Ideally, epidemiologists will identify and define problems. Building on this new knowledge, engineers can pursue countermeasures to solve these problems. Subsequently, both disciplines can then work together to evaluate the real-world effectiveness of the countermeasure(s). To achieve this goal, it is important that motor vehicle injury epidemiology incorporates knowledge of the biomechanics and kinematics characterizing the injury event because this information is fundamental to the development, evaluation, and modification of safety technologies.

This chapter reviews and examines the current state of knowledge of the epidemiology of child MVC injuries and fatalities to provide context for the biomechanical focus of the remaining chapters of this textbook. It begins with a review of the magnitude of the child MVC injury and fatality problem, briefly summarizes existing data sources available for epidemiological analyses, and then describes the risks associated with the various pediatric restraint practices and other factors influencing crash injury and fatalities.

Magnitude of the Problem

With the exception of those under 1 year of age, unintentional injury is the leading cause of death, serious injury, and acquired disability for children and youth, up to 14 years of age (Table 2.1). For children and young adults aged 5–24 years, MVCs are the leading cause of death, representing 63% of unintentional injury deaths in 2007 (Fig. 2.1).

Fatalities represent only the tip of the MVC problem for children. For every 1 fatality, ~18 children are hospitalized and over 400 receive medical treatment for injuries sustained in a crash (CDC NCIPC WISQARS 2010). MVCs are among the top ten leading causes of non fatal injuries treated in hospital Emergency Departments for ages 5–24 years with 876,943 injuries sustained in 2008. Of interest, reductions

2 Epidemiology of Child Motor Vehicle Crash Injuries and Fatalities   35

Table 2.1 Top five leading causes of death by age group, the United States—2007 (CDC NCIPC WISQARS 2010)

<1 year	1–4 years	5–9 years	10–14 years
Congenital anomalies (5,785)	Unintentional injury (1,588)	Unintentional injury (965)	Unintentional injury (1,229)
Short gestation (4,857)	Congenital anomalies (546)	Malignant neoplasms (480)	Malignant neoplasms (479)
SIDS (2,453)	Homicide (398)	Congenital anomalies (196)	Homicide (213)
Pregnancy complications (1,769)	Malignant neoplasms (364)	Homicide (133)	Suicide (180)
Unintentional injury (1,285)	Heart disease (173)	Heart disease (110)	Congenital anomalies (178)

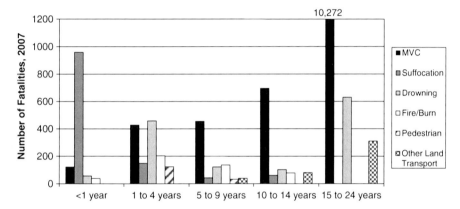

Fig. 2.1 Top five leading causes of unintentional injury death for children, the United States—2007 (CDC NCIPC WISQARS 2010)

in nonfatal injuries due to MVCs from 2007 to 2008 for those children 1–4 years have removed MVC from the top ten list for that age group (CDC NCIPC WISQARS 2010).

The potential exposure of children to MVCs is great. Children travel nearly as much as adults with an average of 3.2 trips per day for children 5–14 years (Santos et al. 2011) (Fig. 2.2).

Through the early 1990s child occupant fatality rates remained relatively stagnant at ~3.5 deaths per 100,000 children (NHTSA 1999b). However, the number of motor vehicle fatalities and serious injuries among children has declined recently. In 2009, 1,314 child occupants aged 0–14 years died in MVCs in the United States representing a 41% decrease from 1996 (NHTSA 1999b) (Fig. 2.3). This shift was more immediate among children ages 0–3 years than among children ages 4–12 years, although the older children also experienced a decline in deaths after 1999 (Nichols et al. 2005). Injuries have seen a similar decline; in 1996, 354,000

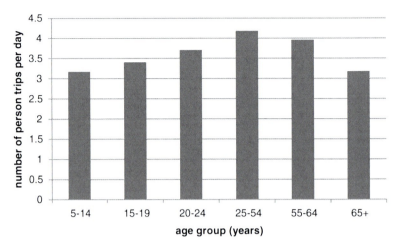

Fig. 2.2 Number of motor vehicle trips per day by age group (Santos et al. 2011)

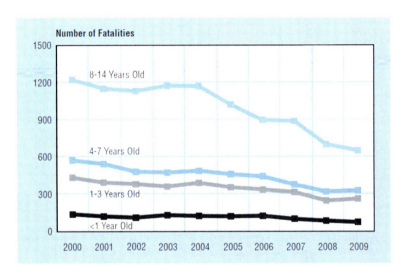

Fig. 2.3 Total traffic fatalities among children aged 14 and under by age group, 2000–2009 (NHTSA 2010b)

occupants aged 0–15 years were injured in MVCs, while in 2009, 179,000 were injured representing a 49% decline. It has been asserted that the mortality and morbidity reductions have been through a combination of increased attention to age-appropriate restraint use and seating position (Arbogast et al. 2004b; Braver et al. 1997; Durbin et al. 2003a, 2005; Elliott et al. 2006), and enhanced laws and enforcement of these laws (Segui-Gomez et al. 2001; Winston et al. 2007).

This burden extends beyond the United States and affects children globally. Worldwide, each year nearly 400,000 people under age 25 sustain fatal injuries on the world's roads—averaging more than 1,000 deaths per day, as noted by The World Health Organization (WHO) in its comprehensive report on Child Injury Prevention with an emphasis on Road Traffic Injuries (WHO 2008). Road traffic injuries are the second leading cause of death for 5–14-year-olds. Approximately one-third of these deaths are to pedestrians, while 65% are to vehicle occupants.

The portion of the text which follows reviews the epidemiologic literature on restraint effectiveness, injury causation scenarios and mechanisms for child occupants and is organized by restraint type. This review was primarily restricted to papers published during the past 10–15 years (i.e., since the mid-1990s) as the focus of this chapter was to establish specifics of the more recent trends and not to delineate the historical variations in pediatric injury patterns in MVCs. Moreover, we have focused our description on the more common injuries sustained by children. We begin with a review of the primary data sources that have been used and/or are available for study of motor vehicle injuries in children.

Primary Data Sources and Methodology

Several public and private databases exist upon which the findings presented throughout this chapter are based. A thorough description of their development, purposes, and content is contained in the Appendix. A brief description of the databases, their uses and limitations are described in Table 2.2.

While there are many sources of data providing information on children in MVCs, no single source provides the sufficient quantity and quality of data to address all issues requiring research in child passenger safety. Existing national surveillance systems such as NASS-GES, Fatality Analysis Reporting System (FARS), Cooperative Crash Injury Study (CCIS) and International Road Traffic and Accident Database (IRTAD), while capable of providing useful information on general trends of child occupant protection, do not collect the child-specific data required to conduct the most relevant analyses such as effectiveness of specific restraint systems or other best practice recommendations or to determine child specific injury mechanisms. Specialized data collection systems such as National Automotive Sampling System – Crashworthiness Data System (NASS-CDS), Special Crash Investigations (SCI), Crash Injury Research and Engineering Network (CIREN), and German In-Depth Accident Study (GIDAS) have the required specificity to identify the nature and body region of serious injuries to child occupants; however, they do not identify an adequate number of children to allow essential analyses to be conducted in a timely fashion. Child-specific data collection systems such as PCPS, CREST, and CHILD have attempted to provide both a sufficient number of cases and depth of child specific data but are difficult to sustain as part of ongoing data collection efforts as they have historically been funded by sole nongovernment sponsors or require consortiums of sponsors. Future progress in child passenger safety research will require better integration of these data sources, as well as the creation of a nationally representative child-focused crash surveillance system that combines the strengths of several existing data collection systems.

Table 2.2 Summary of data sources used in studies of crash injuries to children

Database	Description	Used for	Limitations
Fatality Analysis Reporting System (FARS)	Census of all fatal crashes in the US	Characterizing crashes resulting in fatalities; estimates of restraint system effectiveness	Limited to crashes in which someone died
National Automotive Sampling System—General Estimates System (NASS-GES)	Probability sample of police reported crashes in the United States. Data limited to that contained on the Police Accident Report	Describing national trends in crash characteristics and injuries	Limited detail in data collected (e.g., relies on police report of injury)
National Automotive Sampling System—Crashworthiness Data System (NASS-CDS)	Probability sample of police reported crashes in the United States. Involves a detailed field investigation of the vehicle and scene and abstraction of the occupants' medical records	Describing national trends in crash characteristics and injuries; estimates of restraints system effectiveness; identification of injury sources	Limited number of child occupants identified
Special Crash Investigations (SCI)	Anecdotal dataset of in-depth crash investigations. Inclusion criteria are changed routinely to address emerging traffic safety needs	Identification of new or unique crash circumstances or injury-causing crash scenarios	Nonrepresentative selection of cases. Limited number of child occupant crashes
Crash Injury Research and Engineering Network (CIREN)	A network of trauma centers who collect highly detailed clinical and crash information to determine injury causation	Detailed descriptions of the sources and mechanisms of injury	Limited number of children included. Some bias in the selection of crashes due to hospital-based data collection system

Partners for Child Passenger Safety (PCPS)	Child-specific crash surveillance system that links insurance crash claims to detailed telephone survey data and in-depth crash investigations	Describing trends in crash characteristics and injuries; estimates of restraints system effectiveness	Not nationally representative; limited detail on nature of injuries; some selection bias due to insurance-based case identification. Data collected ended in 2007.
Child Restraint System in Cars (CREST) and Child Led Injury Design (CHILD) projects	In-depth crash investigations conducted by a consortium of European industry, government and academic entities	Case series analysis of in-depth investigations for injury causation	Not a representative sample of crashes
Cooperative Crash Injury Study (CCIS)	In-depth crash investigations of a sample of cases involving newer vehicles which are representative of crashes occurring in the UK	Describing trends in crash characteristics and injuries; estimates of restraint system effectiveness	Limited number of child occupants identified
International Road Traffic and Accident Database (IRTAD)	Traffic fatality data from many countries throughout the world	Comparing traffic fatality rates among countries	Limited crash or injury detail
German In Depth Accident Study (GIDAS)	In-depth crash investigations from two regions in Germany	Case series analysis of in-depth investigations for injury causation	Not a representative sample of crashes involving children

Restraint Use

Restraint Use Policy History and Trends in Restraint Usage

The first US state child occupant restraint law was passed in Tennessee in 1978 (Teret et al. 1986). By 1985, all 50 states and the District of Columbia had passed laws requiring use of child restraints by young children and consequently child restraint use increased dramatically. Beginning in 1995, when children killed by deploying passenger air bags were first reported clinically, attention began to be focused on the unique needs of children in automotive safety. Efforts to ensure appropriate restraint for children have emphasized such issues as improved access to CRS and booster seats, upgraded laws, and educational and media campaigns that recommended that all children under age 13 should ride in the rear seats of vehicles.

In response to evidence of injuries and fatalities to children from deploying passenger air bags, NHTSA initiated a two-pronged response of education and regulation. For a comprehensive summary of the nationwide effort to reduce air-bag-related deaths among children, see Nichols et al. (2005). First, NHTSA, joined by many national organizations, recommended that all child passengers younger than 13 years of age sit in the rear seats of vehicles. Second, in 1997, NHTSA enacted a substantial regulatory change to Federal Motor Vehicle Safety Standard 208, which provided automakers a choice between certifying frontal crash performance for unbelted adults by either rigid barrier tests or sled tests (NHTSA 1997). This change in the standard, in many cases, resulted in the redesign of frontal air bags to reduce the force with which they deploy. These new air bags are generally referred to as "second-generation air bags." The role of the air bag on risks to child occupants is further discussed later in this chapter.

NHTSA began a standardized child passenger safety training and certification program in 1998. Since then, over 35,000 individuals have been certified as child passenger safety technicians. These individuals have participated in thousands of community-based child safety seat clinics and have been a source of information on appropriate restraint guidelines, including the use of booster seats. In addition, several government and industry-sponsored initiatives have drawn significant media attention to the importance of age-appropriate restraint, including the use of booster seats by older children.

Recognizing the importance of laws in both changing restraint behavior and educating the public about recommended restraint practices, beginning in 2000, several states enhanced their child occupant restraint laws through the enactment of booster seat use provisions for older children. While the laws aim to ensure the appropriate use of all forms of child restraints (e.g., child safety seats, belt positioning booster seats, combination seats), the revised laws became generally known as "booster seat laws." Subsequent study of the association of a booster provision in a state child restraint law with changes in child restraint use in that state indicated that

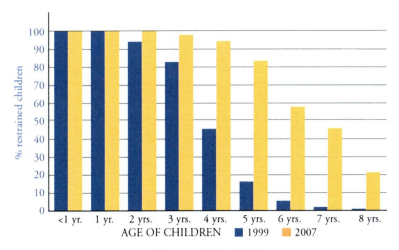

Fig. 2.4 Child restraint use by age: 1999 versus 2007 (CHOP 2008)

booster seat provisions covering children from 4 to 7 years have some effect on all children in this age range. Specifically, children aged 4–5 years in states with booster seat laws were 23% more likely to be reported as appropriately restrained than children in other states, while those aged 6–7 years were twice as likely to be reported as appropriately restrained. For 6–7-year-olds, the effect was much stronger when the law included those aged 4 through 7 years than when it included only those aged 4–5 years (Winston et al. 2007). Booster seat laws have been shown to be related to a decrease in child deaths as well (Farmer et al. 2009).

Due in large part to the increased attention paid to the needs of children in motor vehicle safety beginning in the mid-1990s, the period between 1999 and 2007 witnessed large increases in reported appropriate restraint use (including child safety seats, booster seats, and combination seats) by children ages 4 through 8 years. According to the PCPS data (an insured population), child restraint use for 4–8 year olds in crashes increased to 63% in 2007 from 15% in 1999 (Fig. 2.4). The largest relative increase in CRS use was for the oldest age group (6–8 years of age), yet 57% of these children continued to be inappropriately restrained in seat belts in 2007 (CHOP 2008). The youngest children, 3-year-olds, were primarily restrained by forward facing child restraint systems (FFCRS). Figure 2.5 shows the distribution of child restraint type over time. High back belt-positioning booster seats (BPB) were the most common restraints for the 4–5-year-old children, though the proportion of backless BPB continues to increase. Most appropriately restrained children over age 5 were in belt positioning booster seats, with somewhat more older children in backless, as opposed to high-back boosters (Jermakian et al. 2007b).

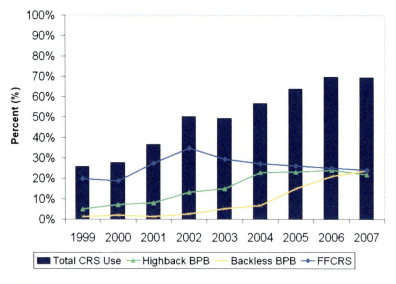

Fig. 2.5 Distribution of child restraint type for 3–8-year-olds: 1999–2007 (CHOP 2008)

NHTSA has conducted an observation study of child restraint use since 2006, the National Survey of the Use of Booster Seats (NSUBS). Survey data are obtained on children 12 years and under in passenger vehicles at a nationwide probability sample of gas stations, day care centers, recreation centers, and restaurants in five fast food chains. Restraint use is observed by trained data collectors prior to or just as the vehicle comes to a stop. Targeting a broader population than PCPS, NSUBS reported in 2009 that 41% of 4–7-year-olds were restrained in booster seats and 14% were in child safety seats (Pickrell and Ye 2010). These percentages have remained relatively flat since 2006. There is an age effect; 68% of 4–5-year-olds were in some form of child restraint; however, this was reduced to 39% for 6–7-year-olds. The variations in the laws and the effect of these differences as discussed above likely influences these percentages. In addition, restraint use for children driven by a belted driver is higher (over 90%) than for those driven by an unbelted driver (NHTSA NCSA 2008; Pickrell and Ye 2010).

Rear-Facing Child Restraints

The American Academy of Pediatrics (AAP) recommends that all infants and toddlers should ride in a rear-facing CRS until they are 2 years of age or until they reach the highest weight or height allowed by the manufacturer of their CRS (Durbin 2011). This recommendation results from the need to support the young child's posterior torso, neck, head, and pelvis and distribute crash forces over the entire body. Developmental considerations put young children at risk for spinal cord injury, including incomplete vertebral ossification and excessive ligamentous laxity (note that developmental anatomy will be discussed in detail in subsequent chapters). The rear-facing child restraint (RFCRS) addresses this risk by supporting the child's head, preventing the relatively large head from loading the proportionately smaller neck. Research on the effectiveness of RFCRS has found them to reduce fatal injury by 71% for infants <1 year of age in passenger cars and by 58% in light trucks (Hertz 1996).

In the US, few children remain rear facing past their first year of age, despite the fact that there are currently many RFCRS that have maximum weight limits beyond 9.2 kg. In fact, in a recent study of NASS data, over 40% of those 0–11 months were in a FFCRS (Henary et al. 2007). In Sweden, children remain rear facing up to the age of 4 years and transition directly from the RFCRS to a booster seat. Swedish researchers have studied the effectiveness of this behavior (Isaksson-Hellman et al. 1997; Jakobsson et al. 2005). They reported that RFCRS reduced the risk of Abbreviated Injury Scale (AIS) Score 2+ injuries by 90%, relative to unrestrained children, reinforcing their policy of children remaining in an RFCRS up to the age of 4 years. In contrast, Australian guidelines recommend rear facing only up to 6 months. In-depth crash research there has revealed no serious neck injuries, in the absence of head contact, among these forward facing children even in very severe frontal impacts (Paine et al. 2003). Of note, top tether use is mandatory in Australia, perhaps influencing these findings.

Recently, Henary et al. (2007) reviewed US crash data to calculate the relative effectiveness of RFCRS compared to FFCRS. These researchers extracted data on crash occupants 0–23 months restrained in a RFCRS or FFCRS from the NASS-CDS system from 1988 to 2003. Across all crash types, children in FFCRS were 76% more likely to be seriously injured than children restrained in RFCRS. When those 12–23 months were analyzed separately, these children were more than five times as likely to be seriously injured when restrained in FFCRS. Of interest, the largest benefits were realized by children in RFCRS involved in side crashes. These authors concluded that for children up to 23 months of age, the RFCRS provides the best protection. The lack of meaningful numbers of children 24 months or older in RFCRS in the US databases prevented extension of these analyses to age groups similar to the Swedish study.

Although the injury risk to children in RFCRS is significantly lower than those restrained in FFCRS (Henary et al. 2007), when injuries occur, they are primarily limited to head injuries. In a review of 31 cases of children restrained in RFCRS from the European CREST project, five sustained AIS3+ injury to the head and four sustained fractures to the extremities (upper and lower) (Lesire et al. 2001). It is important to realize that in European vehicles, it is more common to restrain a rear facing child in the front seat in the absence of a frontal passenger air bag or with the ability to turn the air bag off. As a result, some of the head injuries in this European study were related to the child being positioned in the front seat and having the area of the child restraint containing the child's head contact the dashboard. Of note was the absence of injuries to the neck or spine.

Forward-Facing Child Restraints

The recommendation for FFCRS has been based, in part, on an analysis by Kahane (1986) of laboratory sled tests, observational studies, and police reported crash data from the early 1980s that estimated correctly used CRS reduce the risk of death and injury by ~70% compared with unrestrained children. The engineering tests documented the biomechanical benefits of the CRS in spreading the crash forces over the shoulders and hips and by controlling the excursion of the head and face during a crash. The study further quantified the effectiveness of a partially misused CRS at a 45% reduction in risk of fatality and serious injury.

Estimates of effectiveness based on real-world crashes do vary based on which database was used, the years studied, and the analytical approach taken. Examining the preponderance of evidence, it is difficult to pinpoint a specific numerical value of effectiveness; however, universally, these studies indicated that FFCRS are highly effective at preventing fatal and nonfatal injuries. Two studies on FFCRS fatality effectiveness that are often quoted are Hertz (1996) and Elliott et al. (2006). Using FARS data from 1988 to 1994, Hertz found that, among children between 1 and 4 years of age in passenger cars, those in FFCRS experienced a 54% reduction in deaths compared with unrestrained children. Elliott and colleagues used a more recent dataset to compare the effectiveness of child restraints to seat belts and determined that FFCRS when not seriously misused (e.g., unattached restraint, CRS harness not used, two children restrained with one seat belt) were associated with a 28% reduction in risk for death in children aged 2 through 6 years after adjusting for key confounding variables. When including cases of serious misuse, the effectiveness estimate was slightly lower (21%). Rice and colleagues conducted a matched-cohort study, using FARS data to also estimate the effectiveness of child restraints at reducing the risk of death for children aged 3 and younger (Rice and Anderson 2009). The estimated death risk ratio, comparing children in child restraints to unrestrained children, was 0.33 (95% CI 0.29, 0.37). The authors found similar risk ratios for children in seat belts, suggesting that belts and child restraints were equally as effective at reducing the risk of death when compared to unrestrained children. There was evidence for the superior performance of child restraints for children under age 2 and the authors concluded that parents should be encouraged to use child safety seats instead of seat belts. Several studies have compared the effectiveness of FFCRS at preventing serious injury with effectiveness estimates ranging from a 71 to 82% reduction in serious injury risk in FFCRS compared to children of similar age in seat belts (Arbogast et al. 2004b; Winston et al. 2000; Zaloshnja et al. 2007). A summary table of these effectiveness studies is presented in section "Summary of Restraint Effectiveness Data for Child Occupants."

Historically, the harnesses in FFCRS have varied in design, including T-shields, tray shields and 5-point harnesses. For a thorough description of these differences, see Weber (2002). While sled tests typically show benefits of 5-point harnesses, no study has been able to identify benefits in real-world crash data among the harness types. Designs have evolved to be almost exclusively 5-point harnesses by the mid-2000s.

Although child restraints are very effective in preventing injuries, several authors have reviewed case series of children restrained in child restraints to gain insight into areas of focus for future optimization of these restraints. Sherwood et al. (2003) reviewed detailed police reports of crashes involving 92 fatally injured children (ages 5 and younger) in child restraints to obtain basic crash information and determine the factor most responsible for the fatality. These authors reported that half of the crashes were considered unsurvivable for the child, and 12% of fatalities were judged to result from gross misuse of the child restraint.

In a review of European data, the body regions of injury (defined as AIS3+ plus extremity fractures) for children in FFCRS in decreasing order of prevalence were the extremities, head, and neck (Lesire et al. 2001). Data from the US identified similar trends; the most common body regions of AIS2+ injury were the lower extremity, the face and the head (Arbogast et al. 2002a).

Individual studies have examined how these specific injuries occur. Head injuries sustained by child occupants restrained in child restraints include both contact-induced injuries as well as inertial injuries (Arbogast et al. 2005c; Jakobsson et al. 2005). Injuries such as skull fracture, epidural hematoma, and frontal lobe contusion are contact injuries (Gennarelli 1986, 1993) that are most likely due to excursion of the head and subsequent impact with the vehicle interior. Often, limited initial precrash space or intrusion reduces the space available for the child, thus increasing the likelihood for contact. Another contributing factor to head contact is looseness of both the vehicle seatbelt attaching the child restraint and the child restraint harness—both very common misuses (Bull et al. 1988; Decina and Knoebel 1997; Hummel et al. 1997; Lesire et al. 2001; Muszynski et al. 2005; National Safe Kids Campaign 1999). This laxity has been shown to increase head excursion (Henderson et al. 1994; Hummel et al. 1997). With loose vehicle belt attachment, the CRS is less tightly coupled to the vehicle and does not optimally benefit from the vehicle's energy management. With a loose CRS harness, there is relative movement between the torso of the child and the back of the child seat. While less common than contact-induced injuries to the brain and skull, children in FFCRS also sustain injuries to the brain where there is no evidence of contact to the head. Similar to head excursion, laxity of the child restraint harness and vehicle seat belt has been shown in a series of sled tests with child anthropomorphic test devices (ATDs) to increase resultant head acceleration (Hummel et al. 1997) suggesting contact may not be necessary to produce head injury metrics above suggested thresholds. In a review of over 150 cases of children 1–12 years with head injuries in MVCs, about 60% sustained intracranial injuries without accompanying skull fracture suggesting that at least a portion of these injuries may have occurred due to noncontact mechanisms (Arbogast et al. 2005c). Bohman et al. (2011) in a review of 27 in-depth crash investigations of head injuries sustained by restrained child occupants in frontal crashes have also identified nine cases in which these injuries occurred in the absence of evidence of head contact. The cases without evidence of head/face contact were characterized by high crash severity and accompanied by severe injuries to the thorax and spine. The circumstances under which these potential noncontact head injuries occur deserve further study.

Several researchers reviewed in-depth investigations of crashes from the CIREN database involving children seated in FFCRS with lower extremity injuries to determine the nature of the injuries and the circumstances under which they occurred (Bennett et al. 2006; Jermakian et al. 2007a). Injuries below the knee were the most common and they most often occurred due to interaction with the vehicle seatback in front of the child's seating position. This interaction with the seatback occurred in both frontal and lateral oblique crashes and was exacerbated by possible contributing factors such as intrusion of the front seatback into the child's occupant space or FFCRS misuse resulting in increased excursion of the child during impact.

Concerns have existed about the likelihood of cervical spine injuries in children restrained in FFCRS. Injuries to the cervical spine of young restrained children in MVCs are not common over the spectrum of crash severities. Pediatric cervical spine injuries, however, result in an increased fatality rate due to the fact they more commonly occur in the upper cervical spine compared to cervical spine injuries in adults which occur more often in the lower cervical spine (Brown et al. 2001; Dietrich et al. 1991; Fuchs et al. 1989; Huelke et al. 1991; Kelleher-Walsh et al. 1993; Kokoska et al. 2001; Myers and Winkelstein 1995; Patrick et al. 2000; Tingvall 1987; Vitale et al. 2006; Weber 2002). An analysis of fatal child cervical spine injuries in MVCs revealed an increased prevalence of females among those with these injuries (Stawicki et al. 2009). Many studies have identified a transition in cervical spine injury location that occurs at ~8 or 9 years of age (Elerkay et al. 2000; Finch and Barned 1998; Fuchs et al. 1989; Patrick et al. 2000; Platzer et al. 2007; Zuckerbraun et al. 2004). Children younger than this age demonstrate injuries of the upper cervical spine including "spinal cord injury without radiographic abnormalities" (SCIWORA), while older children sustain injuries to the lower cervical spine. SCIWORA has been documented to occur in about 15–25% of all pediatric cervical spine injuries (Platzer et al. 2007). These injuries are characterized by transient vertebral displacement of the spinal column with subsequent return to normal alignment resulting in a vertebral column that appears radiologically normal; however, injury has occurred to the spinal cord. This injury pattern occurs in the very young as the immature spinal column can stretch up to 5 cm before skeletal or ligamentous rupture (Kokoska et al. 2001). Odent et al. (1999) emphasized the increased frequency of upper cervical spine injuries in the youngest child occupants by reporting 15 cases of odontoid fractures in children less than 6 years—8 of which were children in FFCRS.

A current debate exists on whether severe cervical spine injuries in children in child restraints can occur in the absence of head contact. Huelke et al. (1991) reviewed the literature of case reports and NASS-CDS data from 1980 to 1989 to identify cervical injury without head contact. He concluded that although this injury is rare, the biomechanical characteristics of the immature spine discussed above predispose young children to a noncontact mechanism. Review of Swedish crash data also suggest that noncontact neck injuries with associated basilar skull fractures are possible

Fig. 2.6 Schematic of the LATCH system

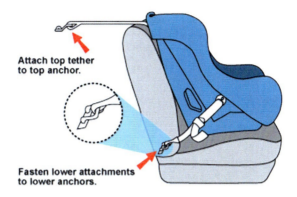

in restrained young child occupants (Jakobssen et al. 2005). In contrast, Australian in-depth case reviews highlight the absence of cases of serious neck injury without head contact to a child in a FFCRS with top tether (Paine et al. 2003). The rare nature of this injury has prevented conclusive resolution to this debate.

Child Restraint Misuse and Novel Attachment Methods to Reduce Misuse

Several studies have highlighted the role misuse plays in how children in child restraints sustain injuries (Arbogast et al. 2004b; Czernowski and Muller 1993; Elliott et al. 2006; Sherwood et al. 2003). In light of this concern, researchers have sought to quantify the frequency and typical modes of CRS misuse (Bulger et al. 2008; Bull et al. 1988; Decina and Knoebel 1997; Decina and Lococo 2005; Eby and Kostyniuk 1999; Margolis et al. 1988, 1992; Morris et al. 2001). Most recently, Decina and Lococo conducted an observational study of over 5,000 children and measured the prevalence of CRS misuse to be 72.6% (Decina and Lococo 2005). In their study, the most common critical misuses were loose harness straps and loose attachment of the CRS to the vehicle by the seat belt.

In response to studies such as these, the method by which a CRS can be attached to the vehicle has been revised in the US. The Lower Anchors and Tethers for Children (LATCH) system was designed to reduce the difficulty associated with installing CRS. This system uses dedicated attachment points in the vehicle rather than using the adult seat belt for child safety seat installation. All vehicles and child restraints manufactured and sold in the United States in September 2002 or later were required to have this anchoring system (NHTSA 1999a). For RFCRS, there are two points of attachment at the base of the child safety seat. These lower anchors buckle into the vehicle at two dedicated attachment points. For FFCRS, a third dedicated attachment point is used for a top tether which is a length of webbing attached on one end near the top of the CRS and on the other end to hardware, such as a ring, bar, or bracket in the vehicle (Fig. 2.6). Most US-based CRS designs incorporate a flexible LATCH lower anchor attachment rather than a rigid lower anchor attachment mechanism (ISOFIX) that is common in Europe and Canada.

Previous research has studied the performance of LATCH (or its rigid lower attachment European counterpart ISOFIX) in laboratory sled test environments (Bilston et al. 2005; Charlton et al. 2004; Sherwood et al. 2004) and documented improved kinematics and reduced ATD injury metrics when compared to the existing seat belt attachment method. Arbogast and Jermakian reviewed cases of CRS attached using the LATCH attachment method and highlighted examples of LATCH misuse; however, to date, no study has evaluated the population-based benefits of this revised attachment system (Arbogast and Jermakian 2007).

At this time, LATCH, however, has not solved the misuse problem. In the first large-scale observation study examining LATCH use and misuse in the United States, data were collected at 66 sites across 7 states in 2005 (Decina and Lococo 2007). The study indicated that many parents purchasing newer vehicles do not update their CRS to take advantage of the available LATCH technology. Approximately one-fifth of the CRS in the vehicles equipped with LATCH did not have tether straps and one-sixth did not have lower attachments. Even when their CRS were LATCH equipped, approximately one-third of the drivers with LATCH-equipped vehicles stated that they could not use LATCH because there were no anchors in their vehicles. Much of the nonuse of lower anchors in this study was related to the fact that the vehicle safety belt was the only method available in the center rear position for installing a CRS. When parents had experience attaching CRS both using the safety belt or lower anchors, three-fourths reported a preference for LATCH, because they found it easier to use, obtained a tighter fit, and felt that the child was more secure.

Booster Seats

It is recommended that children who have outgrown FFCRS (based on the upper weight limit) be restrained in belt-positioning booster seats using the lap and shoulder belts in the rear seat of a vehicle until the vehicle seat belt fits properly—approximately at age 8–12 years. There are two types of belt positioning booster seats, high back and backless or low back. Booster seats raise the child up so that the lap and shoulder belts fit properly. The lap belt should fit low across the child's hips or upper thighs and the shoulder belt should cross the center of the child's shoulder and chest.

Durbin et al. (2003a) published the first real-world study to quantify the benefit of booster seats over seat belts for the young school age child. Using the PCPS dataset, these authors determined that the odds of injury after adjusting for child, crash, driver and vehicle characteristics was 59% lower for 4–7-year-olds in belt positioning booster seats than seat belts. This analysis, conducted on data from 1998 to 2002, was based primarily on children aged 4 and 5 years due to the usage practices during that time period. In the time since then, booster seat use among children 4–8 years of age has seen a threefold increase (CHOP 2008).

As more children, particularly older children, are appropriately restrained in booster seats, the performance of belt-positioning booster seats was revisited (Arbogast et al. 2009a). Arbogast et al. examined a greater percentage of older children; 37% of the study sample using booster seats was 6–8 years of age. After adjusting for potential confounders, children aged 4–8 using belt-positioning booster seats were 45% less likely to sustain AIS2+ injuries than similarly aged children using the vehicle seat belt when considering all crash directions and vehicle model years. Among children restrained in belt positioning booster seats, there was no detectable difference in the risk of injury between the children in backless versus high back boosters.

NHTSA also evaluated the effectiveness of booster seats in preventing injury among 4–8-year-olds, using 1998–2008 data from NASS-CDS and 17 years of combined data from three US States that record the use of booster seats in their reported crash data as a distinct category separate from other types of child safety seats (NHTSA 2010b). NHTSA used a double-pair comparison method, in which each child in a selected vehicle is paired with the adult driver of the vehicle. The risk of injury to a child in a booster seat, relative to the driver, was then compared to the risk of injury to a child in a seat belt, also relative to the driver. The purpose of conducting the analysis this way was to estimate the effect on risk of injury of a single binary factor (in this case booster seats versus adult belts) without having to model the diverse confounding factors or exposure rates that may be affecting injury risk. Instead, the driver of the vehicle is used as a comparison "control" to account for exposure, severity and other confounding factors.

When analyzed collectively the data showed a 14% reduction in overall injuries (from mild to fatal) for children in booster seats relative to children in adult belts. When the analyses were restricted to more severe injuries, sample sizes were insufficient to make reliable inference about the effectiveness of booster seats, though results suggesting benefits of boosters were seen fairly consistently. Unweighted CDS analyses suggested that booster seats were associated with a 45% reduction in $MAIS \geq 2$ injury risk when compared to seat belts for children aged 4–8. This injury reduction estimate was very similar to the results of Arbogast et al. (2009a) noted above, who used a similar definition of injury severity in their analyses.

Rice and colleagues examined the effectiveness of booster seats to reduce the risk of death for children aged 4–8 years, using a matched cohort study of 1996–2006 FARS data (Rice et al. 2009). Estimated death risk ratios for booster seats used with seatbelts, when compared to unrestrained children, were 0.33 (95% CI 0.28–0.40) for children aged 4–5 years and 0.45 (0.31–0.63) for children aged 6–8 years. The estimated risk ratios for seat belt use alone were similar, suggesting that booster seats and seat belts provided similar protection from death in crashes. Attempting to explain the seemingly inconsistent findings that booster seats are more effective than seat belts in reducing the risk of nonfatal injuries, but not fatal injuries, Rice noted that booster seats, which improve seatbelt fit, may not improve collision survivability because the effect of improving seatbelt fit is to lower the probability of injuries to the abdomen and lumbar spine, which can be severe but are much less often fatal than injuries to the head and thorax.

Corden (2005) took a slightly different approach in examining the protection afforded by belt-positioning booster seats by quantifying the decrease in deaths and hospitalizations if all 4–7-year-olds were in booster seats. Using Wisconsin state data from 1998 to 2002, there would be a 57 and 58% reduction in deaths and hospitalizations, respectively, compared to the numbers of deaths and hospitalizations based on restraint practices current at the time.

Of note, in the 1980s and 1990s shield booster seats were common restraints for children of booster seat age. Current recommendations do not advocate the use of shield booster seats in part due to several studies that have highlighted their injury and fatality risks (Edgerton et al. 2004; Whitman et al. 1997). Children in shield booster seats had eight times the risk of sustaining a serious injury than similar age children in FFCRS and four times the risk of sustaining a head injury (Edgerton et al. 2004). These researchers point to the lack of upper torso restraint as a key parameter that leads to suboptimal kinematics, increased head excursion and in several cases, ejection of the child from the shield booster (Whitman et al. 1997).

As stated above, belt-positioning booster seats are designed to better position the occupant such that the vehicle seat belt can provide optimal protection. In practice, this raises the child up and in the case of a high-back belt positioning booster seat moves them forward of the seat bight. The injury patterns in this restraint are related to this positioning. For example, in the Durbin study (2003a), the most common body regions of injury for children in belt positioning booster seats were the head and the face—many of them contact-related injuries. Of note, in this study, children in belt positioning booster seats had no injuries to the abdomen and spine—body regions characterized by the constellation of injuries known as "seat belt syndrome" (SBS) discussed in more detail below.

Review of 108 cases of children restrained in belt-positioning booster seats from the European CREST project shows a pattern of injuries that includes, in decreasing order, lower extremity fractures, AIS3+ injuries to the neck, chest, and abdomen (Lesire et al. 2001). The occurrence of abdominal injuries in children in booster seats in the European literature has been at odds with the US data, which has demonstrated a reduction in abdominal injury risk for booster-seated children (Jermakian et al. 2007b; Trosseille et al. 1997). This may be due, in part, to differences in data collection methodology. The US studies were based on a population-based sample of injured and noninjured children in crashes; however, the European study was based on a convenience sample of children in child restraints who were injured as a result of a crash. These differences in methodology might be responsible for the differences in injury distribution due to potential selection biases associated with convenience samples. Specifically, a convenience sample of injured children does not adequately estimate the appropriate population from which the injured children were drawn—i.e., the number of children in booster seats without abdominal injuries is unknown.

Seat Belts

According to NHTSA, vehicle safety belts are considered to fit correctly when the lap portion of the belt rides low over the hips or thighs and the shoulder portion crosses the sternum and shoulder (Fig. 2.7). Children are usually ready for the adult lap and shoulder belt when they can sit with their backs against the vehicle

Fig. 2.7 Proper positioning of a lap and shoulder belt

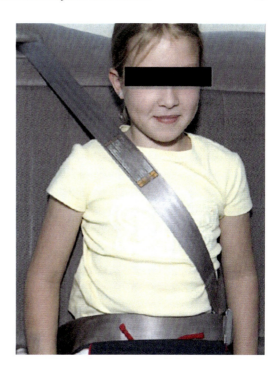

seat back with their knees bent over the vehicle seat edge (typically at ~4′9″) (Klinich et al. 1994).

Using FARS data, NHTSA has evaluated the effectiveness of lap and shoulder belts in the rear rows and found them to be effective in all crash modes for adult and pediatric occupants aged 5 years and older. The estimated fatality reduction was: 77% in rollovers, 42% in side impacts, 29% in frontals, and 31% in rear impacts and other crashes. These findings are applicable to occupants aged 5 years and older and are not specific estimates for children (Morgan 1999).

Two studies have evaluated seat belt effectiveness for children. Chipman et al. (1995) using a database of fatal crashes in Ontario estimated the effectiveness of seat belts for children aged 4–14 years and found that the odds ratio for serious injury or death was 0.60 compared to those without a seat belt. Wisconsin-specific data suggested that 100% seat belt use by children aged 8–15 years would result in 45 fewer deaths and 206 fewer hospitalizations compared to current restraint use rates of 72%. These estimates represent reductions of 45 and 32% for deaths and hospitalizations respectively (Corden 2005).

While seat belts may be recommended for children ages 8 and older, García-Espana and colleagues (2008) have identified an elevated injury risk for these age children compared to their younger counterparts in child restraints. The risk of injury for belted 8–12-year-olds in the rear seat was 1.3% with head and face injuries being most common followed by injuries to the upper extremity and the abdomen. These authors suggest that a systematic approach that includes research, public education, safety regulation, and legislative advocacy needs to be directed toward protection of this age child.

Note that the recommendation for seat belt restraint is a lap and shoulder belt rather than use of a lap belt alone. Lap and shoulder belts have been required in rear outboard positions since 1989. However, it was not until 2005 with a phase-in until 2007 that lap–shoulder belts were required in the center rear seat position. Many manufacturers introduced center rear lap–shoulder belts in advance of this requirement and by model year 2001 most vehicles were equipped as such (Kahane 2004). The benefits associated with this change were evaluated by Arbogast et al. (2004c). These researchers documented that for those children seated in the center rear in seat belts, the presence of a shoulder belt reduced the risk of injury by 81% with the primary benefit seen in reductions in abdominal injury. Parenteau and Viano (2003) previously documented a similar shift in the patterns of injury from lap only belt restraint to lap–shoulder belt. Their study, however, looked at the rear seat as a whole and did not separate out the rear seating positions. Of note, in the García-Espana study, one out of five 8–12-year-olds misused the shoulder portion of the vehicle seat belt suggesting that presence of a shoulder belt does not always lead to proper use (García-Espana and Durbin 2008).

Cases of serious cervical and lumbar spinal cord injury as well as intra-abdominal injuries to children restrained in lap and lap–shoulder belts have been described for many years, resulting in the identification of a so-called seat-belt syndrome (SBS) of injuries to children (Agran et al. 1989, 2007; Garrett and Braunstein 1962; Glassman et al. 1992; Gotschall et al. 1998; Hoy and Cole 1993; Khaewpong et al. 1995; Kulowski and Rost 1956; Lane 1994; Newman and Dalmotas 1993; Sturm et al. 1995; Stylianos and Harris 1990; Tso et al. 1993; Voss et al. 1996). First described by Kulowski and Rost (1956), the term "SBS" was coined by Garrett and Braunstein in 1962 to describe a distinctive pattern of injuries associated with lap belts primarily based on adult crash occupants (Garrett and Braunstein 1962). The first descriptions of SBS in children began appearing in the 1980s as restraint of children became more common with the introduction of rear seat belts (Agran et al. 1989).

The injuries associated with SBS include hip and abdominal contusions (now commonly referred to as the seat belt sign), pelvic (ileal and pubis) fractures, lumbar spine injuries including subluxations and compression fractures of the bodies of L2–L4, and intra-abdominal injuries to both solid organs and hollow viscera. Over the past four decades, a large number of case reports and case series have confirmed a characteristic pattern of injuries, which generally localize to the abdomen and lumbar spine (e.g., Anderson et al. 1991; Arajarvi et al. 1987; Blumenberg 1967; Chance 1948; Chandler et al. 1997; Ciftci et al. 1998; Doersch and Dozier 1968; Hampson et al. 1984; Hendey and Votey 1994; Hingston 1996; Huelke et al. 1995; McCarthy and Lemmon 1996; Moir and Ashcroft 1995; Porter and Zhao 1998; Rogers 1974; Steckler et al. 1969; Sube et al. 1967; Talton et al. 1995; Vandersluis and O'Connor 1987; Vellar et al. 1976; Wagner 1979; Wang et al. 1993; Yarbrough and Hendey 1990).

Intra-abdominal injuries include gastrointestinal tract perforation and small bowel mesenteric tears and perforations (Anderson et al. 1991; Arajarvi et al. 1987;

Blumenberg 1967; Chandler et al. 1997; Ciftci et al. 1998; Doersch and Dozier 1968; Glassman et al. 1992; Gotschall et al. 1998; Hendey and Votey 1994; Hingston 1996; Moir and Ashcroft 1995; Newman and Dalmotas 1993; Porter and Zhao 1998; Sube et al. 1967; Talton et al. 1995; Tso et al. 1993; Vandersluis and O'Connor 1987; Vellar et al. 1976; Wang et al. 1993). These injuries occur more often in children restrained by lap belts but can be sustained by children restrained by lap and shoulder belts as well (Gotschall et al. 1998). Occupants of all ages are at risk of developing SBS, though the poor fit of the belt in younger children likely places them at higher risk than older children. In a large case series of 98 children with SBS treated at Children's National Medical Center in Washington, DC between 1991 and 1997, the mean age of patients was 7.3 (±2.5) years, and 72% of cases were between the ages of 5 and 9 years (Gotschall et al. 1998). Arbogast and colleagues (2004d) further quantified the effect of age on the risk of seat belt related abdominal injuries and highlighted the increased risk for children 4–8 years of age. They were 24.5 times and 2.6 times more likely to sustain an AIS2+ abdominal injury than those 0–3 years and 9–15 years, respectively. Within a specific age group, appropriately restrained children (child restraints or booster seats for those up to 8 years and lap–shoulder belts for those >8 years) were one-third as likely as suboptimally restrained children to suffer an abdominal injury (Nance et al. 2004).

Age influences the abdominal injury pattern as well. Lutz et al. (2003) demonstrated that among restrained children with intra-abdominal injuries, those <8 years of age and restrained by a seat belt were four times more likely to have a hollow visceral than a solid visceral injury when compared with those who were following best practice recommendations for the choice of restraint for their age (in a child restraint or booster seat for those <8 and in a lap–shoulder belt for those 8 years and older).

In addition to abdominal organ injuries, children in seat belts sustain injury to the vessels of the abdominal cavity. Several researchers documented the occurrence of injuries to the abdominal aorta sustained by restrained occupants in MVCs (Anderson et al. 2008; Choit et al. 2006; Dajee et al. 1979; Roth et al. 1997; Swischuk et al. 2007). These injuries often occur in conjunction with bowel injuries and lumbosacral spinal fractures and are due to direct compression between the spine and the seat belt (Randhawa and Menzoian 1990).

In a review of 21 cases of abdominal injury sustained by children restrained in seat belts, researchers identified belt loading directly over the injured organ as the most common mechanism of injury (Arbogast et al. 2007). In these cases, injury occurred in several ways: (1) the child scooting forward prior to the crash on the soft, compressible seat cushion, creating a shallow lap belt angle exacerbated by the forward and downward occupant kinematics of the crash, (2) the child placing the shoulder belt behind his back resulting in a belt geometry that might move the lap belt higher on the abdomen during a crash, and (3) rear seat lap–shoulder systems that have belt geometry that places the lap belt high on the abdomen even when both parts of the seat belt are worn and the child remains seated back against the seat cushion. From these positions, during rapid deceleration, the immaturity of the pediatric pelvis prevents proper anchoring of the lap belt; the belt directly compresses the abdominal contents against the spinal column resulting in mesenteric tears and bowel wall contusions (Leung et al. 1982). Intestinal perforations are likely caused by a sudden increase in intra-luminal pressure, combined with compression of a short segment of bowel by the belt (Tso et al. 1993). Several studies demonstrated that the presence of an abdominal wall contusion significantly increased the likelihood of an intra-abdominal injury including bowel perforation and the need for operative intervention (Chandler et al. 1997; Lutz et al. 2004).

Two types of lumbar spine injuries—compression fractures and Chance fractures—have been described as part of SBS (e.g., Chance 1948; Durbin et al. 2001a; Garrett and Braunstein 1962; Gotschall et al. 1998; Hampson et al. 1984; Hingston 1996; Hoy and Cole 1993; Huelke et al. 1995; Khaewpong et al. 1995; Kulowski and Rost 1956; Moir and Ashcroft 1995; Mulpuri et al. 2007; Newman and Dalmotas 1993; Rogers 1974; Steckler et al. 1969; Sturm et al. 1995; Stylianos and Harris 1990; Subotic et al. 2007; Swischuk et al. 2007; Voss et al. 1996; Yarbrough and Hendey 1990). The mechanisms of these injuries are described in more detail elsewhere in this book.

In addition to the classic abdominal and lumbar spine injuries, young children in seat belts are at increased risk of head and face injuries. These injuries occur more frequently when a child uses a poorly fitting shoulder belt, or the shoulder belt is misused (Arbogast et al. 2002b; Winston et al. 2000). In the extreme, children may place the shoulder belt behind the back or under the arm for comfort, but even young children wearing the shoulder portion may not have adequate torso restraint (Arbogast et al. 2007; Gotschall et al. 1998).

Other researchers have documented additional patterns of injury that, while rare, can occur in children restrained in seat belts. These include diaphragmatic injuries (Shehata and Shabaan 2006), upper cervical spine fractures (Deutsch and Badawy 2008), and iliac wing fractures (Emery 2002). Biomechanical characteristics of the child result in children being at increased risk of these rare injuries; this is discussed in more detail in subsequent chapters of this book.

Summary of Restraint Effectiveness Data for Child Occupants

The majority of the restraint effectiveness research for child occupants has been focused on frontal crashes. Table 2.3 summarizes both the effectiveness for reducing fatalities and that for reducing serious injuries for each restraint best practice recommendation for children as discussed in previous sections. Of note, there is limited research that examines similar analyses for other crash directions; the few studies that exist are summarized later in this chapter.

In order to provide guidance for further optimization of child occupant protection, Table 2.4 provides the distribution of injured body regions stratified by restraint type. In all cases, injuries to the head are the most common injury sustained by children. For children in FFCRS, injuries to the lower extremity, face, and upper extremity are also important. The distribution of body regions of injury between 4- and 7-year-olds in booster seats and 8–15-year-olds in seat belts—both age-appropriate

Table 2.3 Summary of restraint effectiveness data

Restraint	Effectiveness for reducing fatalities	Comparison group	Effectiveness at reducing serious injuries	Comparison group
RFCRS	71% reduction (Hertz 1996)	Unrestrained occupants up to 1 year of age	44% reduction (Henary et al. 2007)	Children aged 0–23 months in FFCRS
			90% reduction (Jakobsson et al. 2005)	Unrestrained occupants up to 4 years of age
FFCRS	71% reduction for correctly used seats (Kahane 1986)	Unrestrained occupants 0–4 years of age	72% reduction (Winston et al. 2000)	Seat belt restrained 2–5-year-olds
	54% reduction (Hertz 1996)	Unrestrained occupants 1–4 years of age	71% reduction (Arbogast et al. 2004b)	Seat belt restrained 1–4-year-olds
	28% reduction (Elliott et al. 2006)	Seat belt restrained 2–6-year-olds	82% reduction (Zaloshnja et al. 2007)	Seat belt restrained 2–3-year-olds
	67% reduction (Rice and Anderson 2009)	Unrestrained children aged 3 and under		
BPB			59% reduction (Durbin et al. 2003a)	Seat belt restrained 4–7-year-olds
			45% reduction (Arbogast et al. 2009a)	Seat belt restrained 4–8-year-olds
			45% reduction in MAIS ≥ 2 injuries (NHTSA 2010)	Seat belt restrained 4–8-year-olds
	67% reduction for children aged 4–5 years (Rice and Anderson 2009)	Unrestrained 4–5-year-olds		
	55% reduction for children aged 6–8 years (Rice and Anderson 2009)	Unrestrained 6–8-year-olds		
Seat belts	29% reduction in frontal impact crashes (Morgan 1999)	Unrestrained occupants aged 5 and older		

Table 2.4 Patterns of AIS2+ injury stratified by restraint type

	RFCRS, 0–11-month-olds	FFCRS, 12–47-month-olds	Belt-positioning booster seats, 4–7-year-olds	Seat belt (lap and lap–shoulder), 4–7-year-olds	Seat belt (lap and lap–shoulder), 8–15-year-olds
Overall AIS2+ injury risk (per 1,000 children in crashes)	2.3	3.0	4.9	16.6	13.6
Head (%)	83.3	56.9	61.1	67.3	62.5
Face (%)	0.0	8.3	7.0	5.8	6.5
Chest (%)	2.4	2.8	5.7	1.3	5.9
Abdomen (%)	2.4	3.3	8.9	17.8	7.1
Neck/spine (%)	0.0	1.7	1.3	0.7	1.6
Upper extremity (%)	7.1	8.3	7.0	4.5	11.0
Lower extremity (%)	4.8	18.8	8.9	2.5	5.4

Data from PCPS from 12/1/98–11/30/07. Limited to model year 1998 and newer vehicles. Differences between restraint types should not be interpreted as statistically significant differences

restraint conditions—is very similar, although the overall injury risk in booster seats is substantially lower. For children between 4 and 7 years old in seat belts, the presence of SBS injuries to the abdomen is highlighted.

Other Factors Associated with Child Occupant Fatality and Injury Risk

Air Bags

Beginning in the 1990s, a portion of the deaths and injuries to children in MVCs were attributed to exposure to deploying passenger air bags. In November 1995, the Morbidity and Mortality Weekly Report issued by the US Centers for Disease Control and Prevention described eight deaths of child occupants involving air-bag deployment that were of special concern because they involved low-speed crashes in which the children otherwise might have survived (CDC 1995). The risk to small occupants from a deploying air bag had been a concern for the automotive industry for several years (Mertz 1988; Kent et al. 2005). As passenger air bags diffused into the market, numerous case reports began appearing in the medical literature describing brain and skull injuries sustained by children in RFCRS and brain and cervical spine injuries sustained by older children often unrestrained or restrained inappropriately in seat belts for their age (CDC 1996; Giguere et al. 1998; Hollands et al. 1996; Huff et al. 1998; Marshall et al. 1998; Willis et al. 1996).

Several researchers reviewed case series of children exposed to deploying passenger air bags to elucidate the mechanisms of injury (Augenstein et al. 1997; Huelke 1997; Kleinberger and Summers 1997; McKay and Jolly 1999; Quinones-Hinojosa et al. 2005; Shkrum et al. 2002; Winston and Reed 1996). For children killed in a RFCRS, the air bag typically deployed into the rear surface of the child restraint often fracturing the plastic shell of the restraint near the child's head causing fatal skull and brain injuries. Older children who were either unrestrained or inappropriately restrained in seat belts for their age were placed in proximity to the deploying air bag due to preimpact braking. In one typical scenario, upon deployment, the air bag causes the neck to go into combined tension and hyperextension loading resulting in a spectrum of injuries to the brain and cervical spine. These include atlanto-occipital fracture, brain stem injuries and diffuse axonal injury of the brain. The largest case series was from NHTSA's Special Crash Investigation program and is summarized in Winston and Reed (1996) and Kleinberger and Summers (1997). Case series of other injuries to child occupants associated with air bag deployment continue to appear in the literature including injuries to the eye (Ball and Bouchard 2001) and upper extremity (Arbogast et al. 2003a).

As a response to growing knowledge of the adverse effects of passenger air bags to child occupants, researchers began quantifying the population-based risks of air bag exposure and the benefits of rear seating. Early evaluations focused on the first generation air bag designs—defined as those designs in vehicles of model year earlier than 1998—and their effect on injuries and fatalities for children (Braver et al. 1997, 1998; Cummings et al. 2002; Kahane 1996; Smith and Cummings 2006. The presence of an air bag uniformly increased the risk to both restrained and unrestrained child occupants. A summary of the findings from the different studies, their age group of focus, and groups of comparison is contained in Table 2.5.

Kuppa et al. evaluated the influence of the air bag on the effectiveness of rear seating using a double-pair comparison study of FARS frontal crash data (Kuppa et al. 2005). Two pairs were analyzed: the first group consists of fatal crashes where a driver and front outboard seat passenger are present and at least one of them was killed; the second group consists of fatal crashes where a driver and a rear outboard seat passenger are present and at least one of them was killed. This analysis considered those vehicles with and without a passenger air bag separately. For restrained children 5 years of age or less, the presence of a passenger air bag increased the benefit, in terms of reduced fatalities, associated with rear seating. For restrained child occupants >8 years of age, the rear seat was still associated with a lower risk of death than the front, but its benefit was less in vehicles with a passenger air bag than in vehicles without. Specifically, the presence of a passenger air bag reduces the ratio of front row risk to rear row risk compared to vehicles without a passenger air bag.

Other researchers studied the impact of passenger air bags on injury risk as well. Based on the PCPS dataset, Durbin and colleagues determined the relative risk of nonfatal injuries to restrained children aged 3–15 years exposed to passenger air bags in frontal impacts compared to those in the front seat in similar crashes with no passenger air bag deployment and reported a 100% increase in injury risk. Exposure to passenger air bags increased the risk of both minor injuries, including facial and chest abrasions, and moderate and more serious injuries, particularly head injuries and upper extremity fractures (Durbin et al. 2003b). Using NASS-CDS data from 1995 to 2002, Newgard et al. reported a trend towards increased risk of serious injury from air bag deployment for children 0–14 years of age compared to those in the front seat with no air bag deployment, although these findings were not statistically significant (Newgard and Lewis 2005).

As described earlier in this chapter, as a result of the many injuries and fatalities to children from deploying passenger air bags, NHTSA revised Federal Motor Vehicle Safety Standard 208, in a way which, in many cases, resulted in the redesign of frontal air bags to reduce the force with which they deploy. These new air bags are generally referred to as "second-generation air bags."

Several studies examined the effect of these design changes on child occupants in real-world crashes. Although the findings were nonsignificant, second-generation air bags reduced the risk of death among right front-seated children 6–12 years of age by 29% compared to no air bag (Olson et al. 2006). For children <6 years of age, both types of air bags increased the risk of death compared to no air bag; however, the increased risk of death associated with air bag deployment was less for second-generation air bags (10%) compared to first generation air bags (66%) adjusted for important crash and occupant parameters including the restraint status of the child and crash direction. Arbogast et al. (2003b, 2005a) quantified the risk of serious

Table 2.5 Summary of studies examining the role of the passenger air bag (PAB) and seat row in determining the risk of death and serious injury to child occupants

Study	Data source	Target group studied	Comparison group	Primary finding
Braver et al. (1997)	FARS 1992–1995	Children <10 years seated in the right front seat in vehicles with a PAB	Children <10 years seated in the right front seat in vehicles with only a driver AB	PAB increased the risk of death by 34%
Braver et al. (1998)	FARS 1988–1995	Children <13 years of age seated in the rear row	Children <13 years of age seated in the front row	In vehicles with a PAB, rear row seating reduced the risk of death by 46%; in vehicles without a PAB, by 35%
Cummings et al. (2002)	FARS 1992–1998 and NASS-GES 1992–1998	Children <13 years of age seated in the right front seat in vehicles in FARS	Children <13 years of age seated in the right front seat in vehicles in NASS-GES	PAB increased the risk of death by 5% for restrained occupants and 37% for unrestrained occupants
Smith and Cummings (2006)	FARS 1990–2001	Children <13 years of age, restrained and seated in the rear row in vehicles with a PAB	Children <13 years of age, restrained and seated in the right front seat in vehicles with a PAB	Rear row seating reduced the risk of death by 38%
Kuppa et al. (2005)	FARS 1993–2003	Children aged 0–12 years seated in the rear outboard seat position	Children aged 0–12 seated in the right front seat position	For restrained occupants in vehicles with a PAB, rear seating reduced the risk of death by 65, 35, and 20% for 0–5, 6–8 and 9–12-year-olds
Durbin et al. (2003b)	PCPS 1998–2001	Restrained children aged 3–15 years exposed to deploying PAB	Restrained children aged 3–15 years seated in the right front seat in vehicles in which there was only a driver AB and it deployed	Twofold increase in risk of AIS2+ injury with PAB exposure
Olson et al. (2006)	FARS 1990–2002	Right front seated children 6–12 years of age in vehicles with a second generation PAB	Right front seated children 6–12 years of age in vehicles with no PAB	Second generation PAB reduced the risk of death by 29% (nonsignificant finding)

(continued)

Table 2.5 (continued)

Study	Data source	Target group studied	Comparison group	Primary finding
Arbogast et al. (2003b), Arbogast (2005a)	PCPS 1998–2002	Restrained children aged 3–15 years exposed to deploying second-generation PAB	Restrained children aged 3–15 years exposed to deploying first-generation PAB	Second generation PAB reduced the risk of AIS2+ injury by 41%
Arbogast et al. (2009b)	PCPS 1998–2007, vehicle model year 1998+	Children aged 0–15 years seated in the front row of vehicles	Children aged 0–15 years seated in the rear row(s) of vehicles	Rear seating reduced the risk of injury by 64% for 0–8-year-olds and 31% for 9–12-year-olds

nonfatal injuries in frontal crashes among belted children in the front seat of vehicles in which second generation passenger air bags deployed, compared with that of belted children in the front seat of vehicles in which first-generation passenger air bags deployed. Serious injuries were reported in 14.9% in the first-generation group versus 9.9% in the second-generation group resulting in a 41% reduction in injury risk. Children in the second-generation group sustained fewer head injuries, including concussions and more serious brain injuries, than in the first-generation group.

Seating Position

In addition to the regulatory changes resulting in the redesign of air bags, a collaborative national response was directed towards encouraging rear seating for those children <13 years of age. In general, this effort has been effective. Data from the FARS from 1990 to 1998 indicated that the proportion of vehicles carrying children in the front declined from 42 to 31% over the 9-year period. By 2007, 95% of infants, 98% of children aged 1–3, and 88% of children aged 4–7 rode in the rear seat (NHTSA 2008). As the use of child restraints, particularly booster seats, by children aged 4–8 years has increased, the proportion of these children sitting in the front seat has declined from 18% in 1999 to 4% in 2007. However, in 2007 approximately one-third of 8- to 12-year-olds were still riding in the front seat (CHOP 2008). Factors associated with child front-seating include a single child occupant alone with the driver, older child age, male or nonparent driver, or lack of a passenger air bag (Agran et al. 1998; Durbin et al. 2004; Ramsey et al. 2000; Wittenberg et al. 1999).

Research on seating position has shown the rear seat to be protective for children regardless of restraint use. Several studies using the FARS dataset, described above in the section on air bags, documented a fatality reduction of 20–65% when the child occupant was seated in the rear rows rather than the front seat, depending on the age of the child and presence of a PAB (Braver et al. 1998; Kuppa et al. 2005; Smith and Cummings 2006). These benefits have been extended to serious injury as well with studies documenting a 40–60% greater risk of injury for children in the front seat compared with children in the rear, after adjusting for crash severity, depending on the restraint status of the child (Durbin et al. 2005; Lennon et al. 2008). Using Utah state-specific data, Berg et al. (2000) reported that children aged 0–14 years were 70% more likely to sustain a serious or fatal injury in the front seat compared to the rear row. An analysis of the CIREN dataset demonstrated that front seat child occupants sustained more severe injuries as measured by an ISS >16 than those seated in the rear rows (Erlich et al. 2006). A recent analysis of PCPS data examined the relative risk of front versus rear row seating in vehicles of model year 1998 and newer and found that rear seating reduced the risk of AIS2+ injury by 64% for those 0–8 years of age and 31% for those 9–12 years of age (Arbogast et al. 2009b). These authors further stratified their data by model year (1998–2002 and 2003+) and demonstrated a crude rear row injury risk that was lower than that of the

Table 2.6 Patterns of AIS2+ injury stratified by seat row

	Front row		Rear row(s)	
	Children 0–7 years in CRS	Children 8–15 years in seat belts	Children 0–7 years in CRS	Children 8–15 years in seat belts
Overall AIS2+ injury risk (per 1,000 children in crashes)	7.5	16.0	3.3	12.1
Head (%)	43.8	57.1	60.0	66.8
Face (%)	12.5	6.4	5.5	6.5
Chest (%)	6.3	7.7	4.2	4.4
Abdomen (%)	12.5	6.9	5.3	7.3
Neck/spine (%)	0.0	1.5	1.5	1.6
Upper extremity (%)	12.5	14.7	9.2	8.2
Lower extremity (%)	12.5	5.6	14.4	5.3

Data from PCPS from 12/1/98-11/30/07. Limited to model year 1998 and newer vehicles. Differences between seat rows should not be interpreted as statistically significant differences

front row for all model year, child age combinations. Durbin and colleagues (2005) examined the combined effect of rear seating and appropriate restraint and found a synergistic effect of the two parameters to provide the best protection for children in crashes. Of interest, the benefits of rear seating for child occupants appeared to extend to side impacts as well with children in the rear 62% less likely to sustain an injury (Durbin et al. 2001b).

Table 2.6 provides the body region specific injury risks stratified by seat row and age/restraint. For all children regardless of seat row or restraint, head injuries are the most common. For front seated children in CRS (including booster seats), other body regions of importance are the face, abdomen, and extremities. It is important to note that few children in CRS are seated in the front row so these body region distributions are based on small numbers. For CRS-restrained children in the rear row(s), except for the extremities, injuries to body regions other than the head are not common. For front seated children in seat belts, injuries are sustained by all body regions except for the neck and spine to which injuries are relatively rare. The body region distribution for seat belt restrained children is very similar between front seated children and rear seated children; however, the overall injury risk is elevated in the front row.

The center rear seat has been generally considered the safest rear seating location for children in child restraints. Lund (2005) used data from the NASS-GES system from 1992 to 2000 to evaluate the effect of seating position on the risk of injury for children in child restraints. They reported children seated in the center rear seat were not at a lower injury risk than children seated at either of the outboard rear seats. In contrast, Kallan et al. (2008) used PCPS data from 1998 to 2006 to evaluate this issue and demonstrated that child occupants restrained in child restraints and seated in the center had an injury risk 43% less than children in child restraints seated in either of the rear outboard positions. These contrasting findings are likely due to how injuries were defined in the two studies. Lund defined injury as any police-reported injury, which includes those of a relatively minor nature. The threshold for injury is higher in Kallan's analysis, resulting in more serious injuries, such as injuries involving internal organs and fractures of the extremity. This more stringent definition of injury facilitates differentiation between the protection afforded by the outboard seat positions and that of the center seat.

Crash Direction

Researchers have described variation in injury and fatality risk by crash direction. For historical data on this topic, the reader is referred to Agran et al. (1989), Henderson et al. (1994), Krafft et al. (1989), Kelleher-Walsh et al. (1993), Langwieder and Hummel (1989), Langwieder et al. (1996) and Tingvall 1987 as examples. A brief review of more recent data follows to set the context for a detailed review of injury causation scenarios and mechanisms by crash direction.

Similar to adult occupants, frontal crashes are the most common crashes experienced by child occupants. As a result, much of the research and restraint evaluation described earlier in this chapter has been focused on this impact direction. However, frontal impacts, side impacts, and rollovers account for similar numbers of child fatalities, and rollovers and side impacts have a greater case fatality rate (Starnes and Eigen 2002). In 2009, there were 164 fatalities for children aged 0–7 years in frontal crashes, compared to 101 in side impacts, and 165 in rollovers with only 50 in rear impacts. For rear-seated children aged 0–7 years, review of FARS data from 1996 to 2005 revealed that the crash mode with the highest fatality risk was rollover crashes (1.37%), followed by right-side (0.47%) and left-side impacts (0.34%) (Viano and Parenteau 2008a) (Fig. 2.8). In general, similar crash direction trends are found for serious injuries with slight variations among the studies due to different years and age groups of study (Erlich et al. 2006; Parenteau and Viano 2003; Viano and Parenteau 2008b).

Side Impacts

Side impact protection for child occupants has received much recent attention attributed to the high fatality and injury rate in this crash direction. Much of this research has focused on quantifying the relative risk among seating positions in side impact crashes. Howard et al. (2004) reported that the risk of fatality for children seated in the near-side position was twice as high as that for children in the center, while the center-seat position had a 40% higher fatality risk than the far-side position. Of interest, near-side crashes on the right side have twice as high fatality rate as near side crashes on the left side. Viano et al. attributed this difference to the dynamics of right side crashes, which often occur as the case vehicle is making a left turn across traffic and likely result in side crashes of increased severity (Viano and Parenteau 2008a).

Howard et al. (2004) extended this work to serious injuries and using the NASS-CDS dataset from 1995 to 2000, quantified that for restrained children 0–12 years

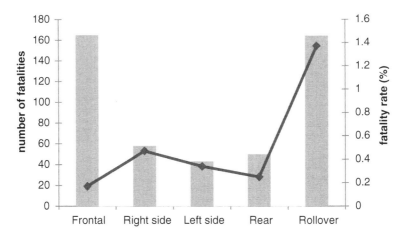

Fig. 2.8 Number of fatalities and fatality risk by crash direction for children aged 0–7. Number of fatalities from FARS 2009 and fatality risk from Viano and Parenteau (2008a)

of age, severe injury as measured by ISS > 16 was greater for children seated nearside (7 per 1,000 children) as compared to those seated in the center (2 per 1,000) or on the far side (1 per 1,000). Similar trends have been delineated for children in child restraints. Injury risk for children in FFCRS nearside to the crash was significantly higher (8.9 injured children per 1,000 crashes) than for children seated on the nonstruck side of the crash (2.1 injured children per 1,000 crashes) (Arbogast et al. 2004a).

Researchers quantified the relative effectiveness of belt-positioning booster seats as compared to seat belts in reducing the risk of injury among 4–8-year-olds in side impact crashes. Children in belt-positioning booster seats were at a 58% reduction in risk of injury than those in seat belts in side impact crashes. Benefits were obtained by reduction of injuries to the head and face as well as the pattern of injuries to the abdomen and spine known as SBS (Arbogast et al. 2005d). Using a broader, more recent sample from the same dataset, Arbogast et al. (2009a) documented that children in side impact crashes benefited the most from booster seats, showing a reduction in injury risk of 68% for near-side impacts and 82% for far-side impacts. In this study, among children restrained in belt positioning booster seats, no difference in side impact effectiveness could be detected between the children in backless versus high back boosters.

Maltese et al. (2005) reviewed cases of seat belt restrained child occupants 4–15 years of age in side impact crashes to explore the role additional occupants play on injury causation. These researchers identified that children who share the rear row with other occupants have a 58% reduced risk of injury in side impact crashes than occupants who sit alone on the rear row.

The limited diffusion of side air bags into the rear seat has prevented many studies of their effectiveness for child occupants. In a study of child occupants from the

PCPS study, Arbogast and Kallan (2007) measured a 10.6% AIS2+ injury rate of children exposed to a deploying side air bag. All injuries were of the AIS2 level and limited to concussions or fractures of the upper extremity. This study did not specifically compare the injury risk of those exposed to a deploying side air bag to those not exposed due to difficulties in identifying nondeployment crashes of similar magnitude to those crashes with deployment.

Many studies have reviewed in depth crash investigations of children restrained in child restraints in side-impact crashes to understand the mechanisms of injury. European studies from the early 1990s reviewed crash data and identified direct contact of the head with the intruding structure or B-pillar as the primary cause of injury or the bullet vehicle (Kamren et al. 1991; Langwieder and Hummel 1989; Langwieder et al. 1996). These researchers identified that most side-impact crashes were forward of direct lateral.

More recent studies confirmed these early findings. Sherwood et al. (2003) studied 14 fatal side-impact crashes involving children in child restraints, and of the six cases with sufficient injury information, head trauma was the cause of fatality. In all fatal side-impact crashes there was intrusion at the child's seating position. In a review of 32 cases of children restrained in FFCRS in 30 side-impact crashes, the most common injuries sustained were to the face, head, and lower extremity (Arbogast et al. 2005b). The absence of chest injuries in this study, a common injury sustained by adults in side impact crashes, and neck/back/spine injuries is of interest. The European-based CREST project reviewed 168 restrained children involved in severe side-impacts and confirmed the head as the most severely injured body region in 62% of the cases. Cervical spine injuries in children in child restraints in side-impact crashes were rare, but when they occurred, often lead to fatality (Lesire et al. 2001). McCray et al. analyzed NASS-CDS cases of 28 children aged 1–3 years in side-impact crashes and described a pattern of injuries that included in decreasing order of frequency: head, torso (including the thorax and the abdomen) and neck. The most common involved physical component associated with these injuries was vehicle interior structures—in particular head contact with the upper door in the area of the windowsill and the intruding seat back. Of interest, the thoracic injuries for children in CRS were primarily lung contusions rather than skeletal rib fractures typically seen in adults in side-impact crashes (McCray et al. 2007).

For children in FFCRS in side-impact crashes, key characteristics that were related to injury were intrusion that entered the child's occupant space or caused an interior part of the vehicle to enter the child's occupant space, forward component of the crash, and the rotation of the CRS, restrained by a seat belt, towards the side of the impact (Arbogast et al. 2005b). One recent study of NASS-CDS data confirmed the role of the longitudinal crash forces in these side-impacts and reported that the most common principal directions of force (PDOF) were 10 and 2 o'clock (McCray et al. 2007).

Several researchers have extended their review of children in side-impact crashes to children restrained in seat belts. In a case series of children treated at a level 1 trauma center that was part of the CIREN network, Orzechowski et al.

(2003) found that seat belt restrained child occupants were 2.8 and 4.8 times as likely to receive an AIS3+ head and chest injury, respectively, in side impacts as compared to frontal impacts. The lower seated height of children increases the risk of head and upper chest contact with the interior door panel or pillars in side-impact crashes. Howard et al. (2004) reported on a two-trauma center database of 28 children 0–12 years of age involved in side impacts and delineated between near-side and far-side injury patterns. Children seated near-side to the crash sustained severe head, trunk, and limb injuries which occurred both in the presence and absence of intrusion. In their study, center-seat and far-side occupants had severe injuries only when unrestrained. Lesire et al. reviewed European data from the CREST crash database, confirmed the significance of head injuries, and documented severe injuries to the chest and abdomen primarily when the child was restrained by a booster seat or was using the adult seat belt (Lesire et al. 2001). The increased frequency of serious head and thoracic injuries to belt restrained children in side impact crashes was reinforced in a review of the CIREN database by Brown et al. (2006). Maltese et al. (2007) in a review of 24 cases of children aged 4–15 restrained in seat belts in side impact crashes found that the majority of the head and face contact points were due to contacts with both interior vehicle structures and structures on the crash partners and were located horizontally within the rear half of the window and vertically from the window sill to the center of the window.

A common injury to adults in side-impact crashes is fracture of the pelvis. In pediatrics, these injuries occur infrequently due to the cartilaginous linkage of the pelvic bones and the increased elasticity of the symphysis pubis and sacroiliac joints. As a result, it has been hypothesized that it takes more force to cause a pediatric pelvic fracture compared to adults (Silber and Flynn 2002). These biomechanical characteristics were evident in a review of pelvic fractures of children in side-impact crashes (Arbogast et al. 2002c). In this study, prepubescent children experienced isolated pubic rami fractures, whereas postpubescent children experienced the more adult-like multiple fractures of the pelvic ring. This distinct injury pattern is directly related to the ossification during puberty of the cartilage connecting the three bones of the pelvis. This will be discussed in more detail in a subsequent chapter.

Other Directions of Impact

Rollover and ejection. Very little research has examined real-world crash data in other crash modes beyond frontal and side impact crashes. Rollover injury and fatality risk in children has received limited attention. Rivara and colleagues (2003) examined the NASS-CDS and FARS datasets from 1993 to 1998 to evaluate the risk associated with rollovers for child occupants 0–15 years of age. They reported about 10% of all children involved in crashes are in a rollover crash with an 11 times increased likelihood of being in a rollover crash in an SUV versus a passenger car. The risk of fatality and injury for children in rollovers was approximately twice that of nonrollover crashes. Using NASS-GES and FARS from a similar time period, other researchers estimated 2.2% of crash-involved children ages 0–12 years were in rollover crashes; however, these crashes represented 28% of the child fatalities. NASS-GES represents a broader dataset than NASS-CDS, which has the requirement of the vehicle being towed as an inclusion criterion; thus the lower estimate of rollover incidence (Howard et al. 2003). Data analyses from the PCPS dataset identified that rollover crashes increased the risk of injury to child occupants 0–15 years of age over three times compared to non-rollover crashes (Daly et al. 2006). More recently, Viano and Parenteau studied NASS and FARS data and found that rollovers accounted for ~60% of the fatalities of children seated in the second row of SUVs. They highlighted the lower belt use and higher likelihood of rollover in SUVs as primary causes of this high percentage (Viano and Parenteau 2008a). Howard et al. (2003) pointed to the role of ejection as a key component of the injury causation scenario for children injured in rollover crashes. Ejection from cargo areas of pickup trucks has also been highlighted as an issue particular to pediatrics (German et al. 2007; Woodward and Bolte 1990).

Rear impacts. Recent interest in child occupant protection in rear-impact crashes has focused on the likelihood of injury to children seated in the rear rows from deforming front seat backs. Jermakian et al. (2008) examined a population-based sample of restrained child occupants, 0–12 years of age, seated in a second row outboard position in rear impact tow-away crashes. For those children with seatback deformation occurring directly in front of them, there was a doubling of the injury risk (4.8 versus 2.1%). The nature of this dataset, however, does not allow the determination of injury causation scenarios. To this end, Viano and Parenteau (2008b) identified 19 cases of children 0–7 years of age involved in rear impact crashes with AIS3+ injuries from the NASS-CDS database from 1997 to 2005. Two-thirds of these cases experienced substantial intrusion into their seating location that caused them to move towards the front seats. The authors identified contact with the front seat back as the source of injury in 10 of the 19 cases; however, they specified that only one of the 19 was injured due to rearward rotation of the front seatback toward the child.

Summary of Injured Body Regions by Direction of Impact

In order to provide guidance for further optimization of child occupant protection, Table 2.7 provides the body region specific injury risks stratified by crash direction and age/restraint. The head is the most common body region injured regardless of crash direction or age/restraint. This is particularly true for near side and rear crashes. The importance of injuries to the extremities in children in CRS is highlighted in frontal and side crashes.

Table 2.7 Patterns of AIS2+ injury stratified by crash direction

	Frontal crashes		Near side crashes		Far side crashes		Rear crashes	
	Children 0–7 years in CRS	Children 8–15 years in seat belts	Children 0–7 years in CRS	Children 8–15 years in seat belts	Children 0–7 years in CRS	Children 8–15 years in seat belts	Children 0–7 years in CRS	Children 8–15 years in seat belts
Overall AIS2+ injury risk (per 1,000 children in crashes)	3.0	12.7	6.7	29.3	3.6	14.6	2.9	8.8
Head (%)	55.7	49.3	60.5	69.6	46.8	59.3	73.0	82.1
Face (%)	8.2	10.4	1.2	4.6	3.2	6.7	5.8	1.0
Chest (%)	4.4	9.0	4.9	4.6	4.8	5.9	3.6	1.8
Abdomen (%)	4.4	8.3	1.2	6.3	3.2	6.7	8.8	6.3
Neck/spine (%)	2.2	1.5	2.5	1.5	1.6	2.2	0.0	1.0
Upper extremity (%)	7.1	14.4	8.6	8.9	27.4	14.4	2.9	3.8
Lower extremity (%)	18.0	7.0	21.0	4.6	12.9	4.8	5.8	4.0

Data from PCPS from 12/1/98-11/30/07. Limited to model year 1998 and newer vehicles. Differences between crash directions should not be interpreted as statistically significant differences

Pedestrians

In 2009, 244 children under 15 years of age died as a pedestrian in the United States, representing one-fifth of the traffic fatalities to this age group. This number represents a 49% decrease in the number of pedestrian fatalities for children since 2000 which is in part attributed to increased attention to improved playing and walking environments for urban children. Incidence rates of pedestrian fatality peak in the young school age child—approximately 6–9 years of age (Dimaggio and Durkin 2002; Miller et al. 2004). Dissection of these numbers reveals a burden dominated by the urban environment with distinct seasonality and time of day effects (Dimaggio and Durkin 2002; Posner et al. 2002, NHTSA 2010a).

The global burden of child pedestrian fatalities is large. Approximately 760,000 pedestrians die each year worldwide with children making up 35% of those fatalities (World Bank 2001). In low and middle income countries, the percentage of pediatric traffic fatalities that are pedestrians outweighs those that are motor vehicle occupants (WHO 2008).

The fatality rates of pedestrians have been documented to vary with vehicle type. NHTSA examined their FARS database from 1997 to 2001 and identified that sport utility vehicles, pickup trucks and large vans have a higher pedestrian fatality rate than passenger cars. These differences were particularly amplified for those <8 years of age in part related to sight lines from these large vehicle types to the younger children (Starnes and Longthorne 2003). This effect has been confirmed by several other researchers (DiMaggio et al. 2006; Henary et al. 2003; Roudsari et al. 2004, 2005; Starnes and Longthorne 2003; Yao et al. 2007).

Child pedestrians have a lower mortality rate than adults and sustain a pattern of injuries that results in a lower ISS and shorter lengths of stay than either young adults or elderly pedestrians (Demetriades et al. 2004; Henary et al. 2003; Peng and Bongard 1999). Understanding the body regions of pedestrian injury requires an understanding of the kinematics of a pedestrian-vehicle impact. Initial impact with the bumper is followed by vehicular hood or windshield impact and then impact with the ground. Injuries associated with each of these three impacts are directly related to pedestrian stature (van Rooij et al. 2004) and thus the injury patterns vary across the pediatric age range. For the smallest children, the hood impact does not necessarily occur and they may be thrown forward or knocked down by the impacting vehicle (Roudsari et al. 2005).

Of note, half of pediatric pedestrian crashes are due to the child darting out between vehicles (Fildes et al. 2004)—rather than at an intersection. As a result, 23% of child pedestrian-vehicle impacts were to the right or left side of the vehicle rather than the front illustrating the importance of considering pedestrian protection when designing the side planes of the vehicle.

Overall for children, at both an AIS2+ and AIS3+ level, the most frequently injured regions in all pediatric age groups were the head/face and the lower extremities (Fig. 2.9). Approximately 15% of child pedestrians with an AIS3+ injury had an injury to the head or face. This percentage was fairly consistent across the pediatric age range. In contrast, the incidence of thoracic injury went down with child age,

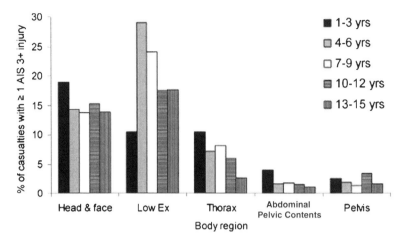

Fig. 2.9 Percentage distribution of pedestrian casualties with at least one AIS3+ injury by body region and age group. The percentages of casualties with ≥1 AIS3+ injury to the upper extremities, spine (lumber or thoracic), or neck (including cervical spine) did not exceed 0.7% in any of the five age groups (reproduced with permission from Ivarsson et al. 2006)

Table 2.8 Top ten priorities for child pedestrian injury mitigation and the associated harm per Fildes et al. (2004)

Top ten priorities for child pedestrian injury mitigation	Harm associated with child pedestrian injury (%)	Harm associated with adult pedestrian injury (%)
Head to hood	16	8
Head to ground	11	8
Leg above the knee to bumper	9	b
Head to front area of vehicle[a]	9	b
Leg below the knee to bumper	8	12
Head to A-pillar	6	7
Pelvis to front area of vehicle[a]	5	4
Head to windshield	4	21
Leg above the knee to front area of vehicle[a]	3	2
Chest to hood	3	4

[a] Front area refers to the grill, lights, and leading edge of the hood
[b] Not in the top ten priorities for adult pedestrian injury mitigation

while lower extremity injury increased with child age; again pointing to the influence of stature in the injury patterns. Of note, when lower extremity injuries occurred, 85% of the injuries were to the femur and most lower extremity fractures are midshaft (Woods et al. 2003).

Fildes et al. (2004) in a review of Australian and German fatal child pedestrian data calculated the HARM or total societal cost for pedestrian events. In this analysis, they identified the top ten body region–vehicle/environment structure impacts that deserve countermeasure development for child pedestrians (Table 2.8).

These areas of emphasis are similar to those highlighted by Roudsari et al. (2005). The harm associated with adult pedestrian injury for these pediatric priorities are also shown in the table for comparison.

Conclusions and Recommendations

While children represent only about 10–15% of the overall traffic fatality burden in the US, MVC remain the leading cause of death and disability for children and young adults and represent close to half of all unintentional injury deaths to children and adolescents. Prevention of the fatalities and disabling injuries associated with MVC is thus a priority for ensuring our children's overall health.

Since the middle 1990s, the safety community has achieved moderate success in reducing this burden. In 2009, 41% fewer children aged 0–14 years died as a result of MVCs than in 1996. These reductions have been achieved, in part, through a combination of increased attention to age-appropriate restraint use and rear seating. These successes should not lead to complacency, however. In 2009, for example, 45% of 4–8-year-olds were not restrained according to best practice recommendations for their age, and one-third of 8–12-year-olds were still seated in the front row.

Even with increases in age-appropriate restraint use and rear seating, the protection of children in motor vehicles must be further optimized. Achieving optimal protection will require knowledge of the unique biomechanical needs of children and youth described in the remaining chapters of this book. Future priorities in child occupant protection must focus on the following issues:

1. *Monitoring trends in child occupant protection through rigorous, child-focused crash surveillance.* Priorities will continue to evolve as child restraint design, restraint use practices, the vehicle fleet, vehicle safety technologies, legislative priorities and regulatory focus continue to evolve. Accurate, timely data are crucial for setting evidence-based priorities. Current crash databases have neither sufficient depth nor breadth of child-specific data to measure important trends, determine effectiveness of occupant protection systems, and identify emerging safety hazards for child occupants.
2. *Optimize the rear seating environment for children and youth.* Children who are too large for current add-on child restraints have an increased injury and fatality risk compared to their younger counterparts. The design and evaluation of a rear row restraint and seating environment that is intrinsic to the vehicle for these children deserves attention.
3. *Develop protection principles for child occupants in side impacts crashes.* Continue to explore the mechanism and causation of injuries sustained by children and youth in side impacts and increase understanding of the timing and nature of a child's interaction with vehicle and restraint components during a typical side impact crash.

4. *Focus epidemiological and injury causation analyses on injuries to the pediatric head and cervical spine.* Head injuries are the most common serious injury sustained by children in MVCs across most age, restraint, and crash direction strata. The exact burden of cervical spine injuries remains unclear in part due to inadequacies of current data collection systems highlighted above. Determination of the prevalence and nature of these injuries, the circumstances under which they occur and the importance of head contact in the causation scenario needs further study.
5. *Assess and evaluate the real-world impact on children and youth of new vehicle safety technologies.* These should include side-impact air bags, advanced frontal air bags, advanced vehicle belt systems such as pretensioners and load limiters, LATCH systems for child restraints, and advanced crash avoidance technologies.
6. *Develop appropriate traffic safety surveillance and intervention programs in the developing world with a particular focus on pedestrians.* Child pedestrian safety outweighs the burden of child occupant safety in many low and middle income countries.

While the safety of children in motor vehicles has improved since the introduction of the first child restraint law in 1978, further advances in child occupant protection will require rigorous collaborative research by epidemiologists and engineers that thoughtfully considers the unique needs of children. Evidence resulting from this research will provide an intellectually sound underpinning to the eventual introduction and adoption of appropriate public policies, improvements in vehicle and restraint design, and adaptation of consumer attitudes and behavior to further increase the protection of children.

Appendix-Data Sources and Methodology

NHTSA Field Crash Databases

To help the traffic safety community identify traffic safety problems, develop and implement vehicle and driver countermeasures, and evaluate motor vehicle safety standards and highway safety initiatives, the National Highway Traffic Safety Administration (NHTSA) developed a variety of field crash databases. Each has a specific purpose and role in the identification of vehicle safety issues and the assessment of countermeasures with the optimal cost/benefit. Many of the analyses described elsewhere in this chapter use these databases as their source of data and thus this description provides context for those studies. In addition, these databases represent resources for researchers to explore topics of motor crash injury for child occupants.

Fatality Analysis Reporting System

The Fatality Analysis Reporting System (FARS) was developed by NHTSA's National Center for Statistics and Analysis (NCSA) in 1975 to assist the traffic safety community in identifying traffic safety problems and evaluating both motor vehicle safety standards and highway safety initiatives. FARS contains data on all fatal traffic crashes within the 50 states, the District of Columbia, and Puerto Rico. To be included in FARS, a crash must involve a motor vehicle traveling on a traffic way customarily open to the public, and result in the death of a person (either an occupant of a vehicle or a nonmotorist such as a pedestrian) within 30 days of the crash event.

FARS case files contain descriptions of each fatal crash reported. Each case has more than 100 coded data elements that characterize the crash, the vehicles, and the people involved. All data elements are reported on the following forms:

- The Accident Form collects information such as the time and location of the crash, the first harmful event, and the numbers of vehicles and people involved.
- The Vehicle and Driver Forms record data on each crash-involved vehicle and driver. Data include the vehicle type, initial and principal impact points, most harmful event, and drivers' license status.
- The Person Form contains data on each person involved in the crash, including age, gender, role in the crash (driver, passenger, nonmotorist), injury severity, and restraint use.

Data are available for every year since FARS was established in 1975 and can be easily queried using a Web-based query system (http://www-fars.nhtsa.dot.gov/Main/index.aspx). To protect individual privacy, no personal information, such as names, addresses, or specific crash locations, is coded. It is important to realize that analyses using the FARS dataset are based on fatal crashes, which may differ in meaningful ways from the general spectrum of crashes on the roads.

National Automotive Sampling System

Established in 1979, the National Automotive Sampling System (NASS) was created as part of a nationwide effort to reduce MVCs, injuries, and deaths on our nation's highways. It is composed of two systems—the Crashworthiness Data System (CDS) and the General Estimates System (GES). Both systems are based on cases selected from a sample of police crash reports. CDS data focus on passenger vehicle crashes, and are used to investigate injury mechanisms to identify potential improvements in vehicle design. GES data focus on the broader overall crash picture, and are used for assessments of the magnitude of a specific traffic safety problem and tracking trends. NASS-CDS and NASS-GES select cases from police crash reports [also known as Police Accident Reports (PARs)] at police agencies within randomly selected areas of the country. Data from GES and CDS are weighted in such a manner that statistical analyses using these weighted data can result in national estimates of the traffic safety topic being studied.

For NASS-CDS, field research teams located at Primary Sampling Units (PSU's) across the country study ~5,000 crashes involving passenger cars, light trucks, vans, and utility vehicles each year. About 200 of these crashes involve a child occupant. Trained crash investigators obtain data from crash sites, studying evidence such as skid marks, fluid spills, broken glass, and bent guard rails. They locate the vehicles involved, photograph them, document the crash damage using sophisticated measuring procedures, and identify interior locations struck by the occupants. These researchers follow up on their on-site investigations by interviewing crash occupants and reviewing medical records to determine the nature and severity of injuries. The data collected by the PSU's are quality controlled by one of two NASS Zone Centers which have the responsibility for coordinating and supervising the activities of the field offices, keeping field offices informed regarding changes in functional and administrative procedures, sharing ideas and concepts throughout the system regarding new techniques, procedures, and components found on vehicles and updating field offices regarding changes in system hardware and software.

NASS-GES data comes from a nationally representative sample of police reported MVCs of all types, from minor to fatal. For a crash to be eligible for the GES sample it must involve at least one motor vehicle traveling on a traffic way, the result must be property damage, injury, or death and a PAR must be completed. These accident reports are chosen from 60 areas that reflect the geography, roadway mileage, population, and traffic density of the US GES data collectors make weekly visits to ~400 police jurisdictions in the 60 areas across the United States, where they randomly sample about 50,000 PARs each year. About 10,000 of these crashes involve a child occupant. The data collectors obtain copies of the PARs and send them to a central contractor for coding. No other data are collected beyond the selected PARs. Trained data entry personnel interpret and code ~90 data elements data directly from the PARs into an electronic data file. An annual publication, Traffic Safety Facts, is produced with GES data for nonfatal crashes, combined with information on fatal crashes from FARS. NASS-GES provides broad estimates of child crash exposures but with little data specific to children, especially on important parameters such as injury and restraint use. NASS datasets can be downloaded for analyses from NHTSA's Web site (http://www.nhtsa.gov/NCSA).

Special Crash Investigations

Since 1972, NCSA's Special Crash Investigations (SCI) Program has provided NHTSA with an in depth and detailed level of crash investigation data. The data collected ranges from data maintained in routine police and insurance crash reports to data from special reports by professional crash investigation teams. Hundreds of data elements relevant to the vehicle, occupants, injury mechanisms, roadway, and safety systems involved are collected for each of the more than 200 crashes designated for study annually.

SCI cases are intended to serve as an anecdotal dataset useful for examining special crash circumstances or outcomes from an engineering perspective. To this end, the inclusion criteria are changed routinely to address emerging traffic safety needs. The benefit of this program lies in its ability to locate unique real-world crashes anywhere in the country, and conduct in depth clinical investigations in a timely manner, which can be used by the automotive safety community to improve the performance of its state-of-the-art safety systems. Individual and select groups of cases have triggered both individual companies and the industry as a whole to improve the safety performance of motor vehicles, including passenger cars, light trucks, or school buses.

The SCI program's flexibility allows for detailed investigation of any newly emerging technologies related to automotive safety. A number of incidents involving alternative fuel vehicles, passenger side air bag deployments, vehicle-to-pedestrian impacts, and child restraints have been investigated. A focus of the SCI program in the 1990s was investigation of cases of serious injuries and fatalities to children exposed to deploying air bags. Summary tables of these cases as well as any other cases the SCI program has investigated are available on NHTSA's Web site (http://www-nass.nhtsa.dot.gov/BIN/logon.exe/airmislogon), as are copies of completed SCI reports.

Crash Injury Research and Engineering Network

In order to maximize the integration of crash data collected by engineers and crash reconstructionists with detailed injury and radiological information collected by clinical teams at hospitals, NHTSA has funded hospital-related studies since the 1980s. In 1991, NHTSA initiated the Highway Traffic Injuries Studies. Over the next several years, research projects were funded at four Level One Trauma Centers to collect detailed injury information on motor vehicle occupants involved in crashes. In 1996, the Crash Injury Research and Engineering Network (CIREN) was developed in response to the need for a uniform centralized data system when three additional Level I Trauma Hospitals were added to the network as part of a settlement agreement with General Motors.

CIREN is a sponsor-led multicenter research program involving a collaboration of clinicians and engineers in academia, industry, and government pursuing in-depth studies of crashes, injuries, and treatments to improve processes and outcomes. Its mission is to improve the prevention, treatment, and rehabilitation of MVC injuries to reduce deaths, disabilities, and human and economic costs.

There are not any CIREN centers focused specifically on children; however, all sites enroll children who present to their trauma centers. These child occupants are screened using criteria pertaining to specific vehicle and crash characteristics as well as the injury and restraint status of the child. Once a subject meets the criteria for enrollment and consents to participation in the study, an in-depth investigation is initiated to collect detailed crash and injury information. Traumatology experts review the occupant's radiology and clinical data for the location and type of the injury. Crash investigators investigate the crash vehicles and the crash scene to determine the severity of the impact and the physical evidence of occupant's interaction within the crash environment. Mechanical and biomechanical engineers experienced in the field of crash testing and biomechanics research evaluate each case to determine the role of the vehicle's design and the level of interaction with the occupant. Together, these multidisciplinary teams of engineers and clinicians review the cases to assess injury causation scenarios and injury mechanisms. These data are publicly available and can be accessed by http://nhtsa-nrdapps.nhtsa.dot.gov/bin/cirenfilter.dll.

Partners for Child Passenger Safety

In 1998, researchers at Children's Hospital of Philadelphia and State Farm Insurance Companies responded to a need for a crash surveillance system focused exclusively on children and created the Partners for Child Passenger Safety (PCPS) Study. Based on many of the design specifications of the NHTSA databases, PCPS data have been used to conduct many of the child-focused analyses that are discussed in this chapter. In place from 1998 to 2007, PCPS consisted of a large-scale, child-specific crash surveillance system: insurance claims from State Farm functioned as the source of subjects, with telephone survey and on-site crash investigations serving as the primary sources of data. Durbin et al. (2001c) described the study methods in detail.

Briefly, passenger vehicles qualifying for inclusion were State Farm-insured, model year 1990 or newer, and involved in a crash with ≥1 child occupant <16 years of age. Qualifying crashes were limited to those that occurred in 16 states and the District of Columbia, representing three large regions of the United States. A stratified cluster sample was designed to select passenger vehicles (the unit of sampling) for the conduct of a telephone survey with the driver. Probability sampling was based on two criteria: whether the vehicle was towed from the scene and the level of medical treatment received by the child passenger(s). For a subset of crashes of specific interest, in-depth crash investigations were conducted using a similar methodology to the SCI program.

International Data Sources

Several countries outside the United States have championed in-depth crash investigation programs to understand child occupant injuries. Examples of these programs are featured in several studies highlighted elsewhere in this chapter. A partial list of these programs includes the following:

- Child Restraint System in Cars (CREST) and Child Led Injury Design (CHILD)—European collaborative research projects (http://www.childincarsafety.org)
  - The CREST project (1999–2000) and the CHILD project (2002–2006) were European Union funded research efforts focused on child safety that involved a consortium of European industry, government and academic entities. As part of these projects, over 800 in-depth crash investigations involving child motor vehicle occupants were conducted. A subset of these in-depth investigations was reconstructed as full-scale vehicle to vehicle crash tests. A third project in this series, Child Advanced Safety Project for European Roads (CASPER) was started in 2009.
- Cooperative Crash Injury Study (CCIS)—United Kingdom (http://www.lboro.ac.uk/research/esri/vehicle-road-safety/projects/ccis.htm)
  - Based at Loughborough University, the CCIS study started in 1983 and investigates more than 1,200 crashes annually. The study selects a sample of cases involving newer vehicles which are representative of crashes occurring in the UK. Both adult and pediatric motor vehicle occupants are studied.
- German In Depth Accident Study (GIDAS)—Germany (http://www.gidas.org/en/home)
  - Officially started in 1999, GIDAS collects crash and medical data through in-depth investigation from ~2,000 cases annually in two cities in Germany—Hanover and Dresden. Cases are chosen based on a statistical sampling plan from crashes reported to police, rescue services and fire departments in the targeted areas. Both adult and pediatric motor vehicle occupants are studied.
- International Road Traffic and Accident Database (IRTAD) (http://www.irtad.net)
  - IRTAD, started in 1988, is a database that houses traffic fatality data from many countries throughout the world. It is maintained by the International Traffic Forum and the Organization for Economic Co-operation and Development. At present the following countries are included: Australia, Austria, Belgium, Canada, the Czech Republic, Denmark, Finland, France, Germany, Greece, Iceland, Ireland, Italy, Japan, Korea (South Korea), Luxembourg, the Netherlands, New Zealand, Norway, Poland, Portugal, Sweden, Switzerland, Spain, Slovakia, Slovenia, Turkey, Hungary, the United States, the UK. It serves as an important resource in comparing road safety metrics between various developed countries.
- Volvo's statistical crash database—Sweden—(Jakobssen et al. 2005)
  - Volvo has compiled a crash database from crashes involving their vehicles identified through a Swedish Insurance company. Vehicle damage information is sent to Volvo's in-depth crash investigation team and injury data is gathered from medical records. Several publications have focused on the child occupants that are in this privately held database.

Each of these research efforts has varied objectives and data collection methods, which make it difficult to merge the data for analysis. They do however represent rich sources of detailed data on child occupant injury.

References

Agran P, Winn D, Dunkle D (1989) Injuries among 4 to 9 year old restrained motor vehicle occupants by seat location and crash impact site. Am J Dis Child 143:1317–1321

Agran PF, Anderson CL, Winn DG (1998) Factors associated with restraint use of children in fatal crashes. Pediatrics 102:E39

Agran P, Dunkle D, Winn D (2007) Injuries to a sample of seat belted children evaluated and treated in a hospital emergency room. J Trauma 27:58–64

Anderson P, Rivara F, Maier R et al (1991) The epidemiology of seatbelt-associated injuries. J Trauma 31:60–67

Anderson SA, Day M, Chen MK et al (2008) Traumatic aortic injuries in the pediatric population. J Pediatr Surg 43:1077–1081

Arajarvi E, Santavirta S, Tolonen J (1987) Abdominal injuries sustained in severe traffic accidents by seat belt wearers. J Trauma 27:393–397

Arbogast KB, Jermakian JS (2007) Field use patterns and performance of child restraints secured by lower anchors and tethers for children (LATCH). Accid Anal Prev 39:530–535

Arbogast KB, Kallan MJ (2007) The exposure of children to deploying side airbags: an initial field assessment. Annu Proc Assoc Adv Automot Med 51:245–259

Arbogast KB, Cornejo RA, Kallan MJ et al (2002a) Injuries to children in forward facing child restraints. Annu Proc Assoc Adv Automot Med 46:213–230

Arbogast KB, Durbin DR, Kallan MJ et al (2002b) The role of restraint and seat position in pediatric facial fractures. J Trauma 52:693–698

Arbogast KB, Mari-Gowda S, Kallan MJ et al (2002c) Pediatric pelvic fractures in side impact collisions. Stapp Car Crash J 46:285–296

Arbogast KB, DeNardo MB, Xavier AM et al (2003a) Upper extremity fractures in restrained children exposed to passenger airbags. SAE World Congress and Exhibition

Arbogast KB, Durbin DR, Kallan MJ et al (2003b) Effect of vehicle type on the performance of second generation airbags for child occupants. Annu Proc Assoc Adv Automot Med 47:85–99

Arbogast KB, Chen I, Durbin DR et al (2004a) Injury risks for children in child restraint systems in side impact crashes. Conference of the International Research Council on Biomechanics of Injury (IRCOBI)

Arbogast KB, Durbin DR, Cornejo RA et al (2004b) An evaluation of the effectiveness of forward facing child restraint systems. Accid Anal Prev 36:585–589

Arbogast KB, Durbin DR, Kallan MJ et al (2004c) Evaluation of pediatric use patterns and performance of lap shoulder belt systems in the center rear. Annu Proc Assoc Adv Automot Med 48:57–72

Arbogast KB, Mong DA, Marigowda S et al (2004d) Evaluating pediatric abdominal injuries. Stapp Car Crash J 48:479–494

Arbogast KB, Durbin DR, Kallan MJ et al (2005a) Injury risk to restrained children exposed to deployed first- and second-generation airbags in frontal crashes. Arch Pediatr Adolesc Med 159:342–346

Arbogast KB, Ghati Y, Menon RA et al (2005b) Field investigation of child restraints in side impact crashes. Traffic Inj Prev 6:351–360

Arbogast KB, Jermakian JS, Ghati Y et al (2005c) Patterns and predictors of pediatric head injury. Conference of the International Research Council on Biomechanics of Injury (IRCOBI)

Arbogast KB, Kallan MJ, Durbin DR (2005d) Effectiveness of high back and backless belt-positioning booster seats in side impact crashes. Annu Proc Assoc Adv Automot Med 49:201–213

Arbogast KB, Kent RW, Menon RA et al (2007) Mechanisms of abdominal organ injury in seat belt-restrained children. J Trauma 62:1473–1480

Arbogast KB, Jermakian JS, Kallan MJ et al (2009a) Effectiveness of belt positioning booster seats: an updated assessment. Pediatrics 124:1281–1286

Arbogast KB, Kallan MJ, Durbin DR (2009b) Front versus rear seat injury risk for child passengers: evaluation of newer model year vehicles. Traffic Inj Prev 10:297–301

Augenstein J, Perdeck E, Williamson J et al (1997) Air bag induced injury mechanisms for infants in rear facing child restraints. Proceedings of the 41st Stapp Car Crash Conference

Ball DC, Bouchard CS (2001) Ocular morbidity associated with air bag deployment: a report of seven cases and a review of the literature. Cornea 20:159–163

Bennett TD, Kaufman R, Schiff M et al (2006) Crash analysis of lower extremity injuries in children restrained in forward-facing car seats during front and rear impacts. J Trauma 61:592–597

Berg M, Cook L, Corneli H et al (2000) Effect of seating position and restraint use on injuries to children in motor crashes. Pediatrics 105:831–835

Bilston L, Brown J, Kelly P (2005) Improved protection for children in forward-facing restraints during side impacts. Traffic Inj Prev 6:135–146

Blumenberg R (1967) The seat belt syndrome: sigmoid colon perforation. Ann Surg 165: 637–639

Bohman K, Arbogast KB, Bostom O (2011) Head injury causation scenarios for belted, rear-seated children in frontal impact. Traffic Inj Prev 12:62–70

Braver ER, Ferguson SA, Greene MA et al (1997) Reductions in deaths in frontal crashes among right front passengers in vehicles equipped with passenger airbags. JAMA 278:1437–1439

Braver E, Whitfield R, Ferguson S (1998) Seating positions and children's risk of dying in motor vehicle crashes. Inj Prev 4:181–187

Brown RL, Brunn MA, Garcia VF (2001) Cervical spine injuries in children: a review of 103 patients treated consecutively at a level I pediatric trauma center. J Pediatr Surg 36: 1107–1114

Brown JK, Jing Y, Wang S et al (2006) Patterns of severe injury in pediatric car crash victims: Crash Injury Research Engineering Network database. J Pediatr Surg 41:363–367

Bulger EM, Kaufman R, Mock C (2008) Childhood crash injury patterns associated with restraint misuse: implications for field triage. Prehosp Disaster Med 23:9–15

Bull MJ, Stroup KB et al (1988) Misuse of car safety seats. Pediatrics 81:98–101

Centers for Disease Control and Prevention (1996) Update: fatal air bag related injuries to children – United States 1993-1996. MMWR 45:1073–1076

Centers for Disease Control and Prevention (1995) Air-bag-associated fatal injuries to infants and children riding in front passenger seats – United States. MMWR 44:845–847

Centers for Disease Control and Prevention National Center for Injury Prevention and Control Web-based Injury Statistics Query and Reporting System. http://www.cdc.gov/injury/wisqars. Accessed 2010

Chance GQ (1948) Note on a type of flexion fracture of the spine. Br J Radiol 21:452

Chandler CF, Lane JS, Waxman KS (1997) Seatbelt sign following blunt trauma is associated with increased incidence of abdominal injury. Am Surg 63:885–888

Charlton J, Fildes B, Laemmle R et al (2004) A preliminary evaluation of child restraints and anchorage systems for an Australian car. Annu Proc Assoc Adv Automot Med 48:73–86

Children's Hospital of Philadelphia (2008) PCPS Fact and Trend Report

Chipman ML, Liu J, Hu X (1995) The effectiveness of safety belts in preventing fatalities and major injuries among school-aged children. Annu Proc Assoc Adv Automot Med 39:133–145

Choit RL, Tredwell SJ, Leblanc JG et al (2006) Abdominal aortic injuries associated with Chance fractures in pediatric patients. J Pediatr Surg 41:1184–1190

Ciftci AO, Tanyel FC, Salman AB et al (1998) Perforation due to blunt abdominal trauma. Pediatr Surg Int 13:259–264

Corden T (2005) Analysis of booster seat and seat belt use: How many Wisconsin childhood deaths and hospitalizations could have been prevented 1998-2002? Wis Med J 104:42–45

Cummings P, Koepsell T, Rivara F et al (2002) Air bags and passenger fatality according to passenger age and restraint use. Epidemiology 13:525–532

Czernowski W, Muller M (1993) Misuse mode and effects analysis—an approach to predict and quantify misuse of child restraint systems. Accid Anal Prev 25:323–333

Dajee H, Richardson IW, Iype MO (1979) Seat belt aorta: acute dissection and thrombosis of the abdominal aorta. Surgery 85:263–267

Daly L, Kallan MJ, Arbogast KB et al (2006) Risk of injury to child passengers in sport utility vehicles. Pediatrics 117:9–14
Decina LE, Knoebel KY (1997) Child safety seat misuse patterns in four states. Accid Anal Prev 29:125–132
Decina LE, Lococo KH (2005) Child restraint system use and misuse in six states. Accid Anal Prev 37:583–590
Decina LE, Lococo KH (2007) Observed LATCH use and misuse characteristics of child restraint systems in seven states. J Safety Res 38:273–281
Demetriades D, Murray J, Martin M et al (2004) Pedestrians injured by automobiles: relationship of age to injury type and severity. J Am Coll Surg 199:382–387
Deutsch RJ, Badawy MK (2008) Pediatric cervical spine fracture caused by an adult 3-point seatbelt. Pediatr Emerg Care 24:105–108
Dietrich AM, Ginn-Pease ME, Bartkowski HM et al (1991) Pediatric cervical spine fractures: predominantly subtle presentation. J Pediatr Surg 26:995–1000
DiMaggio C, Durkin M (2002) Child pedestrian injury in an urban setting: descriptive epidemiology. Acad Emerg Med 9:54–62
DiMaggio C, Durkin M, Richardson LD (2006) The association of light trucks and vans with paediatric pedestrians deaths. Int J Inj Contr Saf Promot 13:95–99
Doersch K, Dozier W (1968) The seat belt syndrome: the seat belt sign, intestinal and mesenteric injuries. Am J Surg 116:831–833
Durbin DR (2011) American Academy of Pediatrics Committee on Injury, Violence, and Poison Prevention. Policy Statement – Child Passenger Safety. Pediatrics 127(4):788–793
Durbin DR, Arbogast KB, Moll EK (2001a) Seat belt syndrome in children: a case report and review of the literature. Pediatr Emerg Care 17:474–477
Durbin DR, Bhatia E, Holmes J et al (2001c) Partners for child passenger safety: a unique child-specific crash surveillance system. Accid Anal Prev 33:407–412
Durbin DR, Elliott MR, Arbogast KB (2001b) The effect of seating position on risk of injury for children in side impact collisions. Annu Proc Assoc Adv Automot Med 45:61–72
Durbin DR, Elliott MR, Winston FK (2003a) Belt-positioning booster seats and reduction in risk of injury among children in vehicle crashes. JAMA 289:2835–2840
Durbin DR, Kallan MJ, Elliott MR et al (2003b) Risk of injury to restrained children from passenger airbags. Traffic Inj Prev 4:58–63
Durbin DR, Chen I, Elliott M et al (2004) Factors associated with front row seating of children in motor vehicle crashes. Epidemiology 15:345–349
Durbin DR, Chen IG, Smith R (2005) Effects of seating position and appropriate restraint use on the risk of injury to children in motor vehicle crashes. Pediatrics 115:e305–e309
Eby DW, Kostyniuk LP (1999) A statewide analysis of child safety seat use and misuse in Michigan. Accid Anal Prev 31:555–566
Edgerton EA, Orzechowski KA, Eichelberger MR (2004) Not all child safety seats are created equal: the potential dangers of shield booster seats. Pediatrics 113:e153–e158
Elerkay MA, Theodore N, Adams M et al (2000) Pediatric cervical spine injuries: report of 102 cases and review of the literature. J Neurosurg 92:12–17
Elliott MR, Kallan MJ et al (2006) Effectiveness of child safety seats vs seat belts in reducing risk for death in children in passenger vehicle crashes. Arch Pediatr Adolesc Med 160:617–621
Emery KH (2002) Lap belt iliac wing fracture: a predictor of bowel injury in children. Pediatr Radiol 32:892–895
Erlich PF, Bown JK, Sochor MR et al (2006) Factors influencing pediatric Injury Severity Score and Glasgow Coma Scale in pediatric automobile crashes: results from the Crash Injury Research Engineering Network. J Pediatr Surg 41:1854–1858
Farmer P, Howard A, Rothman L et al (2009) Booster seat laws and child fatalities: a case control study. Inj Prev 15:348–350
Fildes B, Gabler H, Otte D et al (2004) Pedestrian impact priorities using real-world crash data and harm. Conference of the International Research Council on Biomechanics of Injury (IRCOBI)

Finch GD, Barned MJ (1998) Major cervical spine injuries in children and adolescents. J Pediatr Orthop 18:811–814

Fuchs S, Barthel M et al (1989) Cervical spine fractures sustained by young children in forward facing car seats. Pediatrics 84:348–354

García-Espana JF, Durbin DR (2008) Injuries to belted older children in motor vehicle crashes. Accid Anal Prev 40:2024–2028

Garrett JW, Braunstein PW (1962) The seat belt syndrome. J Trauma 2:220–238

Gennarelli TA (1986) Mechanisms and pathophysiology of cerebral concussion. J Head Trauma Rehabil 1:23–29

Gennarelli TA (1993) Mechanisms of brain injury. J Emerg Med 11:5–11

German JW, Klugh A III, Skirboll SL (2007) Cargo areas of pickup trucks: an avoidable mechanisms for neurological injuries in children. J Neurosurg 106:368–371

Giguere JF, St-Vil D, Turmel A et al (1998) Airbags and children: a spectrum of C-spine injuries. J Pediatr Surg 33:811–816

Glassman S, Johnson J, Holt R (1992) Seatbelt injuries in children. J Trauma 33:882–886

Gotschall C, Better A, Bulas D et al (1998) Injuries to children restrained in 2- and 3-point belts. Annu Proc Assoc Adv Automot Med 42:29–44

Hampson S, Coombs R, Hemingway A (1984) Case reports: fractures of the upper thoracic spine: an addition to the "seatbelt" syndrome. Br J Radiol 57:1033–1034

Henary B, Crandall J, Bhalla K et al (2003) Child and adult pedestrian impact: the influence of vehicle type on injury severity. Annu Proc Assoc Adv Automot Med 47:105–126

Henary B, Sherwood CP, Crandall JR et al (2007) Car seats for children: rear facing for best protection. Inj Prev 13:398–402

Henderson M, Brown J, Paine M (1994) Injuries to restrained children. Annu Proc Assoc Adv Automot Med 38:75–87

Hendey G, Votey S (1994) Injuries in restrained motor vehicle accident victims. Ann Emerg Med 24:77–84

Hertz E (1996) Revised estimates of child restraint effectiveness. NHTSA Research Note. www-nrd.nhtsa.dot.gov/Pubs/96855.pdf. Accessed 10 July 2012

Hingston GR (1996) Lap seat belt injuries. NZ Med J 109:301–302

Hollands CM, Winston FK, Stafford PW et al (1996) Severe head injury caused by air bag deployment. J Trauma 41:920–922

Howard A, McKeag AM, Rothman L et al (2003) Ejections of young children in motor vehicle crashes. J Trauma 55:126–129

Howard A, Rothman L, McKeag A et al (2004) Children in side-impact motor vehicle crashes: seating positions and injury mechanisms. J Trauma 56:1276–1285

Hoy GA, Cole WG (1993) The paediatric cervical seat belt syndrome. Injury 24:297–299

Huelke DF (1997) Children as front seat passengers exposed to air bag deployments. Proceedings of the 41st Stapp Car Crash conference

Huelke D, Mackay G et al (1991) A review of cervical fractures and fracture-dislocations without head impacts sustained by restrained occupants. Accid Anal Prev 25:731–743

Huelke DF, Mackay GM, Morris A (1995) Vertebral column injuries and lap-shoulder belts. J Trauma 38:547–556

Huff GF, Bagwell SP, Bachman D (1998) Airbag injuries in infants and children: a case report and review of the literature. Pediatrics 102:e2

Hummel T, Langwieder K et al (1997) Injury risks, misuse rates and the effect of misuse depending on the kind of child restraint system. Proceedings of the 41st Stapp Car Crash conference

Isaksson-Hellman I, Jakobosson L, Gustafsson C et al (1997) Trends and effects of child restraint systems based on Volvo's Swedish accident database. Proceedings of the 41st Stapp Car Crash conference

Ivarsson BJ, Crandall JR, Okamoto M (2006) Influence of age-related stature on the frequency of body region injury and overall injury severity in child pedestrian casualties. Traffic Inj Prev 7:290–298

Jakobsson L, Isaksson-Hellman I, Lundell B (2005) Safety for the growing child: experiences from Swedish accident data. 19th International technical conference on the enhanced safety of vehicles (ESV), Washington, DC

Jermakian JS, Locey CM, Haughey LJ et al (2007a) Lower extremity injuries in children seated in forward facing child restraint systems. Traffic Inj Prev 8:171–179

Jermakian JS, Kallan MJ, Arbogast KB (2007b) Abdominal injury risk for children in belt-positioning booster seats. 20th International technical conference on the enhanced safety of vehicles (ESV), Lyon, France

Jermakian JS, Arbogast KB, Durbin DR (2008) Injury risk for rear impacts: role of the front seat occupant. Ann Adv Automot Med 52:109–116

Kahane CJ (1986) An evaluation of child passenger safety: the effectiveness and benefits of safety seats. NHTSA DOT HS 806 890

Kahane C (1996) Fatality reduction by air bags: analysis of accident data through early 1996. NHTSA DOT HS 808 470

Kahane CJ (2004) Lives saved by the FMVSS and other vehicle safety technologies, 1960-2002-Passenger cars and light trucks – With a review of 19 FMVSS and their effectiveness in reducing fatalities, injuries and crashes. NHTSA DOT HS 809 833

Kallan MJ, Durbin DR, Arbogast KB (2008) Seating patterns and corresponding risk of injury among 0- to 3-year-old children in child safety seats. Pediatrics 21:e1342–e1347

Kamren B, Kullgren A, Lie A et al (1991) Side protection and child restraints—accident data and laboratory test including new test methods. 13th International technical conference on experimental safety vehicles (ESV), Paris, France

Kelleher-Walsh B, Walsh M et al (1993) Trauma to children in forward-facing car seats. Presented at the child occupant protection symposium, Warrendale, PA, August 1993

Kent RW, Viano DC, Crandall JR (2005) The field performance of frontal air bags: a review of the literature. Traffic Inj Prev 6:1–23

Khaewpong N, Nguyen T, Bents F et al (1995) Injury severity of restrained children in motor vehicle crashes. Proceedings of the 39th Stapp Car Crash conference

Kleinberger M, Summers L (1997) Mechanisms of injuries for adults and children resulting from airbag interaction. Annu Proc Assoc Adv Automot Med 41:405–420

Klinich KD, Pritz HB, Beebe MS et al (1994) Survey of older children in automotive child restraints. Proceedings of the 38th Stapp Car Crash conference

Kokoska ER, Keller MS, Rallo MC et al (2001) Characteristics of pediatric cervical spine injuries. J Pediatr Surg 36:100–105

Krafft M, Nygren C, Tingvall C (1989) Rear seat occupant protection: a study of children and adults in the rear seat of cars in relation to restraint use and car characteristics. 12th International technical conference on experimental safety vehicles (ESV), Gothenberg, Sweden

Kulowski K, Rost W (1956) Intra-abdominal injury from safety belts in auto accidents. Arch Surg 73:970–971

Kuppa S, Saunders J, Fessahaie O (2005) Rear seat occupant protection in frontal crashes. 19th International technical conference on the enhanced safety of vehicles (ESV), Washington, DC

Lane J (1994) The seat belt syndrome in children. Accid Anal Prev 26:813–820

Langwieder K, Hummel TH (1989) Children in cars: the injury risk and the influence of child protection systems. 12th International technical conference on experimental safety vehicles (ESV), Gothenberg, Sweden

Langwieder K, Hell W, Willson H (1996) Performance of child restraint systems in real-life lateral collisions. Proceedings of the 40th Stapp Car Crash conference

Lennon A, Siskind V, Haworth N (2008) Rear seat safer: seating position, restraint use and injuries in children in traffic crashes in Victoria, Australia. Accid Anal Prev 40:829–834

Lesire P, Grant R, Hummel T (2001) The CREST project accident data base. 17th International technical conference on the enhanced safety of vehicles (ESV), Amsterdam, The Netherlands

Leung Y, Tarriere C, Lestrelin D et al (1982) Submarining injuries of 3 pt. belted occupants in frontal collisions—description, mechanisms, and protection. Proceedings of the 26th Stapp Car Crash conference

Lund UJ (2005) The effect of seating location on the injury of properly restrained children in child safety seats. Accid Anal Prev 37:435–439

Lutz N, Arbogast KB, Cornejo RA et al (2003) Suboptimal restraint affects the pattern of abdominal injuries in children involved in motor vehicle crashes. J Pediatr Surg 38:919–923

Lutz N, Nance ML, Arbogast KB et al (2004) Incidence and significance of abdominal wall bruising in restrained children involved in motor vehicle crashes. J Pediatr Surg 39:972–975

Maltese MR, Chen IG, Arbogast KB (2005) Effect of increased rear row occupancy on injury to seat belt restrained children in side impact crashes. Annu Proc Assoc Adv Automot Med 49:229–243

Maltese MR, Locey CM, Jermakian JS et al (2007) Injury causation scenarios in belt-restrained nearside child occupants. Stapp Car Crash J 51:299–311

Margolis LH, Molnar LJ, Wagenaar AC (1988) Recognizing the common problem of child automobile restraint misuse. Pediatrics 81:717–720

Margolis LH, Wagenaar AC, Molnar LJ (1992) Use and misuse of automobile child restraint devices. Am J Dis Child 146:361–366

Marshall KW, Koch BL, Egelhoff JC (1998) Air bag related deaths and serious injuries in children: injury patterns and imaging findings. Am J Neuroradiol 19:1599–1607

McCarthy M, Lemmon G (1996) Traumatic lumbar hernia: a seat belt injury: case report. J Trauma 40:121–122

McCray L, Scarboro M, Brewer J (2007) Injuries to children one to three years old in side impact crashes. 20th International technical conference on the enhanced safety of vehicles conference (ESV), Lyon, France

McKay MP, Jolly BT (1999) A retrospective review of air bag deaths. Acad Emerg Med 6:708–714

Mertz H (1988) Restraint performance of the 1973–76 GM air cushion restraint system. SAE 880400, SP-736

Miller TR, Zaloshnja E, Lawrence BA et al (2004) Pedestrian and pedalcyclist injury costs in the United States by age and injury severity. Annu Proc Assoc Adv Automot Med 48:265–284

Moir JS, Ashcroft GP (1995) Lap seat-belts: still trouble after all these years. JR Coll Surg Edinb 40:139–141

Morgan C (1999) Effectiveness of lap/shoulder belts in the back outboard seating positions. NHTSA DOT HS 808 945

Morris SD, Arbogast KB, Durbin DR et al (2001) Misuse of booster seats. Inj Prev 6:281–284

Mulpuri K, Jawadi A, Perdios A et al (2007) Outcome analysis of chance fractures of the skeletally immature spine. Spine 32:E702–E707

Muszynski CA, Yoganandan N, Pintar FA et al (2005) Risk of pediatric head injury after motor vehicle accidents. J Neurosurg 102:374–379

Myers BS, Winkelstein BA (1995) Epidemiology, classification, mechanism, and tolerance of human cervical spine injuries. Crit Rev Biomed Eng 23:307–409

Nance ML, Lutz N, Arbogast KB et al (2004) Optimal restraint reduces the risk of abdominal injury in children involved in motor vehicle crashes. Ann Surg 239:127–131

National Highway Traffic Safety Administration (2010b) Booster seat effectiveness estimates based on CDS and state data. NHTSA DOT HS 811 338

National Highway Traffic Safety Administration (1997) Federal motor vehicle safety standards: occupant crash protection, FMVSS 208. Fed Regist 62:12960–12975

National Highway Traffic Safety Administration (1999a) Federal motor vehicle safety standards, child restraint systems, child restraint anchorage systems. Final Rule Fed Regist 64:10785–10850

National Highway Traffic Safety Administration (1999b) Traffic safety facts. NHTSA DOT HS 809 100

National Highway Traffic Safety Administration (2008) Child restraint use in 2007 – overall results. NHTSA DOT HS 810 931

National Highway Traffic Safety Administration (2010a) Traffic safety facts: 2009 data children. NHTSA DOT HS 811 387

National Safe Kids Campaign (1999) Child passengers at risk in America: a national study of car seat misuse. http://www.safekids.org/buckleup/study.html. Accessed 2010

Newgard CD, Lewis RJ (2005) Effects of child age and body size on serious injury from passenger air bag presence in motor vehicle crashes. Pediatrics 115:1579–1585

Newman JA, Dalmotas D (1993) Atlanto-occipital fracture dislocation in lap-belt restrained children. Presented at the child occupant protection symposium, Warrendale, PA, August 1993

Nichols JL, Glassbrenner D, Compton RP (2005) The impact of a nationwide effort to reduce airbag-related deaths among children: an examination of fatality trends among younger and older age groups. J Safety Res 36:309–320

Odent T, Langlais J, Glorion C et al (1999) Fractures of the odontoid process: a report of 15 cases in children younger than 6 years. J Pediatr Orthop 19:51–54

Olson C, Cummings P, Rivara F (2006) Association of first and second generation air bags with front occupant death in car crashes: a matched cohort study. Am J Epidemiol 164:161–169

Orzechowski KM, Edgerton EA, Bulas DI et al (2003) Patterns of injury to restrained children in side impact motor vehicle crashes: the side impact syndrome. J Trauma 54:1094–1101

Paine M, Griffiths M, Brown J et al (2003) Protecting children in car crashes: the Australian experience. 18th International technical conference on the enhanced safety of vehicles (ESV), Nagoya, Japan

Parenteau C, Viano D (2003) Field data analyses of rear occupant injuries, Part II: children, toddlers, and infants. SAE World Congress and Exhibition

Patrick DA, Bensard DD, Moore EE et al (2000) Cervical spine trauma in the injured child: a tragic injury with potential for salvageable functional outcome. J Pediatr Surg 35:1571–1575

Peng RY, Bongard FS (1999) Pedestrian versus motor vehicle accidents: an analysis of 5,000 patients. J Am Coll Surg 189:343–348

Pickrell TM, Ye TJ (2010) The 2009 national survey of the use of booster seats. NHTSA DOT HS 811 377

Platzer P, Jaindl M, Thalhammer G et al (2007) Cervical spine injuries in pediatric patients. J Trauma 62:389–396

Porter RS, Zhao N (1998) Patterns of injury in belted and unbelted individuals presenting to a trauma center after motor vehicle crash: seat belt syndrome revisited. Ann Emerg Med 32:418–424

Posner JC, Liao E, Winston FK et al (2002) Exposure to traffic among urban children injured as pedestrians. Inj Prev 8:231–235

Quinones-Hinojosa A, Jun P, Manley GT et al (2005) Airbag deployment and improperly restrained children: a lethal combination. J Trauma 59:729–733

Ramsey A, Simpson E, Rivara FP (2000) Booster seat use and reasons for nonuse. Pediatrics 106:e20

Randhawa MPS, Menzoian JO (1990) Seat belt aorta. Ann Vasc Surg 4:370–377

Rice TM, Anderson CL (2009) The effectiveness of child restraint systems for children aged 3 years or younger during motor vehicle collisions: 1996 to 2005. Am J Public Health 99:252–257

Rice TM, Anderson CL, Lee AS (2009) The association between booster seat use and risk of death among motor vehicle occupants aged 4-8: a matched cohort study. Inj Prev 15:379–383

Rivara FP, Cummings P, Mock C (2003) Injuries and death of children in rollover motor vehicle crashes in the United States. Inj Prev 9:76–80

Rogers L (1974) Injuries peculiar to traffic accidents: seat belt syndrome, laryngeal fracture, hangman's fracture. Tex Med 70:77–83

Roth SM, Wheeler JR, Gregory RT et al (1997) Blunt injury of the abdominal aorta: a review. J Trauma 42:748–755

Roudsari BS, Mock CN, Kaufman R et al (2004) Pedestrian crashes: higher injury severity for light truck vehicles compared with passenger cars. Inj Prev 10:154–158

Roudsari BS, Mock CN, Kaufman R (2005) An evaluation of the association between vehicle type and the source and severity of pedestrian injuries. Traffic Inj Prev 6:185–192

Santos A, McGuckin N, Nakamoto HY et al (2011) Summary of travel trends 2009 National Household Transportation Survey. USDOT Federal Highway Administration FHWA-PL-11-022

Segui-Gomez M, Wittenberg E, Glass R et al (2001) Where children sit in cars: the impact of Rhode Island's new legislation. Am J Public Health 91:311–313

Shehata SMK, Shabaan BS (2006) Diaphragmatic injuries in children after blunt abdominal trauma. J Pediatr Surg 41:1727–1731

Sherwood CP, Ferguson SA et al (2003) Factors leading to crash fatalities to children in child restraints. Annu Proc Assoc Adv Automot Med 7:343–359

Sherwood CP, Abdelilah Y, Crandall JR et al (2004) The performance of various rear facing child restraint systems in a frontal crash. Annu Proc Assoc Adv Automot Med 487:303–321

Shkrum MJ, McClafferty NES et al (2002) Driver and front seat passenger fatalities associated with air bag deployment: Part 2: a review of injury patterns and investigative issues. J Forensic Sci 47:1–6

Silber J, Flynn J (2002) Changing patterns of pediatric pelvic fractures with skeletal maturation: implications for classification and management. J Pediatr Orthop 22:22–26

Smith KM, Cummings P (2006) Passenger seating position and the risk of passenger death in traffic crashes: a matched cohort study. Inj Prev 12:83–86

Starnes M, Eigen AM (2002) Fatalities and injuries to 0–8 year old passenger vehicle occupants based on impact attributes. NHTSA DOT HS 809 410

Starnes M, Longthorne A (2003) Child pedestrian fatality rate by striking vehicle body type: a comparison of passenger cars, sport utility vehicles, pickups, and vans. NHTSA DOT HS 809 640

Stawicki SP, Holmes JH, Kallan MJ et al (2009) Cervical spine injury and motor vehicle crash characteristics: an analysis of child occupant fatalities in a unique, linked national dataset. Injury 40:864–867

Steckler R, Epstein J, Epstein B (1969) Seat belt trauma to the lumbar spine: an unusual manifestation of the seat belt syndrome. J Trauma 9:508–513

Sturm PF, Glass RBJ, Sivit CJ et al (1995) Lumbar compression fractures secondary to lap-belt use in children. J Pediatr Orthop 15:521–523

Stylianos S, Harris B (1990) Seatbelt use and patterns of central nervous system injury in children. Pediatr Emerg Care 6:4–5

Sube J, Ziperman H, McIver W (1967) Seat belt trauma to the abdomen. Am J Surg 113:346–350

Subotic U, Holland-Cunz S, Bardenheuer M et al (2007) Chance fracture – a rare injuries in pediatric patients? Eur J Pediatr Surg 17:207–209

Swischuk LE, Jadhav SP, Chung DH (2007) Aortic injury with chance fracture in a child. Emerg Radiol 14:431–433

Talton DS, Craig MH, Hauser CJ et al (1995) Major gastroenteric injuries from blunt trauma. Am Surg 61:69–73

Teret SP, Jones AS, Williams AF et al (1986) Child restraint laws: an analysis of gaps in coverage. Am J Public Health 76:31–34

Tingvall C (1987) Children in cars. Some aspects of the safety of children as car passengers in road traffic accidents. Acta Paediatr Scand Suppl 339:1–35

Trosseille X, Chamouard F, Tarriere C (1997) Abdominal injury risk to children and its prevention. Conference of the International Research Council on Biomechanics of Injury (IRCOBI)

Tso E, Beaver B, Halter JA (1993) Abdominal injuries in restraint pediatric passengers. J Pediatr Surg 28:915–919

van Rooij L, Meissner M, Bhalla B et al (2004) A comparative evaluation of pedestrian kinematics and injury prediction for adults and children upon impact with a passenger car. SAE World Congress and Exhibition

Vandersluis R, O'Connor HMC (1987) The seat-belt syndrome. CMAJ 137:1023–1024

Vellar ID, Vellar DJ, Mullany CJ (1976) Rupture of the bowel due to road trauma: the emergence of the "seat belt syndrome". Med J Aust 1:694–696

Viano DC, Parenteau CS (2008a) Fatalities of children 0-7 years old in the second row. Traffic Inj Prev 9:231–237

Viano DC, Parenteau CS (2008b) Field accident data analysis of 2nd row children and individual case reviews. SAE Government/Industry Meeting, Washington, DC, May 2008, Paper 2008-01-1851

Vitale MG, Goss JM, Matsumoto H et al (2006) Epidemiology of pediatric spinal cord injury in the United States years 1997 and 2000. J Pediatr Orthop 26:745–749

Voss L, Cole PA, D'Amato C (1996) Pediatric chance fractures from lapbelts: unique case report of three in one accident. J Orthop Trauma 10:421–428

Wagner AC (1979) Disruption of abdominal wall musculature: unusual feature of seat belt syndrome. AJR 133:753–754

Wang S-F, Tiu C-M, Chou Y-H et al (1993) Obstructive intestinal herniation due to improper use of a seat belt: a case report. Pediatr Radiol 23:200–201

Weber K (2002) Child passenger protection. In: Nahum A, Melvin J (eds) Accidental injury: biomechanics and prevention. Springer, New York

Whitman GR, Brown KA, Cantor A et al (1997) Booster with shield child restraint case studies. Proceedings of the 41st Stapp Car Crash conference

Willis BL, Smith JL, Vernon DD et al (1996) Fatal air bag mediated craniocervical trauma in a child. Pediatr Neurosurg 24:323–327

Winston F, Reed R (1996) Airbags and children: results of a National Highway Traffic Administration special investigation into actual crashes. Proceedings of the 40th Stapp Car Crash conference

Winston FK, Durbin DR, Kallan MJ et al (2000) The danger of premature graduation to seat belts for young children. Pediatrics 105:1179–1183

Winston FK, Kallan MJ, Elliott MR et al (2007) Effect of booster seat laws on appropriate restraint use by children 4 to 7 years old involved in crashes. Arch Pediatr Adolesc Med 161:270–275

Wittenberg E, Nelson T, Graham J (1999) The effect of passenger airbags on child seating behavior in motor vehicles. Pediatrics 104;1247–1250

Woods W, Sherwood C, Ivarsson J et al (2003) A review of pediatric injuries at a level I trauma center. 18th International technical conference on the enhanced safety of vehicles (ESV), Nagoya, Japan

Woodward GA, Bolte RG (1990) Children riding in the back of pickup trucks: a neglected safety issue. Pediatrics 86:683–691

World Bank (2001) Road safety. http://www.worldbank.org/transport/roads/safety.htm. Accessed 19 Sep 2008

World Health Organization (2008) World report on child injury prevention. http://whqlibdoc.who.int/publications/2008/9789241563574_eng.pdf. Accessed 2010

Yao J, Yang J, Otte D (2007) Head injuries in child pedestrian accidents – in-depth case analysis and reconstructions. Traffic Inj Prev 8:94–100

Yarbrough BE, Hendey GW (1990) Hangman's fracture resulting from improper seat belt use. South Med J 83:843–845

Zaloshnja E, Miller TR, Hendri D (2007) Effectiveness of child safety seats vs safety belts for children aged 2 to 3 years. Arch Pediatr Adolesc Med 161:65–68

Zuckerbraun BS, Morrison K, Gaines B et al (2004) Effect of age on cervical spine injuries in children after motor vehicle collisions: effectiveness of restraint devices. J Pediatr Surg 39:483–486

Chapter 3
Experimental Injury Biomechanics of the Pediatric Extremities and Pelvis

Johan Ivarsson, Masayoshi Okamoto, and Yukou Takahashi

Introduction

The paucity of pediatric postmortem human subjects (pediatric PMHS) for biomechanical research has led to the development of biofidelity requirements and injury assessment reference values (IARVs) for pediatric anthropomorphic test devices (ATDs) through geometrical scaling of adult PMHS data. Geometrical scaling relies on the assumption of geometrical similarity between the adult and child and does not account for any differences in tissue material properties. Attempts have been made to improve the accuracy of the scaled responses and IARVs by also accounting for the difference in Young's modulus between adult and pediatric bone (Irwin and Mertz 1997; van Ratingen et al. 1997; Mertz et al. 2001; Ivarsson et al. 2004a, b). However, the development of more biofidelic pediatric ATDs and accurate IARVs requires access to validation data that do not rely on the assumptions and simplifications associated with scaling. Access to accurate data from testing of pediatric tissues and anatomical structures would also facilitate the development of computational models for simulation of the response and injury of pediatric subjects in various loading situations.

The current chapter provides a detailed summary of the sparse experimental data that exist for tissues and anatomical structures of the pediatric pelvis and lower and upper extremities. The chapter contains two main sections: material properties and

---

J. Ivarsson, Ph.D. (✉)
Biomechanics Practice, Exponent, Inc.,
23445 N 19th Avenue, Phoenix, AZ 85027, USA
e-mail: jivarsson@exponent.com

M. Okamoto • Y. Takahashi
Honda R&D Co. Ltd, 4633 Shimotakanezawa, Tochigi, Japan

structural properties. The section on material properties summarizes available experimental data on the tissue level for the pediatric pelvis and extremities. The tissues for which relevant data are available are cortical (compact) bone, trabecular (cancellous, spongy) bone, growth plate (epiphyseal plate, physis) cartilage, tendon, ligament, and articular cartilage. The studies reviewed are further subgrouped into studies that have used pediatric human tissue and those that have utilized animal tissue intended to represent human pediatric tissue. While the majority of the tissue properties presented are mechanical properties such as Young's modulus and ultimate stress, physical properties such as tissue density and porosity have also, when available, been included. Studies on soft tissue structures such as tendon and ligament do not always provide data on the tissue level but instead present data pertaining to the structure such as tensile stiffness and ultimate tensile force. While such properties technically are structural rather than material, they have been included here for completeness.

The section on structural properties summarizes available experimental data on the loading response and tolerance of the pediatric pelvis and extremities. The anatomical structures for which relevant data are available are the pelvis, femur, tibia, fibula, humerus, radius, and ulna. No data from animal models have been included in this section given the anatomical differences between a human child and an immature animal.

The studies reviewed in this chapter are limited to those that have utilized biological models of human pediatric structures or tissues. It should, however, be mentioned that any adult data from studies that also include immature data generally are included in the review given that adult and pediatric data obtained under identical experimental conditions provide the opportunity to compare the difference in response and tolerance of immature and mature tissues and structures. Results from testing of pediatric and adult anatomical structures under identical experimental conditions further allow for evaluation of traditional scaling. The many studies that rely upon geometrical scaling of adult data, either human or animal, have not been included in the review, nor have studies that utilize nonbiological models such as computational models or pediatric ATDs.

The definition of "pediatric" has intentionally been left ambiguous since, for example, the degree to which tissue from an 18-year-old differs from the tissue of an older population of donors has to be considered as valuable information given the general lack of pediatric biomechanical data.

Finally, it is important to emphasize that the juvenile development of animals is different from the development of the pediatric human and that correlating the age of an immature animal to an equivalent human pediatric developmental age therefore is very difficult. Previous investigators have attempted to correlate animal age and human age based on osseous development (Ching et al. 2001), size, anatomical arrangement, and other developmental factors (Kent et al. 2006), but the lack of a direct correlation between immature animal data and pediatric human data must be considered in any interpretation of experimental results from an animal model of the human child.

Material Properties

Cortical Bone

Cortical (compact) bone is one of the two main types of osseous tissue and makes up approximately 80% of the total weight of the human skeleton (the other main type is trabecular bone). It is a tightly packed tissue that composes the outer cortical layer of all bones in the skeleton as well as the hollow shaft (diaphysis) of the long bones of the upper and lower extremities.

Human Subject Studies

Tensile Loading

The majority of studies that have used tissue from pediatric PMHS to determine the material properties of pediatric cortical bone have utilized tensile loading. One of the earliest tensile loading studies is the one by Hirsch and Evans (1965). They conducted quasi-static tensile tests using specimens machined from femora harvested at the autopsy of nine pediatric subjects ranging in age from newborn to 14 years. After machining, the test specimens were stored frozen at −20°C and were allowed to thaw prior to testing. The smallest specimens used were approximately 25 mm long, 1 mm thick, and 3–4 mm wide (with a reduced middle region of length 5 mm and width 1.3 mm). The long axis of the specimens was in the axial direction of the intact femora. The moist specimens were loaded in the direction of the long axis at a constant rate of 1 mm/min at a temperature of 20–22°C and a humidity of 65%. The tensile strain was measured by a strain gauge attached to the reduced middle region of the specimen. Ultimate force was successfully recorded in all 16 tests, whereas incorrect strain measurements due to faulty strain gauge attachment were said to have occurred in four of the tests. The results are summarized in Table 3.1.

Hirsch and Evans grouped the results into an "infant" group (including the data from the specimens from the donors up to and including 6 months of age) and a "child" group (including the data from the four specimens from the 14-year-old male donor) and observed that both the ultimate stress and Young's modulus were higher in the "child" group than in the "infant" group, whereas the ultimate strain did not exhibit any consistent trend. These observations were confirmed by the authors of the current chapter using one-tailed t-tests for comparing the means of the ultimate stress ($p<0.0001$) and Young's modulus ($p<0.0001$) between the "infant" and "child" groups and a two-tailed t-test for comparing the means of ultimate strain ($p=0.374$) between the same groups (the results from the four tests for which the strain measurements were incorrect were excluded from the t-tests of ultimate strain and elastic modulus).

Table 3.1 Results from the quasi-static tensile tests of specimens of pediatric femoral cortical bone by Hirsch and Evans (1965)

Age	Sex	Specimen #	A (mm^2)	F_u (N)	σ_u (MPa)	ε_u (%)	E (GPa)
Newborn	Female	22-1	1.14	147.2	129.1	1.90	12.57
1 day	Male	23-1	1.68	161.9	96.3	2.20	8.86
		23-2	1.65	142.2	86.2	2.00	6.96
1 day	Female	3-1[a]	1.54	142.2	92.4	0.50	16.69
2 days	Male	7-1[a]	2.20	122.6	55.7	1.00	7.25
2 days	Unknown	20-1	1.97	191.3	97.2	3.18	7.85
1 week	Female	9-1[a]	1.45	161.9	111.6	–	–
		9-2	1.36	157.0	115.4	1.60	12.89
2 months and 5 days	Male	21-1[a]	1.73	103.0	59.5	1.00	7.88
6 months	Unknown	10-1	1.30	149.6	115.1	2.31	8.91
		10-2	1.46	161.9	110.9	1.76	11.42
		10-3	1.48	159.4	107.7	2.90	8.04
14 years	Male	12-1	1.30	215.8	166.0	1.86	14.26
		12-2	1.37	274.7	200.5	1.72	22.30
		12-3	1.22	220.7	180.9	1.41	31.89
		12-4	1.70	245.3	144.2	2.68	19.62

A: cross-sectional area, F_u: ultimate force, σ_u: ultimate stress, ε_u: ultimate strain, E: Young's modulus
[a] Faulty strain gauge attachments precluded accurate estimates of ultimate strain and elastic modulus

Vinz (1969, 1970, 1972) conducted quasi-static tensile tests using 198 cortical specimens obtained from the femora of 48 PMHS of predominantly male gender and various age. Prior to testing, the bone material was stored for 48 h in Formalin and up to 4 days in NaCl solution. Vinz grouped the results by donor age into seven different age groups and reported, in addition to Young's modulus and ultimate stress and strain, physical tissue properties including density and volume and weight percentages of water in fresh cortical tissue. The results for the different age groups are shown in Table 3.2 and suggest that tensile strength and Young's modulus increase, while ultimate strain decreases with age during childhood.

Yamada (1970) reported on the results from Ko's (1953) tensile tests of standardized specimens of wet femoral cortical bone harvested from 36 Japanese PMHS ranging in age from the second to eight decade of life. The results from the tests are shown in Table 3.3 and indicate that bone strength increases from youth to the third or fourth decade of life and decreases thereafter, whereas bone ductility seems to decrease monotonically with age.

Wall et al. (1974, 1979) conducted quasi-static tensile tests with cortical specimens from fresh femora obtained at autopsy of PMHS ranging in age from 13 to 97 years. The specimens were 50 mm long, 2 mm thick, and 5 mm wide with a central section of length 20 mm in which the width was narrowed to 2 mm. The specimens were cut such that the long axis was in the direction of the long axis of the intact femora. The moist specimens were loaded up to fracture in the direction of the long axis at a constant rate of 1 mm/min. The tests were carried out at body temperature. Only data from the specimens that fractured in the central section

Table 3.2 Mean values (SD) of the results from the quasi-static tensile tests of cortical specimens of human cadaveric femora by Vinz (1969, 1970, 1972)

Age span	N	n	ρ (kg/m³)	ρ' (kg/m³)	q (volume%)	p (weight%)	σ_u (MPa)	E (MPa)	ε_u (%)	σ_u' (MPa)	E' (MPa)
0–2 weeks	13	29	1,781 (26)	2,265 (21)	38.2 (3.1)	19.8 (1.5)	54.9 (7.1)	10,497 (2,060)	0.63 (0.24)	89.0 (15.6)	16,971 (4,218)
3–5 months	5	13	1,727 (110)	2,244 (27)	41.5 (10.0)	20.5 (1.1)	59.4 (2.8)	11,380 (2,943)	0.82 (0.25)	101.7 (22.6)	19,424 (8,437)
7–11 months	5	16	1,832 (24)	2,262 (26)	34.1 (3.3)	19.4 (1.9)	76.5 (7.5)	18,247 (5,101)	0.73 (0.16)	116.2 (17.2)	27,664 (9,221)
1.5–2.2 years	6	25	1,877 (52)	2,269 (21)	30.9 (5.2)	16.9 (1.3)	89.3 (8.8)	22,465 (6,965)	0.59 (0.21)	129.5 (22.6)	32,471 (12,655)
4–13 years	5	36	1,948 (32)	2,304 (17)	27.3 (3.4)	14.2 (1.0)	101.7 (13.3)	33,256 (6,671)	0.55 (0.18)	139.8 (25.0)	45,715 (10,497)
18–40 years	9	60	2,000 (20)	2,332 (11)	24.9 (2.1)	12.1 (0.6)	103.5 (12.3)	35,316 (8,044)	0.46 (0.10)	137.8 (20.3)	47,088 (11,772)
70–85 years	5	31	1,968 (22)	2,337 (8)	27.6 (2.2)	12.0 (0.7)	82.7 (13.3)	22,465 (5,690)	0.47 (0.13)	114.3 (21.9)	31,000 (8,829)

N: number of PMHS from which the specimens were obtained, n: number of specimens tested within the age group, ρ: density, ρ': density of dried tissue, q: volume percentage of water in fresh cortical tissue = $(\rho' - \rho)/(\rho' - 1)$, p: weight percentage of water in fresh tissue, σ_u': ultimate stress of homogenous (cavity free) tissue = $\sigma_u/(1-q)$, E': Young's modulus of homogenous (cavity free) tissue = $E/(1-q)$

Table 3.3 Mean values (SD) of the results from the quasi-static tensile tests of wet specimens of human femoral cortical bone by Ko (1953) and reported on by Yamada (1970)

Age span (years)	σ_u (MPa)	ε_u (%)
10–19	113.8 (1.5)	1.48 (–)
20–29	122.6 (1.0)	1.44 (0.007)
30–39	119.7 (1.9)	1.38 (0.014)
40–49	111.8 (2.5)	1.31 (0.027)
50–59	93.2 (1.4)	1.28 (0.015)
60–69	86.3 (2.4)	1.26 (0.05)
70–79	86.3 (2.4)	1.26 (0.05)

σ_u: ultimate stress, ε_u: ultimate strain

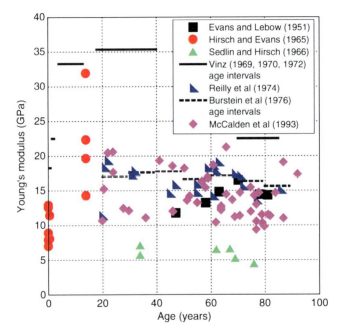

Fig. 3.1 Young's modulus of human femoral cortical bone as a function of donor age. All data were obtained in quasi-static testing with tension applied in the direction of the long bone axis. The specimens used by Vinz (1969, 1970, 1972) were stored in formalin and NaCl solution prior to testing, which most likely affected the properties

(where the stress was considered to be purely tensile) were included. The mean ± SD of the tissue density and ultimate stress for the four specimens tested from a 13-year-old female subject were 1,933 ± 23 kg/m^3 and 82.8 ± 9.8 MPa, respectively. The corresponding results for the specimens obtained from the only additional pediatric subject (a 15-year-old male) were 1,936 ± 22 kg/m^3 and 73.8 ± 7.3 MPa.

Figure 3.1 shows published data on Young's modulus of pediatric and adult human femoral cortical bone as a function of donor age. All data are from studies that subjected specimens to quasi-static tension in the direction of the long bone axis. Neglecting the results from the studies by Vinz (who preserved the specimens

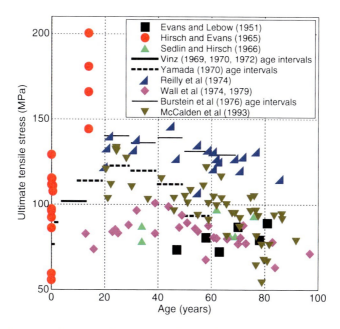

Fig. 3.2 Ultimate tensile stress of human femoral cortical bone as a function of donor age. All data were obtained in quasi-static testing with tension applied in the direction of the long bone axis. The specimens used by Vinz (1969, 1970, 1972) were stored in formalin and NaCl solution prior to testing, which most likely affected the properties

prior to testing), the majority of data points fall within the approximate range of 10–20 GPa. The exceptions are the study by Hirsch and Evans (1965), for which some of the specimens obtained from the infant subjects (age ≤ 6 months) had a modulus below 10 GPa and two of the specimens from the 14-year-old subject for which the modulus exceeded 20 GPa, and the study by Sedlin and Hirsch (1966) in which the authors reported a modulus in the approximate range of 4.4–7.1 GPa for the 34- to 76-year-old subjects. As shown in Fig. 3.1, the current available data in the pediatric age range are too sparse to allow for an accurate estimate of how Young's modulus of cortical bone in tension (in the axial direction) varies with age during maturation.

Figure 3.2 shows published data on the ultimate tensile stress of pediatric and adult human femoral cortical bone as a function of donor age. Again, all data are from studies that subjected specimens to quasi-static tension in the direction of the long bone axis. The majority of data points fall within the approximate range of 75–140 MPa, and there appears to be a trend towards a decrease in ultimate strength with age from early adulthood to senescence. The existing pediatric data are too sparse and show too much variation for subjects of similar age to allow for an accurate estimate of how the ultimate tensile stress of cortical bone (in the axial direction) varies with age during maturation. It is, however, worth noticing that three of

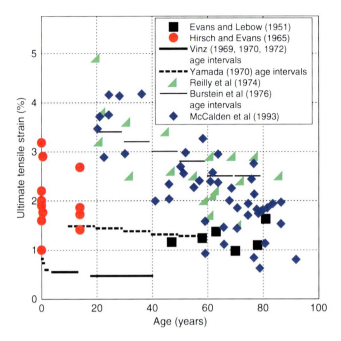

Fig. 3.3 Ultimate tensile strain of human femoral cortical bone as a function of donor age. All data were obtained in quasi-static testing with tension applied in the direction of the long bone axis. The specimens used by Vinz (1969, 1970, 1972) were stored in formalin and NaCl solution prior to testing, which most likely affected the properties

the four specimens from the 14-year-old subject tested by Hirsch and Evans (1965) were substantially stronger than any other specimen tested in the other studies. It is unfortunately not possible to say whether this subject had unusually strong bones or whether the high strength recorded for these specimens is due to inaccurate force and/or specimen cross-sectional area measurements. However, given the relatively low average strengths recorded for the specimens from one 13-year-old female and one 15-year-old male tested by Wall et al. (1974, 1979) and the medium strength reported by Yamada (1970) for the 10–20 years age interval, there is no reason to believe that the high strength reported by Hirsch and Evans (1965) reflects the normal strength of cortical bone at age 14 years.

Figure 3.3 shows published data on the ultimate tensile strain of pediatric and adult human femoral cortical bone as a function of donor age. As in Figs. 3.1 and 3.2, all data are from studies that subjected specimens to quasi-static tension in the direction of the long bone axis. As shown, there is a clear trend of a decreasing ultimate strain with age from early adulthood through the eight decade of life. As for Young's modulus and ultimate tensile stress, the data in the pediatric age range are too sparse to allow for an accurate estimate of how the ultimate tensile strain of cortical bone (in the axial direction) varies with age during maturation.

Table 3.4 Mean values (SD) of the results from the quasi-static three-point bending tests of wet specimens of human femoral cortical bone by Tsuda (1957) and reported on by Yamada (1970)

Age span (years)	σ_u (MPa)	ε_u (%)
10–19	151.1 (–)	8.6 (–)
20–29	173.6 (10.8)	7.5 (0.41)
30–39	173.6 (10.8)	6.6 (0.54)
40–49	161.9 (20.6)	6.2 (0.70)
50–59	154.0 (19.6)	6.2 (0.70)
60–69	139.3 (28.4)	5.3 (0.45)
70–79	139.3 (28.4)	5.3 (0.45)

σ_u: ultimate bending stress, ε_u: ultimate axial strain

Bending Loading

Two studies have determined the material properties of human pediatric cortical bone from bending tests with small specimens cut from femora. Yamada (1970) reported on the results from Tsuda's (1957) quasi-static three-point bending tests of standardized specimens of wet femoral cortical bone harvested from 28 Japanese PMHS ranging in age from the second to eight decade of life. The results are shown in Table 3.4 and indicate that bone strength increases from youth to the third or fourth decade of life and decreases thereafter, and that bone ductility decreases monotonically with age.

Currey and Butler (1975) conducted quasi-static three-point bending tests on wet cortical specimens obtained from the femoral mid-shaft of PMHS in the age range 2–48 years. The specimens were 23–26 mm long, 3 mm wide, and 2 mm thick and were cut such that the long axis was in the direction of the long axis of the intact femora, and the height was in the radial direction relative to the intact bone. The specimens were deflected at a rate of 5 mm/min and oriented such that the part of the specimen nearest the periosteal surface was in compression, and the part closer to the endosteal side was in tension. All tests were conducted at room temperature and with a span length of 20 mm. The variables recorded were ultimate bending strength, Young's modulus, fracture energy density, midspan deflection at fracture, and ash content. For specimens that did not break cleanly, the authors calculated the deflection and energy density to the point on the force–deflection curve at which the force had decreased to 50% of the maximum recorded force. The results from the tests appear in Table 3.5 and Figs. 3.4 and 3.5. Using the mean values reported for each pediatric donor (2–17 years) except the 6-year-old who had suffered from diabetes (according to Currey and Butler, the diabetes might have affected the bone properties), the results indicate that the ultimate bending strength ($p<0.001$), Young's modulus ($p<0.001$), and ash content ($p=0.013$) increase, while fracture energy density ($p<0.03$) and deflection to fracture ($p<0.005$) decrease with age during maturation. Comparing the mean values from the pediatric specimens (2–17 years) to the corresponding mean values reported for the adult specimens (26–48 years), ultimate bending strength ($p<0.001$), Young's modulus ($p<0.0001$), and ash content ($p<0.0003$) were all higher for the specimens from the adult donors, while fracture energy density ($p<0.0003$) and deflection to fracture ($p<0.0001$) were higher for the pediatric specimens.

Table 3.5 Mean values (SD) of the results from the quasi-static three-point bending tests of wet specimens of human femoral cortical bone by Currey and Butler (1975). The 6-year-old subject had suffered from diabetes that, according to Currey and Butler, might have affected the bone properties

Age (years)	Sex	N	σ_u (MPa)	E^a (GPa)	W_u (kJ/m^2)	δ_u (mm)	Ash (%)
2	Female	4	157.8 (41.8)	8.22 (2.82)	21.5 (10.0)	1.91 (0.62)	59.98 (1.04)
2.5	Male	4	168.8 (5.4)	9.15 (1.46)	19.1 (4.0)	1.91 (0.56)	60.57 (2.10)
3	Male	4	157.0 (24.0)	7.92 (1.78)	19.4 (10.6)	2.12 (0.88)	62.12 (0.34)
3.5	Male	8	150.0 (25.7)	9.71 (1.90)	16.0 (5.7)	1.69 (0.25)	61.04 (0.87)
4	Female	10	176.8 (17.4)	9.85 (1.80)	19.7 (5.1)	1.76 (0.32)	61.55 (0.28)
6	Male	5	207.2 (32.0)	13.75 (2.30)	21.6 (6.0)	1.79 (0.25)	63.08 (1.11)
8	Female	11	190.4 (15.9)	12.27 (1.06)	16.3 (2.7)	1.50 (0.23)	63.20 (0.58)
13	Male	7	185.7 (20.1)	11.77 (1.16)	17.1 (5.0)	1.63 (0.37)	63.02 (0.50)
14	Male	6	183.5 (22.3)	11.45 (1.37)	15.2 (5.4)	1.58 (0.34)	62.61 (0.53)
16	Male	9	204.7 (14.1)	12.08 (1.50)	18.1 (3.3)	1.56 (0.27)	61.35 (0.82)
17	Male	12	194.2 (15.9)	12.21 (1.91)	16.2 (4.5)	1.50 (0.31)	65.02 (0.74)
26	Male	10	206.4 (20.2)	14.38 (2.06)	15.2 (2.8)	1.33 (0.13)	63.70 (0.76)
28	Male	6	195.0 (15.9)	13.35 (1.47)	11.6 (2.4)	1.19 (0.22)	65.18 (0.86)
32	Male	8	206.4 (6.2)	14.8 (0.91)	15.6 (1.4)	1.32 (0.11)	64.79 (0.59)
39	Male	6	188.0 (15.7)	14.17 (0.93)	9.6 (1.7)	1.01 (0.12)	64.52 (1.02)
44	Male	7	225.4 (8.5)	16.20 (1.27)	13.7 (3.2)	1.11 (0.21)	65.98 (0.32)
46	Female	6	218.8 (19.4)	15.38 (1.03)	12.0 (2.7)	1.13 (0.20)	64.60 (0.49)
48	Male	6	220.8 (8.6)	15.43 (0.96)	14.9 (1.5)	1.26 (0.12)	64.58 (1.07)

^{a}The mean values and standard deviations for Young's modulus reported by Currey and Butler (1975) are one order of magnitude greater than given in this table. Considering that Currey (2001) cites the 1975 publication and gives exactly the values shown in this table, it is assumed that the (abnormally high) values for Young's modulus reported by Currey and Butler (1975) are just a result of a typographical error

N: number of samples tested from each subject, σ_u: ultimate bending stress, E: Young's modulus, W_u: fracture energy density, δ_u: ultimate deflection

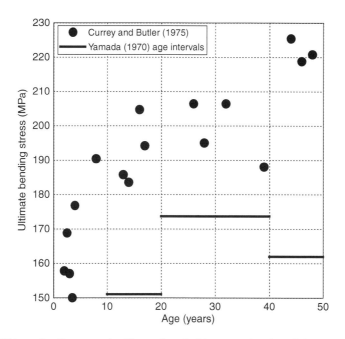

Fig. 3.4 Ultimate bending strength of femoral cortical bone as a function of donor age. The data were obtained in quasi-static three-point bending tests by Tsuda (1957) [reported on by Yamada (1970)] and Currey and Butler (1975). The data point from the 6-year-old diabetic subject tested by Currey and Butler is not plotted

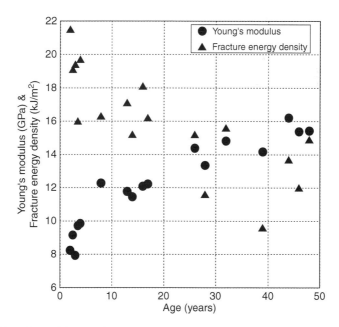

Fig. 3.5 Young's modulus and fracture energy density of femoral cortical bone as a function donor age. The data were obtained in quasi-static three-point bending tests by Currey and Butler (1975). The data for the 6-year-old diabetic subject are not plotted

It should be emphasized that bending as a loading mode for determining the material properties of bone has been criticized for ignoring the plastic behavior of bone. Burstein et al. (1972) estimated that application of linear beam theory to force–deflection data recorded in a three-point bending test would over predict the ultimate tensile stress of bone by a factor 1.56 for a square cross section and by a factor of 2.1 for a circular cross-section. This is most likely the explanation to why the ultimate bending strength reported by both Yamada (1970) and Currey and Butler (1975) is substantially higher than the majority of the data on ultimate tensile stress presented in Fig. 3.2. Comparing Yamada's (1970) results on ultimate stress of femoral cortical bone in tension and bending for each individual age interval, the latter is 33–65% higher.

Bending has also been used in two studies investigating how the energy absorption characteristics of human cortical bone change with age from early childhood through the ninth and tenth decades of life (Currey 1979; Currey et al. 1996). Currey (1979) conducted dynamic impact testing with wet cortical specimens obtained from the left or right femur of 39 PMHS, ranging in age from 2 to 96 years, to study the effect of age on the impact energy absorption of bone. Nine of the subjects were pediatric, ranging in age from 2 to 17 years. Six specimens of length 32 mm, approximate width 3 mm, and approximate depth 2 mm were cut from the midshaft section of each femur. Using a Hounsfield plastics impact tester, the specimens were impacted and loaded to failure in four-point bending. The distance between the outer supports was 25 mm, and the distance between the inner points on the pendulum impacting the specimen was 7 mm. All specimens were impacted such that the subperiosteal surface was loaded in compression and the subendosteal surface in tension. Other properties measured for some of the test specimens included ash content, dried defatted specimen density, bone material density, calcium content, and porosity (the volume proportion of the bone section occupied by cavities).

The energy absorbed demonstrated a marked decrease with age with the youngest specimens showing a mean value of approximately 28 kJ/m^2 and the 90-year-old specimens less than 10 kJ/m^2. Porosity was found to decline from about 9% at age 3 years to about 3% in young adulthood and then increase slowly with age, reaching approximately 12% in the tenth decade of life. Bone material density was found to increase rapidly with age during the first two decades of life and slower thereafter. The increase in material density with age, which Currey stated was most likely due to the decreasing rate of bone reconstruction with age, was, however, relatively modest with the youngest specimens demonstrating an approximate material density of 1,980 kg/m^3 and the oldest about 2,100 kg/m^3. Calcium content demonstrated a steady increase with age ($p<0.005$). Based on the findings, Currey concluded that at least 40% of the variation in impact energy absorption of bone with age can be attributed to the increasing mineralization of bone with age. The strong connection between impact energy absorption and mineralization was in turn attributed to the negative effect that mineral content has on the ability of bone to undergo plastic deformation before a crack develops. This negative effect of mineralization on impact energy absorption was found to be intensified by the high porosity in old bone. The same was, however, not true for young bone for which impact energy absorption was found to be more or less independent of porosity. Although the experiments did not provide any information on the relative contribution to the total absorbed energy of events that occurred prior to and after fracture initiation, Currey hypothesized that young bone, apart from showing more plastic deformation prior to fracture initiation, also may have a more energetically expensive fracture process than older bones. This hypothesis was based on the difference in fracture surface characteristics between the old and the young test specimens with the younger specimens demonstrating a surface that appeared torn and had more pullouts than the older specimens.

Currey et al. (1996) conducted both quasi-static and dynamic bending tests with wet femoral cortical specimens obtained from nine PMHS, ranging in age from 4 to 82 years. Three of the subjects were pediatric (4, 6, and 10 years old). A total of 88 specimens were prepared for testing. The 44 quasi-static test specimens were taken just proximal to midshaft and had a breadth of 5 mm, a gauge length of 25 mm, and a depth varying between 2.5 and 5 mm. Material was removed from the middle of the specimens, leaving a waist with a height of 2.2 mm and a base of approximately 4.5 mm. All quasi-static tests were performed in three-point bending at a deflection rate of 0.1 mm/min and with the specimens oriented such that the side closest to the original periosteal side of the shaft was in tension. The 44 dynamic test specimens were taken just distal to mid-shaft and had an overall length of 40 mm, a gauge length of 35 mm, a breadth of 3 mm, and a depth of 2 mm. The dynamic tests were conducted using the Hounsfield plastic impact tester previously used by Currey (1979). Following testing, a small sample was taken from each specimen for subsequent ashing.

In agreement with the findings of Currey (1979), the energy absorbed in the impact tests demonstrated a marked decrease with age. The energy absorbed by the 4-year-old specimens was approximately 36 kJ/m^2, while the corresponding number for the 82-year-old specimens was about 7 kJ/m^2. Age had a similar effect on the work of fracture recorded in the quasi-static tests with the numbers ranging from approximately 4.9 kJ/m^2 at age 4 years to about 1.9 kJ/m^2 at age 82 years. Ash content increased regularly with age in the femora from the four youngest subjects (4–25 years) but showed no clear trend thereafter. Regression analyses demonstrated that age, as opposed to ash content, explained slightly more of the variation in work of fracture (60% vs. 53%) and absorbed impact energy (57% vs. 52%). According to Currey et al., this suggested that there are features associated with age, other than mineral content, which decrease the mechanical quality of bone.

Animal Studies

Tensile Loading

Torzilli et al. (1982) conducted tensile tests with specimens of femoral and tibial bone harvested from seven immature (16–48 weeks old) and two adult (120 weeks old) pure-breed Labrador Retrievers. The immature animals were all from four litters of puppies, totaling 20 animals, from two bitches and one stud. The bones were stored frozen at $-25°C$ and milled into dog-bone shaped test specimens with a quadratic cross section and a total length of 7.5 mm. All specimens were aligned parallel to the longitudinal axis of the intact femora and tibiae. The specimens were tested wet at an approximate strain rate of 0.08/s and demonstrated approximately bi-linear stress–strain behavior. The yield point was determined from the intersection of the best fit straight lines drawn through the elastic and plastic portions of the stress–strain curve. Due to this geometric construction, the reported yield stress was slightly higher than the true yield stress. For specimens which the plastic portion of the stress–strain curve was not linear, the slope was determined from the best straight line fit of the initial linear region. The results from the tests appear in Table 3.6. Using the data from the immature specimens (16–48 weeks), Torzilli et al. utilized linear regression to evaluate the change in tissue properties during maturation. Yield strain and failure strain were independent of age for both tibial and femoral bone and did not differ from the corresponding values obtained for the adult specimens. Yield stress, failure stress, and Young's modulus for tibial bone increased with age during maturation ($p<0.05$). The same trend with age during maturation was observed for the corresponding properties of femoral bone, but the confidence level was slightly lower (90–95%). The plastic modulus was found to increase with age during maturation for both tibial and femoral bone ($p<0.05$).

Compression Loading

Torzilli et al. (1982) also conducted compressive tests with specimens of tibial and femoral bone obtained from the same nine pure-breed Labrador Retrievers used in the previously described tensile tests. Storage of the bones, machining of test specimens, rate of deformation, definition of the yield point, and assessment of the change in tissue properties during maturation were identical to what previously have been described for the corresponding tensile experiments. All but one sample demonstrated a decreasing stress with increasing strain after yield, resulting in a negative plastic modulus and a failure stress that was lower than the yield stress. The results appear in Table 3.7. Yield strain and failure strain for the immature bones were found to be independent of age for both tibial and femoral bone and did not differ from the corresponding values obtained for the adult specimens. Yield stress, failure stress, and Young's modulus for tibial bone were all found to increase with age during maturation ($p<0.05$). The same trend with age during maturation was observed for the corresponding properties of femoral bone, but the confidence level was slightly lower (90–95%). The plastic modulus did not demonstrate significant dependence on age during maturation.

Discussion

The preceding summary of experimental studies on the material properties of pediatric cortical bone reveals that valuable and useful data for this particular tissue exist.

Table 3.6 Mean values (SD) of the results from the quasi-static tensile tests of femoral and tibial bone specimens from Labrador Retrievers by Torzilli et al. (1982)

Bone	Age (weeks)	N	σ_y (MPa)	ε_y (%)	σ_f (MPa)	ε_f (%)	E (GPa)	P_{mod} (GPa)
Femur	16	3	60 (5)	0.9 (0.2)	79 (8)	5.1 (0.7)	7.1 (1.8)	0.459 (0.025)
	24	1	89	1.6	131	11.4	5.6	0.430
	24	4	103 (21)	1.1 (0.4)	137 (17)	6.1 (0.9)	11.1 (5.3)	0.684 (0.213)
	40	3	100 (19)	1.5 (0.6)	147 (19)	7.3 (0.8)	7.5 (2.2)	0.818 (0.038)
	48	4	87 (37)	1.3 (0.8)	153 (19)	5.8 (1.0)	7.4 (3.7)	1.500 (0.507)
	120	12	123 (11)	0.8 (0.2)	156 (19)	4.3 (1.5)	15.6 (2.8)	1.200 (0.769)
Tibia	16	1	64	1.0	87	6.2	6.2	0.455
	20	3	79 (5)	1.4 (0.1)	117 (6)	6.8 (0.4)	5.9 (0.9)	0.705 (0.042)
	24	7	91 (6)	1.0 (0.3)	130 (11)	5.9 (1.0)	10.0 (3.4)	0.817 (0.120)
	24	12	103 (11)	1.4 (0.6)	150 (16)	7.1 (1.3)	8.2 (2.7)	0.840 (0.129)
	40	7	122 (14)	0.9 (0.1)	184 (25)	6.7 (1.5)	13.3 (2.1)	1.050 (0.134)
	48	8	136 (9)	1.0 (0.3)	195 (16)	6.2 (1.0)	14.5 (4.2)	1.100 (0.113)
	120	27	124 (12)	0.8 (0.3)	185 (15)	6.0 (0.9)	15.7 (3.5)	1.200 (0.139)
	120	18	137 (17)	1.0 (0.4)	197 (20)	6.3 (1.0)	14.1 (3.7)	1.200 (0.244)

N: number of specimens from each bone, σ_y: yield stress, ε_y: yield strain, σ_f: failure stress, ε_f: failure strain, E: Young's modulus, P_{mod}: plastic modulus

Table 3.7 Mean values (SD) of the results from the quasi-static compression tests of femoral and tibial bone specimens from Labrador Retrievers by Torzilli et al. (1982)

Bone	Age (weeks)	N	σ_y (MPa)	ε_y (%)	σ_f (MPa)	ε_f (%)	E (GPa)	P_{mod} (GPa)
Femur	16	3	94 (52)	0.9 (0.4)	104 (41)	7.0 (1.7)	9.2 (4.5)	0.14 (0.15)
	24	2	133 (80)	1.6 (0.3)	99 (4)	8.2 (0.0)	7.5 (5.1)	−0.49 (1.30)
	24	6	185 (48)	1.2 (0.2)	114 (30)	8.7 (0.2)	15.5 (3.5)	−0.81 (0.86)
	40	3	202 (63)	1.0 (0.1)	178 (27)	8.1 (0.3)	16.6 (2.3)	−0.43 (0.48)
	48	3	205 (20)	1.2 (0.3)	67 (79)	7.5 (0.1)	17.6 (4.9)	−2.20 (1.50)
	120	12	229 (16)	1.1 (0.2)	143 (52)	5.6 (0.2)	21.0 (3.9)	−2.00 (0.91)
Tibia	16	3	72 (7)	0.6 (0.0)	69 (21)	7.3 (0.2)	12.1 (1.8)	−0.19 (0.47)
	20	8	118 (15)	1.0 (0.3)	107 (11)	7.5 (0.2)	12.3 (3.8)	−0.17 (0.10)
	24	9	103 (20)	1.1 (0.1)	101 (21)	7.8 (0.2)	8.7 (2.2)	−0.06 (0.04)
	24	11	139 (14)	1.3 (0.3)	125 (11)	7.6 (0.2)	11.5 (2.0)	−0.25 (0.17)
	40	8	178 (29)	1.2 (0.3)	160 (20)	7.1 (1.1)	15.9 (6.0)	−0.21 (0.76)
	48	9	205 (15)	1.0 (0.2)	148 (21)	7.4 (0.5)	21.9 (5.1)	−0.91 (0.50)
	120	27	198 (25)	1.2 (0.3)	148 (34)	7.2 (0.6)	16.5 (4.3)	−0.83 (0.62)
	120	18	225 (23)	1.2 (0.2)	168 (25)	7.5 (0.2)	18.7 (4.0)	−0.88 (0.48)

N: number of specimens from each bone, σ_y: yield stress, ε_y: yield strain, σ_f: failure stress, ε_f: failure strain, *E*: Young's modulus, P_{mod}: plastic modulus

The study by Currey and Butler (1975) provides a strong indication that the strength and stiffness of cortical bone increase with age during maturation, and that adult cortical bone is stronger and stiffer than pediatric. It further suggests that the fracture energy density of cortical bone decreases with age during maturation and consequently is lower in mature than in immature cortical bone. The more recent studies by Currey (Currey 1979; Currey et al. 1996) show that impact energy absorption of cortical bone decreases during maturation, which is in agreement with his previously published finding of a declining fracture energy density with age during maturation (Currey and Butler 1975). While the results from Currey's studies, and possibly also the previously reviewed studies by Hirsch and Evans (1965) and Vinz (1969, 1970, 1972), provide enough information to conclude that some material properties, such as strength, stiffness (Young's modulus), and energy absorption, increase or decrease with age during maturation and are lower or higher in immature than in mature cortical bone, there is unfortunately not enough data available to allow for determining accurate ranges for various cortical bone material parameters for specific ages during maturation. This paucity is clearly illustrated in the graphs in Figs. 3.1, 3.2, and 3.3 which show fairly extensive data for the >20 years age interval but very few data points in the 0–20 years age interval.

Finally, it is essential to comment on the study utilizing cortical bone specimens from Labrador Retrievers by Torzilli et al. (1982). While the results of this study in general appear to support those of Currey and Butler (1975), i.e., strength and stiffness of cortical bone increase with age during maturation, and adult cortical bone is stronger and stiffer than pediatric, the extensive numerical results from this study are unfortunately not directly applicable to human. This is mainly for two reasons. First, the juvenile development of a Labrador Retriever is different from the development of the pediatric human, and it is consequently very difficult to correlate the ages of the immature dogs utilized by Torzilli et al. to corresponding human developmental ages. Second, the composition of cortical bone is species dependent, and the cortical bone in a child and a Labrador Retriever of equivalent developmental ages do therefore not necessarily demonstrate the same, or even similar, material properties (Aerssens et al. 1998). For the purpose of adding to the knowledge of the material properties human pediatric cortical bone, the study by Torzilli et al. (1982) consequently does little else than strengthening the previously discussed indications of the study by Currey and Butler (1975).

Trabecular Bone

Trabecular (spongy, cancellous) bone is one of the two main types of osseous tissue and makes up approximately 20% of the total weight of the human skeleton (the other main type is cortical bone). It consists of an irregular latticework of thin bone plates referred to as trabeculae. Trabecular bone makes up the interior structure of short, flat, and irregularly shaped bones as well as the interior of the long bone epiphyses.

Table 3.8 Mean values of the results from Lindahl's (1976) compression testing of dried samples of pediatric (14–19 years) human vertebral and tibial trabecular bone

Entity	Lumbar vertebrae		Proximal tibia	
	Females	Males	Females	Males
Ultimate stress (MPa)	5.78	7.36	5.04	6.23
Yield stress (MPa)	4.57	6.37	3.38	5.53
Ultimate strain (%)	10.25	9.61	12.85	15.09
Yield strain (%)	6.52	6.97	6.17	12.33
Young's modulus (MPa)	71.8	85.9	49.4	45.4

Human Subject Studies

There are only a handful of experimental studies that have reported on results from tests on pediatric human trabecular bone, and all of them have utilized compression as the only loading mode. One early study is the one by Weaver and Chalmers (1966). They conducted compression tests on cubes of human trabecular bone (side length 12.7 mm) obtained from the central portion of lumbar vertebrae and from the inferior portion of the calcaneal tuberosity. The sides of the cubes were oriented in accordance with the major trabecular structure. The specimens were obtained within 8 h of death and were immediately stored at −20°C. Prior to testing, the specimens were allowed to thaw for 3 h at room temperature. The ultimate stress for the lumbar vertebra specimens obtained from the five pediatric subjects (males aged 10, 14, 15, 15, and 16 years) was 3.06 ± 0.52 MPa (mean ± SD) and was significantly lower than the corresponding 4.40 ± 0.98 MPa recorded for the 22 specimens from male subjects in the age range 18–50 years ($p=0.007$, two tailed t-test). The ultimate stress recorded for the calcaneal tuberosity specimens from the pediatric subjects (males aged 10, 14, 15, and 15 years) was 4.09 ± 1.25 MPa and was not significantly different from the corresponding 3.47 ± 1.79 MPa recorded for the 20 specimens from male subjects in the age range 18–50 years ($p=0.52$, two tailed t-test).

Lindahl (1976) conducted quasi-static compression tests on rectangular specimens of trabecular bone obtained from the lowermost 2–4 lumbar vertebrae and the proximal tibia of 60 PMHS aged 14–89 years. Prior to testing, the trabecular bone material was dried and stored in air at 3–5°C and conditioned for 2 months at $65 \pm 3\%$ relative humidity and 20 ± 1.0°C. The specimens were loaded in the axial direction at a rate of 0.05 mm/min. The average values obtained for the specimens from the pediatric subjects (14–19 years) are shown in Table 3.8. It is important to point out that the drying of the specimens in Lindahl's study most likely had a substantial influence on the material properties. Cowin et al. (1987) emphasized the importance of keeping bone samples for mechanical testing moistened and stated that the properties of fresh bone tissue can vary in a matter of minutes if the bone is allowed to dry.

Mosekilde et al. (1987) conducted quasi-static compression tests on cylindrical specimens (diameter 7 mm, length 5 mm) of trabecular bone obtained from the central part of the first lumbar vertebra of 42 PMHS (27 females and 15 males) aged

15–87 years. Samples were obtained both in the horizontal and vertical directions. Prior to testing, the frozen samples (−20°C) were thawed slowly to +20°C. The specimens were loaded at a constant compression rate of 2 mm/min. The maximum stress, maximum stiffness, and energy absorption capacity showed a significant decrease ($p<0.001$) with age, whereas the maximum strain increased with age in both the vertical ($p<0.001$) and horizontal ($p<0.06$) directions. The vertical direction maximum stress, energy absorption, maximum stiffness, and maximum strain of the specimen from the only pediatric subject (a 15-year-old female) were approximately 4.6 MPa, 0.85 mJ/mm^2, 126 MPa, and 7.8%, respectively. The corresponding horizontal direction values were approximately 2.5 MPa, 0.52 mJ/mm^2, 68 MPa, and 7.7%.

Ding et al. (1997) and Ding (2000) conducted quasi-static compression tests on cylindrical specimens (diameter 7.5 mm, height 7.5 mm) of tibial trabecular bone obtained from the medial and lateral condyles of 31 Caucasian PMHS (5 females, 31 males) aged 16–83 years. All specimens were stored frozen in physiological saline at −20°C until testing. The orientation of the specimens was such that the longitudinal axis of the tibia corresponded to that of the cylinder and thus to the direction of compression. Young's modulus was obtained from nondestructive testing (395 tests) of the moist specimens, while ultimate stress, ultimate strain, and energy absorption to failure were obtained from subsequent destructive testing (183 tests). The authors constructed scatter plots showing recorded values of Young's modulus, ultimate stress, ultimate strain, energy absorption to failure, tissue density, apparent density, mineral concentration, apparent ash density, collagen concentration, and collagen density as functions of donor age and fitted linear functions or second degree polynomials to the data. The Young's modulus, ultimate stress, ultimate strain, energy absorption to failure, and apparent density for the specimens obtained from the only pediatric subject (a 16-year-old male) were approximately 600 MPa, 9.04 MPa, 2.77%, 131.3 kJ/m^3, and 0.54 g/cm^3, respectively, and were within the corresponding ranges of values recorded for the specimens from the young adult (≤40 years) donors in the same study.

Animal Studies

Nafei et al. (2000) conducted quasi-static nondestructive and destructive compression testing with cubes (side length 4.2 mm) of wet subchondral trabecular bone from the center of the medial tibial condyle of skeletally immature lambs aged 3, 6, and 9 months and skeletally mature sheep aged 36 and 80 months (the authors referred to a pilot study that had shown that lambs reach skeletal maturity around the 15th month of life). All animals were of the same breed, came from the same farm, and had a known birth date, gender, and body weight. A bone age of 1 month in sheep was said to roughly correspond to that of 1 year in human. There were samples from ten animals in each age group. Several physical properties were measured for each sample including tissue density, apparent density, and bone mineral content (BMC) (Table 3.9). Statistical analyses demonstrated that the tissue density, apparent

Table 3.9 Mean values (SE) of the physical properties measured in the nondestructive compression tests of wet subchondral trabecular bone samples from the center of the medial tibial condyle of skeletally immature lambs (3, 6, and 9 months) and skeletally mature sheep (36 and 80 months) by Nafei et al. (2000). $N=10$ animals in each age group

	Age (months)				
Property	3	6	9	36	80
ρ (g/cm^3)	2.15 (0.01)	2.24 (0.01)	2.25 (0.01)	2.27 (0.01)	2.26 (0.01)
ρ_a (g/cm^3)	0.405 (0.01)	0.448 (0.03)	0.429 (0.02)	0.563 (0.04)	0.665 (0.04)
BMC (%)	61.7 (0.7)	65.6 (0.7)	67.0 (0.4)	66.4 (0.3)	67.6 (0.8)

ρ: tissue density, ρ_a: apparent density, *BMC*: bone mineral content

Table 3.10 Mean values (SE) of the mechanical properties measured in the nondestructive compression tests of wet subchondral trabecular bone samples from the center of the medial tibial condyle of skeletally immature lambs (3, 6, and 9 months) and skeletally mature sheep (36 and 80 months) by Nafei et al. (2000). $N=10$ animals in each age group

	Age (months)				
Property	3	6	9	36	80
W_v (kJ/m^3)					
z	1.49 (0.15)	0.76 (0.19)	0.51 (0.04)	0.30 (0.05)	0.21 (0.04)
y	0.82 (0.11)	0.41 (0.07)	0.27 (0.02)	0.24 (0.08)	0.24 (0.04)
x	0.41 (0.07)	0.31 (0.04)	0.26 (0.03)	0.22 (0.06)	0.20 (0.04)
W_e (kJ/m^3)					
z	4.44 (0.35)	3.82 (0.34)	5.16 (0.42)	5.19 (0.57)	7.62 (0.96)
y	2.79 (0.21)	2.89 (0.26)	3.24 (0.42)	3.49 (0.51)	4.88 (1.00)
x	2.86 (0.19)	2.81 (0.24)	3.45 (0.51)	3.79 (0.51)	5.54 (0.91)
E (MPa)					
z	543 (62)	601 (71)	822 (111)	938 (176)	1,550 (275)
y	389 (43)	402 (54)	474 (50)	562 (81)	890 (172)
x	369 (29)	413 (44)	534 (42)	528 (82)	732 (124)

z: axial direction, y: anteroposterior direction, x: lateromedial direction, W_v: viscoelastic energy absorption, W_e: elastic energy absorption, E: Young's modulus

density, and BMC were all significantly lower for the immature than for the mature animals ($p<0.05$). Pairwise comparisons between age groups showed that tissue density and BMC were significantly lower in the 3-month old animals than in the animals from the four other age groups.

Each sample was nondestructively compressed in each of the three orthogonal directions to approximately 0.6% strain at a constant strain rate of 0.001/s. Properties recorded included Young's modulus, viscoelastic energy absorption (the difference between the energy absorbed during loading and the energy released during unloading), and elastic energy absorption (the energy released during unloading) (Table 3.10). The specimens were allowed to recover for at least 24 h between tests. The Young's modulus correlated positively with increasing age for all three test

Table 3.11 Mean values (SE) of the mechanical properties measured in the axial destructive tests of wet subchondral trabecular bone samples from the center of the medial tibial condyle of skeletally immature lambs (3, 6, and 9 months) and skeletally mature sheep (36 and 80 months) by Nafei et al. (2000). $N=10$ animals in each age group

Property	Age (months)				
	3	6	9	36	80
E (MPa)	646 (60)	599 (91)	764 (116)	869 (224)	1,514 (296)
σ_u (MPa)	11.6 (0.8)	14.3 (1.5)	14.0 (1.7)	16.6 (3.4)	26.3 (4.7)
ε_u (%)	3.8 (0.18)	3.69 (0.16)	3.47 (0.14)	3.05 (0.14)	2.78 (0.20)
W_u (kJ/m³)	250 (26)	295 (33)	277 (28)	409 (65)	492 (65)

E: Young's modulus, σ_u: ultimate stress, ε_u: ultimate strain, W_u: energy absorption to failure

directions. Considering only the immature age groups, the Young's modulus showed a positive correlation with age in the axial ($p=0.025$), anteroposterior ($p=0.005$), and mediolateral ($p=0.225$) directions. The viscoelastic energy absorption decreased with increasing age over the entire age span (3–80 months), but the dependence on age was strongest during maturation with a 50% reduction from age 3 to 6 months in both the axial and mediolateral directions. Comparing the immature and mature groups, the elastic energy absorption in the immature group was 30% lower in the axial direction ($p=0.017$), 35% lower in the anteroposterior direction ($p=0.003$), and 30% lower in the mediolateral direction ($p=0.08$).

Following the nondestructive testing, each sample was compressed to failure in the axial direction at a constant rate of 0.001/s. Properties recorded included Young's modulus, ultimate stress, ultimate strain, and energy absorption to failure (Table 3.11). In agreement with the findings from the nondestructive axial direction testing, the Young's modulus in the mature group was almost double that of the immature group ($p=0.017$). Further analysis of the Young's modulus in the different age groups indicated that bone stiffness increases slowly during skeletal maturation, but that a further significant increase occurs after skeletal maturity. The ultimate stress demonstrated an age dependence similar to that of the Young's modulus and was 38% lower in the immature than in the mature group ($p=0.017$). The ultimate strain was found to demonstrate an age dependence that was almost exactly the opposite to that of the ultimate stress and was 20% higher in the immature than in the mature group ($p=0.001$). The energy absorption to failure showed relatively modest variation during skeletal maturation but increased with age from age 9 to 80 months. Comparing the mature and immature groups, the absorption was 39% lower in the immature group ($p=0.001$).

Nafei et al. (2000) emphasized that the numerical values were not directly applicable to human, but that the patterns of variations and the influential factors probably applied to human bone. They concluded that age has a significant influence on the physical and mechanical properties of trabecular bone, and that skeletally immature trabecular bone is less dense, weaker, less stiff, more deformable, and has higher shock-absorptive qualities than skeletally mature trabecular bone.

Discussion

Combining the results from the few studies that have tested samples of human pediatric cortical bone, there are still only a handful of data points available, and with the exception of one 10-year-old subject utilized by Weaver and Chalmers (1966), all specimens were harvested from subjects 14 years of age or older. Consequently, the existing human pediatric data do not provide any information on how the material properties depend on age during maturation.

In contrast to the human subject studies, the study utilizing trabecular bone samples from lamb and sheep by Nafei et al. (2000) reported on the results from a large number of specimens from animals at various developmental stages. Nafei et al.'s findings that strength and stiffness increase during maturation and thus are higher in skeletally mature than in skeletally immature animals are in agreement with the previously reported findings for cortical bone from human (Currey and Butler 1975) and dog (Torzilli et al. 1982). However, in contrast to the energy absorption to failure for human cortical bone, which according to Currey and Butler (1975) decreases with age during maturation and is lower in mature than in immature bone, Nafei et al.'s results indicate that the energy absorption characteristics for trabecular bone show modest variation during, at least, part of the skeletal maturation process but then increase with age. As previously emphasized in the discussion of the cortical dog bone study by Torzilli et al. (1982) and also pointed out by the authors themselves, the numerical results from the study by Nafei et al. are not directly applicable to human. This is for reasons associated with differences in trabecular bone composition (and thus material properties) between human and ovine (Aerssens et al. 1998), as well as the difficulties associated with correlating the developmental age between the two species. However, for the purpose of studying the influence of age on the material properties of human trabecular bone during maturation, the study by Nafei et al. provides patterns of variations and identifies influential factors that, according to the authors themselves, probably are applicable to human trabecular bone development.

Growth Plate Cartilage

The growth plate (epiphyseal plate, physis) is a region of cartilage present between the epiphysis and metaphysis in immature long bones. Its function is to produce longitudinal growth of the diaphysis by endochondral ossification. It begins as a planar structure but becomes contoured in response to biomechanical stresses allowing the characteristic shape of each bone to develop. There are small projections of cartilage extending into the metaphysis called mammillary projections providing some resistance to shear and torsional forces, as this region of the bone is a potential source of weakness. The growth plate contains different zones that may be defined according to function or histology and are described differently by different sources. Eventually, the growth plate cartilage cells cease to divide and the cartilage is replaced by bone.

The growth plate typically fails in torsion but is also susceptible to crushing or shearing forces. Many classification systems exist for these injuries, but the most widely used is the system devised by Salter and Harris (1963), which defines the following five types of growth plate injury:

- Type I (8%): Separation of the epiphysis and metaphysis through the growth plate. This type of fracture is more commonly seen in infants and younger children following shearing, torsion, or avulsion causing separation through the growth plate. The hypertrophic zone is cleaved leaving the growing cells on the epiphysis in continuity with their blood supply, giving a good prognosis.
- Type II (72%): A fracture extending into the metaphysis. The fracture line runs through the hypertrophic zone of the growth plate and out through the metaphysis. Again, the blood supply to the growing cell layer is preserved, giving a good prognosis.
- Type III (6.5%): A split of the epiphysis extending into the joint. This is an intraarticular fracture of the epiphysis with extension through the hypertrophic layer. The prognosis is usually good but as the fracture line traverses the growing cell layer, growth disturbance may occur following injury. The greater the fragment displacement, the more likely this is to occur.
- Type IV (12%): A vertical split through the epiphysis, growth plate, and metaphysis. This type of fracture is most common at the lower end of the humerus. Again, the incidence of growth disruption is related to the degree of blood supply disruption.
- Type V (<1%): Compression of the growth plate. This is the rarest injury type and the one most likely to lead to growth arrest. The growth plate is crushed, damaging the reserve and proliferative cell layers. There is minimal displacement, and this injury is often diagnosed in retrospect once growth disturbance is recognized.

Human Subject Studies

Chung et al. (1976) conducted quasi-static shear tests on 25 pairs of hips obtained from pediatric PMHS ranging in age from 5 days to 15 years and 10 months to determine the shear strength and modes of failure of the capital femoral epiphyseal plate (Fig. 3.6). A custom made jig with clamps to hold the femoral neck, trochanter, and the proximal part of the femoral shaft was used to hold the specimens during testing. One specimen of each pair was tested after removal of the perichondrial fibrocartilaginous complex (PFC) surrounding the epiphyseal plate, while the contralateral specimen was tested with the PFC intact. The load was applied in the anteroposterior direction exactly over the secondary center of ossification, with the line of travel of the loading ram parallel to the plane of the epiphyseal plate. As such, the specimens were assumed to fail in shear. The failure mode, shear failure load, cross-sectional area, and calculated shear failure stress from the tests in which the PFC was left intact and was removed are listed in Tables 3.12 and 3.13, respectively.

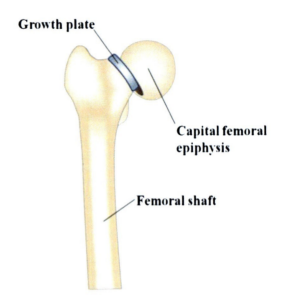

Fig. 3.6 Drawing illustrating the proximal femur and the location of the capital femoral epiphyseal plate (growth plate). As shown, relative motion (slip) has here occurred between the capital femoral epiphysis and growth plate. This condition is normally referred to as slipped capital femoral epiphysis (SCFE)

Of the 25 matched pairs of hips tested, there were 18 pairs for which failure occurred by means of epiphyseal slipping in both specimens. Using the results for the 18 specimens of those matched pairs that had been tested with the PFC intact, Chung et al. utilized linear regression to estimate the following function predicting the average shear strength as a function of age:

$$\tau_f = 0.644 + 0.054 \times age \quad (r = 0.79),$$

where τ_f is the average shear failure stress in MPa on the cross section made up of the capital epiphyseal plate and the surrounding PFC, and *age* is the child age in years. It is important to emphasize that this function does not predict the average shear failure stress of the capital epiphyseal plate but instead the average shear failure stress of the combined structure consisting of the capital epiphyseal plate and the surrounding PFC.

Chung et al. (1976) also used the results from the matched pairs for which failure occurred by means of epiphyseal slipping in both specimens to study the contribution of the PFC to the shear resistance of the PFC-epiphyseal plate combination. For each of those matched pairs, the ratio between the shear failure loads recorded for the specimen with and the specimen without the PFC was determined. This ratio was high for the matched pairs from the youngest subjects but approached one for the matched pairs from subjects older than 3 years. This led the authors to conclude that the contribution of the PFC to the overall shear resistance is relatively low in children older than 3 years.

Table 3.12 Results from Chung et al.'s (1976) shear tests of human pediatric hips for which the PFC was left intact

Subject	Specimen #	Age (years, months)	Race	Sex	Failure mode	F_u (N)	A (mm²)	τ_f (MPa)
1	1	5 days	Black	F	High metaphys. fract.	60.2	154	0.39
2	3	0, 3	White	F	Low metaphys. fract.	99.0	201	0.49
3	5	0, 6	White	F	Greenstick bending	120.7	291	0.42
4	7	0, 7	White	M	Plate fracture	75.5	–	–
5	9	1, 1	White	M	Greenstick bending	132.4	497	0.27
6	11	1, 2	Black	M	Salter I	320.8	485	0.66
7	13	1, 6	White	M	Salter I	285.5	377	0.76
8	15	1, 6	Black	M	Salter I	395.3	518	0.76
9	17	1, 7	White	M	Neck fracture	171.7	419	0.41
10	19	2, 0	Black	M	Salter I	857.4	611	1.40
11	21	2, 0	White	M	Greenstick bending	196.2	442	0.44
12	23	2, 6	White	F	Greenstick bending	366.5	358	1.02
13	25	2, 9	White	F	Neck fracture	470.9	541	0.87
14	27	3, 0	White	F	Punch fracture head	544.5	536	1.02
15	29	3, 6	Black	F	Punch fracture head	461.1	675	0.68
16	31	4, 0	Black	F	Greenstick bending	892.7	917	0.97
17	33	5, 5	White	F	Salter I	686.7	653	1.05
18	35	6, 0	White	M	Neck fracture	696.5	704	0.99
19	37	8, 8	Black	F	Salter II	1,404.8	987	1.42
20	39	9, 10	White	M	Neck fracture	1,520.6	1,182	1.29
21	41	10, 9	White	M	Salter I	1,672.6	1,696	0.98
22	43	12, 0	White	F	Neck fracture	1,697.1	1,382	1.23
23	45	12, 4	White	F	Salter II	912.3	1,454	0.63
24	47	13, 2	White	F	Salter II	1,334.2	1,146	1.16
25	49	15, 10	Black	F	Intertronch. fracture	2,060.1	–	–

Salter I: crack through the growth plate–metaphyseal bone junction without a significant portion of the metaphysis attached to the growth plate, *Salter II*: crack through the epiphyseal–metaphyseal bone junction leaving a significant metaphyseal beak attached to the growth plate, F_f: shear failure force, A: cross-sectional area of the PFC–epiphyseal plate combination, τ_f: shear failure stress (F_f/A)

Table 3.13 Results from Chung et al.'s (1976) shear tests of human pediatric hips for which the PFC was removed prior to testing

Subject	Specimen #	Age (years, months)	Race	Sex	Failure mode	F_u (N)	A (mm^2)	τ_f (MPa)
1	2	5 days	Black	F	Salter I	8.9	41	0.22
2	4	0, 3	White	F	Salter I	40.1	74	0.54
3	6	0, 6	White	F	Salter I	80.2	140	0.58
4	8	0, 7	White	M	Salter I	11.1	71	0.16
5	10	1, 1	White	M	Salter I	58.9	288	0.20
6	12	1, 2	Black	M	Salter I	134.4	210	0.64
7	14	1, 6	White	M	Salter I	120.4	216	0.56
8	16	1, 6	Black	M	Salter I	132.4	295	0.45
9	18	1, 7	White	M	Salter I	54.0	157	0.34
10	20	2, 0	Black	M	Salter I	441.5	338	1.31
11	22	2, 0	White	M	Salter I	127.0	209	0.61
12	24	2, 6	White	F	Salter I	281.0	201	1.40
13	26	2, 9	White	F	Salter I	431.6	268	1.61
14	28	3, 0	White	F	Salter I	323.7	300	1.08
15	30	3, 6	Black	F	Salter I	423.8	408	1.04
16	32	4, 0	Black	F	Salter I	642.1	449	1.43
17	34	5, 5	White	F	Salter I	407.1	395	1.03
18	36	6, 0	White	M	Salter I	539.6	459	1.18
19	38	8, 8	Black	F	Salter II	1,092.4	675	1.62
20	40	9, 10	White	M	Salter II	1,353.8	865	1.57
21	42	10, 9	White	M	Salter I	1,447.0	1,154	1.25
22	44	12, 0	White	F	Salter II	1,255.7		–
23	46	12, 4	White	F	Salter II	382.6	1,088	0.35
24	48	13, 2	White	F	Salter II	–	897	–
25	50	15, 10	Black	F	Punch fracture	2,786.0		–

The shear failure stress data given in the last column were not presented by Chung et al. but were calculated by the authors of the current chapter using the shear failure forces and cross-sectional areas recorded by Chung et al.

Salter I: crack through the growth plate–metaphyseal bone junction without a significant portion of the metaphysis attached to the growth plate, *Salter II*: crack through the epiphyseal–metaphyseal bone junction leaving a significant metaphyseal beak attached to the growth plate, F_f: shear failure force, A: cross-sectional area of the epiphyseal plate, τ_f: shear failure stress (F_f/A)

Chung et al. (1976) did not determine the shear failure stress for the specimens tested after the PFC had been removed. However, in order to provide data on the individual shear strength of epiphyseal plate cartilage, the authors of the current chapter estimated this property using the shear failure loads and cross-sectional areas recorded by Chung et al. in the tests of the specimens for which the PFC had been removed (Table 3.13). Figure 3.7 shows the calculated shear failure stress as a function of donor age for all specimens with and without the PFC that failed at the level of the epiphyseal plate. Linear functions have been fitted to the data from the specimens with and without the PFC and indicate that the shear failure stress for

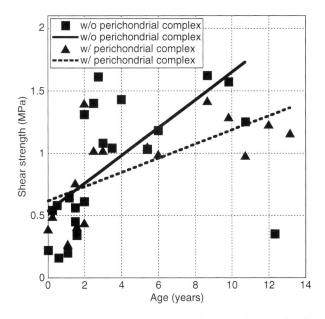

Fig. 3.7 Shear failure stress of the human capital femoral growth plate as a function of donor age. The data were obtained in quasi-static shear tests of 25 pairs of human hips by Chung et al. (1976). Only data for the specimens that failed at the level of the epiphyseal plate are shown. The regression line fitted to the data from the specimens without the PFC has the function $\tau_f = 0.53 + 0.111 \times \text{age}$ ($r = 0.703$) and does not account for the data point from the 12 years and 4 months old specimen considered to be an outlier. The corresponding function fitted to the data from the specimens tested with the PFC intact is $\tau_f = 0.61 + 0.057 \times \text{age}$ ($r = 0.680$)

the PFC-epiphyseal plate combination and the epiphyseal plate cartilage is fairly similar[1].

Williams et al. (2001) conducted tensile tests with 40 bone–growth plate–bone samples (0.51 mm × 2.29 mm × 15 mm) cut from two femoral heads removed from two male cerebral palsy patients aged 8 and 14 years. The two patients had required resection of the femoral head due to chronic and painful hip dislocation. The average growth plate thickness for the 40 samples was 1.35 mm. All samples were distracted to failure in uniaxial tension at a strain rate of 0.003/s, but only one sample from the 8-year-old and seven samples from the 14-year-old failed through the growth plate cartilage. For those eight samples, the mean ± SD ultimate stress, ultimate strain, and

---

[1] The linear function in Figure 3.7 that is fitted to the data from the specimens tested with the PFC intact is slightly different from the linear function fitted to the same data by Chung et al (1976). This difference is explained by the fact that Chung et al. used the average shear failure stress for specimens from subjects of the same age, whereas the function shown in Figure 3.7 was fitted to all individual data points. For the same reason, the correlation coefficients of the two functions (0.68 vs. 0.79) are also different.

tangent modulus were 0.98 ± 0.29 MPa, 31 ± 7%, and 4.26 ± 1.22 MPa, respectively. Williams et al. emphasized that in vivo loading of the capital femoral growth plate in cerebral palsy patients with chronic painful hip dislocation would be expected to be abnormal, and that it therefore was questionable whether the growth plate samples tested were normal.

Animal Studies

Morscher et al. (1965) conducted tensile tests on specimens of proximal tibia obtained from rats aged 25–85 days. Sexual maturation in the rat was said to begin at an age of 40 days. Prior to testing, the proximal ends of the tibiae were dissected free from all soft tissues including muscles, ligaments, and periosteum. The recorded values of ultimate tensile strength of the proximal tibial growth plate increased from approximately 0.6 MPa for the youngest specimens to 1.25 MPa in the oldest specimens (75 days old male rats and 85 days old female rats).

Bright et al. (1974) harvested the 560 tibiae from 280 rats aged 25–65 days and dissected them free of all soft tissue. Each tibia was positioned in a holding device and loaded at the proximal epiphysis in a direction perpendicular to the proximal growth plate. This loading configuration imposed a combination of shear and bending moment in the proximal tibia and caused failure to occur through the growth plate (the loading configuration was said to simulate clinical growth plate trauma). The growth plate failure properties determined included nominal shear failure stress (the ratio of applied force at failure and growth plate area) and maximum tensile stress due to the imposed bending moment. The nominal shear stress at failure varied from approximately 0.43 MPa in the youngest specimens to 1.6 MPa in the oldest (female) specimens. A similar trend with age was recorded for the tensile stress at failure which varied from approximately 3.9 MPa in the youngest specimens to 11.8 MPa in the oldest (female) specimens. Bright et al. hypothesized that transverse loading cause shear cracks to start developing within the growth plate (along the planes of highest shear stress) when the growth plate structure has absorbed about 50% of its failure energy. If the transverse load is released at this point, the shear cracks remain and cause a permanent reduction of the transverse loading strength of the growth plate. However, if the applied transverse load instead continues to increase, a secondary crack occurs at the outermost fiber of the growth plate due to the tension imposed by the applied bending moment. This secondary crack then propagates and causes failure of the growth plate.

Amamilo et al. (1985) subjected the proximal tibiae from 110 rats aged 25–65 days to axial and perpendicular loading to failure at a rate of 10 mm/min to investigate the strength of the epiphysis–metaphysis junction (growth plate) with and without the surrounding periosteum present. The age range 25–65 days was said to be the period of rapid growth and sexual maturation in the rat. For the specimens without periosteum, the failure stress recorded in axial and perpendicular loading did not increase monotonically with age. The ultimate tensile stress in the axial tests with specimens without periosteum was in the approximate range of 0.4–0.9 MPa. The corresponding range of shear stress recorded in the perpendicular tests was 0.4–0.7 MPa. Amamilo et al. concluded that the periosteum plays an important role in protecting the growth plate because removal of the periosteum resulted in a loss of perichondrial stability. This is in agreement with Morscher (1968) who demonstrated that the failure load of the growth plate with intact periosteum is almost twice that required when the periosteum is absent.

Rudicel et al. (1985) obtained 40 femora from 20 male immature New Zealand white rabbits aged 2, 4, 6, 10, and 14 weeks (four pairs of femora in each age group) and subjected them to quasi-static (0.5 mm/s) shear loading to failure in the major plane of the capital femoral growth plate. Complete failure was defined as separation of the capital femur from the adjacent metaphysis. The failure shear stress was found to increase monotonically with age from approximately 1.4 MPa at age 2 weeks to about 3.2 MPa at age 6 weeks after which the rate of increase leveled off. The approximate failure shear stress for the 10 and 14 week old specimens was 3.8 and 4.1 MPa, respectively. Rudicel et al. also noted that the zone of the growth plate through which shear fracture occurred progressed from the columnar zone in the youngest specimens to the zone of hypertrophy in the oldest specimens.

Guse et al. (1989) performed tensile tests (strain rate 0.25 mm/s) on specimens of distal femoral growth plate removed from 16 skeletally immature rabbits of various age and weight. The femora from 12 rabbits were tested with the perichondrium–periosteum complex intact, while the femora from the four remaining rabbits were tested after the complex had been removed. For the femora without the perichondrium/periosteum complex, the recorded values of ultimate tensile strength were in the range of 0.8 to 2.5 MPa and tended to increase with age and body weight. The authors stated that the ultimate tensile properties of the perichondrial ring are not tested until the stiffer and less elastic growth plate has failed and consequently, that the tensile strength of the entire growth plate complex is mainly provided by the growth plate itself and not by the surrounding perichondrium/periosteum complex. This is agreement with Bright et al. (1974) but contradicts the findings of Morscher (1968) and Amamilo et al. (1985).

Cohen et al. (1992) determined the tensile properties of distal femoral growth plates from 12-month-old cows using uniformly prepared bone–growth plate–bone specimens obtained from predetermined anatomical sites on the physis. The mean ultimate stress and strain varied with location on the physis and were in the ranges of 2.16–4.10 MPa and 11.7–16.0%, respectively. The constitutive behavior of the growth plate was described by a function of the form $\sigma = A(e^{B\varepsilon} - 1)$ in which the parameters A and B were determined by nonlinear regression. The tangent modulus at 75% of the ultimate strain was chosen to represent the stiffness of the growth plate in tension and varied from 23.5 MPa (posterior/medial region) to 48.6 MPa (anterior region).

Fujii et al. (2000) performed tensile tests on epiphysis–growth plate–metaphysis specimens from radii and ulnae removed from 8-week-old rabbits. Prior to testing, the periosteum and perichondrial ring were removed. All specimens fractured according to Salter-Harris I. The mean ultimate stress recorded for the radial and ulnar growth plates were 1.05 and 1.03 MPa, respectively.

Williams et al. (2001) conducted tensile tests on 40 bone–growth plate–bone samples cut from tibiae of three 12–18 month old heifers and three 5-month-old male calves. The average growth plate thickness of the 21 heifer samples and 19 calf samples were 0.65 and 0.50 mm, respectively. All specimens were loaded in uniaxial tension to failure at a strain rate of 0.008/s. Compared to the calf samples, the heifer samples failed at 34% higher tensile stress (1.82 ± 0.55 MPa vs. 1.36 ± 0.53 MPa, $p<0.001$) and 65% greater tensile strain ($38 \pm 13\%$ vs. $23 \pm 6\%$, $p<0.01$). The tangent modulus was, however, 10% higher for the samples from the calves (7.52 ± 2.15 MPa vs. 6.89 ± 1.96 MPa, $p=0.04$).

Discussion

The study by Chung et al. (1976) provides relatively thorough data for the shear strength of human growth plate cartilage and indicates that this strength increases with age during maturation. Further testing is required to characterize the stiffness as well as the response and failure threshold in loading perpendicular to the growth plate.

While the animal studies are not directly applicable to human due to potential differences in growth plate cartilage composition between species as well as the difficulty associated with correlating the developmental ages between man and animal, they provide several interesting indications. First, five of the six animal studies that reported on the influence of age on the properties concluded that the tensile strength and/or shear strength of the growth plate cartilage increase with age during maturation, a finding in support of Chung et al.'s (1976) human data. Second, the three reviewed animal studies that reported on shear strength reported ranges similar to those determined for the human specimens by Chung et al. (1976), suggesting that these species (rat and rabbit), and possibly also other species, may be appropriate for animal models for studying the response and tolerance of human growth plate cartilage to various types of loading.

Tendon

Tendons attach muscles to bone and transmit the forces generated by muscle contraction to effect the desired movement at the joint. They are dense, regularly arranged tissues containing a parallel arrangement of collagen fibrils aligned along the axis of the tensile forces they experience. The insertion to bone contains fibrocartilage which helps to dissipate the forces at this interface. Several experimental studies have investigated the tensile properties of tendons as functions of age from young adulthood to senescence and reported only minor effects (Hubbard and Soutas-Little 1984; Johnson et al. 1994; Flahiff et al. 1995). The methodology and findings from the studies that have investigated the effects of age during maturation are summarized below.

Human Subject Studies

Rollhäuser (1950) conducted tensile testing with tendon specimens obtained from fresh PMHS and amputated extremities. In total, specimens were obtained from 43 adult and infant subjects. Five to eight tendons were obtained from each subject. Given the maximal tensile force of the testing apparatus (approximately 392 N), Rollhäuser selected either small tendons with cross-sectional areas not exceeding 3 mm^2 or larger tendons that were split longitudinally into individual thinner strands. The ultimate strength and elongation for the infant specimens were in the approximate ranges 29–44 MPa and 14–18%, respectively. The corresponding ranges for the adult specimens were 42–113 MPa and 10–12.5%. Young's modulus was found to be about 1 GPa for the adult tendon and 350 MPa for the infant tendon. Rollhäuser attributed the lower strength of the infant tendon to its higher water content but also due to the micellae of the adult tendon having better alignment with the load path.

Stucke (1950) investigated the mechanical properties of the human Achilles tendon using specimens obtained from pediatric and adult PMHS. Ultimate stress and strain were measured for two pediatric specimens (13 and 15 years). The youngest specimen demonstrated ultimate stress and strain of 52 MPa and 3.5%, respectively. The corresponding values recorded for the specimen from the 15-year-old subject was 44 MPa and 6.9%. The highest mechanical performance was said to occur for specimens from subjects in their third decade of life for which the average ultimate stress and strain were 46 MPa and 7.34%, respectively. According to Elliot (1965), the relatively low values of ultimate strength reported by Stucke (1950) was possibly due to the methodology used, which involved the excision of two wedge-shaped segments from the midsection of the specimens in order to ensure that rupture would occur there and not at the end clamp.

Blanton and Biggs (1970) investigated the tensile strength of human tendons harvested from the upper and lower extremities of 20 embalmed fetuses, two unembalmed fetuses, seven unembalmed adult amputated limbs, and five embalmed adult PMHS. One of the unembalmed fetuses was a full-term stillborn and the other one a premature stillborn. Tendons selected for tensile strength testing had to be at least 25.4 mm long and of "fairly" uniform diameter. The cross-sectional area of each selected tendon was estimated by projecting the tendon cross section onto a screen at ×30 magnifications and using a planimeter read to the nearest one-hundredth square inch. Each tendon selected for testing was mounted in a custom designed apparatus and subjected to quasi-static tension to failure. All specimens ruptured "fairly" evenly and usually in the middle between the two gripping jaws of the test apparatus. Means, standard deviations, and ranges of ultimate tensile strength of all the fetal and adult tendon specimens are shown in Table 3.14. The embalmed specimens had higher ultimate strength than the unembalmed specimens. Considering the unembalmed specimens only, the mean ultimate strength of fetal specimens was somewhat higher than the corresponding adult value. This contradicts the previously summarized findings of Rollhäuser (1950) as well as Elliot (1965) who stated that the tensile strength of human tendon increases with age from birth to adulthood.

Table 3.14 Ultimate tensile strength of fetal and adult human tendons as reported by Blanton and Biggs (1970)

Group	Number of specimens	Mean ± SD (MPa)	Range (MPa)
Fetal embalmed	149	68.8 ± 56.1	9.8–477.1
Fetal unembalmed	11	25.8 ± 32.9	7.1–128.9
Adult embalmed	43	118.3 ± 39.5	55.6–190.5
Adult unembalmed	38	20.1 ± 9.9	9.3–48.3

Table 3.15 Mean ± SD of ultimate tensile strength and ultimate tensile strain for calcaneal tendinous tissue as reported by Yamada (1970)

Age group (years)	σ_u (MPa)	ε_u (%)
0–9	52.0	11.0
10–19	54.9	10.0
20–29	54.9 ± 0.88	9.9 ± 0.11
30–59	54.9 ± 0.88	9.5 ± 0.14
60–69	52.0 ± 1.37	9.1 ± 0.23
70–79	43.2 ± 1.67	9.1 ± 0.23

σ_u: ultimate tensile strength, ε_u: ultimate tensile strain

Yamada (1970) reported on the tensile properties of calcaneal tendinous tissue obtained from 42 persons and 19 fetuses. Means and standard deviations of ultimate tensile strength and ultimate tensile strain for various age groups are shown in Table 3.15. The tensile strength appears to increase from the first decade of life and then remains constant from the second decade until senescence after which it decreases. The ultimate tensile strain appears to decrease monotonically from the first to the seventh and eight decades of life.

Kubo et al. (2001) investigated the growth changes in the elastic properties of human tendon structures in vivo. Nine younger boys (10.8 ± 0.9 years), nine elder boys (14.8 ± 0.3 years), and 14 young adult males (24.7 ± 1.6 years) performed isometric knee joint extension, while the elongation of the tendon and aponeurosis of the vastus lateralis muscle was measured noninvasively using a B-mode ultrasonic apparatus. The mean ± SD of maximum voluntary contraction recorded in the young boys group, elder boys group, and young adult male group was 114.1 ± 13.4, 164.9 ± 20.7, and 234.0 ± 38.9 Nm, respectively. The corresponding results for the tensile compliance of the tendon structures were 41 ± 9, 29 ± 11, and 18 ± 3 μm/N and indicate that the tensile stiffness of tendon structures increases during maturation. Material properties were not reported.

Animal Studies

Shadwick (1990) conducted tensile tests on digital flexor and extensor tendons removed from the front feet of two newborn and nine adult (age 4–5 months) pigs.

Table 3.16 Mean ± SD or range of the mechanical properties of digital extensor and flexor tendons from newborn and mature pigs as determined by Shadwick (1990). The results for newborn extensor and flexor tendons were indistinguishable and have therefore been combined

Age and tendon type	E (GPa)	σ_u (MPa)	ε_u (%)	Hysteresis (%)	$W_{3\%}$ (J/kg)
Newborn extensor and flexor	0.16 ± 0.02	<20	~12–15	24.5 ± 3.5	7
Mature extensor	0.76 ± 0.12	40–50	n/a	17.5 ± 5.2	164
Mature flexor	1.66 ± 0.16	80–90	6.8–8.5[a]	9.2 ± 2.4	415

[a]Based on the results from two specimens
E: Young's modulus (the gradient of the linear portion of the stress–strain curve), σ_u: ultimate tensile strength, ε_u: strain at σ_u, $W_{3\%}$: strain energy density recovered elastically from 3% strain

Tendon cross-sectional areas were estimated by dividing the wet weight of the tendon by its length and density. The tendons were subjected to either quasi-static extension at a constant elongation rate of 5 mm/min or to sinusoidal loading at frequencies of 0.055 or 2.2 Hz. In the quasi-static extension tests, the change in length was monitored using a video dimension analyzer that recorded the displacement of two surface markers glued onto the central section of the specimen. The cyclic tests utilized an electronic extensometer to measure the length changes at the central portion of the specimen.

The flexor and extensor tendons obtained from the newborn animals demonstrated indistinguishable mechanical properties. Following the initial toe region associated with straightening of collagen fibrils, the stress–strain curve was linear with a gradient (Young's modulus) that averaged 0.16 GPa. The mechanical properties of the tendons from the adult animals were highly dependent on tendon type with the flexor tendons demonstrating approximately twice the tensile strength and elastic modulus but only half the energy dissipation of the extensor tendons. A summary of the mechanical properties determined from the tests is shown in Table 3.16. The newborn tendons demonstrated lower stiffness, ultimate strength, and energy dissipation but higher ultimate strain and hysteresis than the mature tendons. According to Shadwick, the major findings of the study were that the mechanical properties of the pig digital tendons change dramatically from birth to maturity, and that these changes are significantly greater for the major load bearing flexors than for the less stressed extensors.

Haut et al. (1992) conducted tensile failure experiments on patella–patellar tendon–tibia specimens removed from 14 male and 13 female medium and large breed dogs aged 6–180 months. Each specimen was cyclically loaded (preconditioned) between 90 and 180 N at 1 Hz for 20 cycles. This was immediately followed by extraction to failure at a nominal strain rate of 1/s. Sixteen of the 27 tested specimens failed by avulsion fracture of bone from the patella. Neither the failure loads nor the mode of failure was a function of animal age. The tensile modulus was calculated as the slope of the linear part of the stress–strain curve. A positive trend but no significant age dependence ($p \geq 0.05$) was found for the tensile modulus that averaged 474 ± 101 MPa. For the 11 specimens that failed by tendon rupture, ultimate stress, ultimate strain, and failure strain energy density did not vary with animal age, being 122 ± 25.6 MPa, 32.3 ± 7.8%, and 25.7 ± 8.7 J/cm^3, respectively. Total

Table 3.17 Mean ± SEM of the tensile properties of immature and mature rabbit Achilles tendon as determined by Nakagawa et al. (1996)

Age group	σ_u (MPa)	E (MPa)	ε_u (%)
Immature (mean age 3 weeks)	23.9 ± 3.8	281.0 ± 104.6	15.7 ± 2.9
Young adult (8–10 months)	67.3 ± 4.2	618.0 ± 87.0	16.3 ± 2.7
Old (4–5 years)	66.7 ± 5.7	530.5 ± 91.0	16.3 ± 1.8

E: Young's modulus (the gradient of the linear portion of the stress–strain curve), σ_u: ultimate tensile strength, ε_u: strain at σ_u

collagen concentration decreased significantly with age, whereas the insoluble fraction of collagen increased significantly with age.

Nakagawa et al. (1996) conducted tensile tests on specimens of Achilles tendons removed from four immature (mean age 3 weeks), four young adult (age range 8–10 months), and five old (age range 4–5 years) Japanese white rabbits. The cross-sectional area of each tendon specimen was estimated as the average of the cross-sectional area measured at the proximal, middle, and distal regions using an area micrometer. Following ten preloading cycles, the specimens were extended to failure at a strain rate of 1/s. The change in length was monitored using a video dimension analyzer that recorded the displacement of two parallel lines drawn at the mid-substance of the tendon perpendicular to its long axis.

All specimens failed mid-substance. The tensile strength of the immature specimens was significantly ($p < 0.05$) lower than the corresponding strength of the specimens from the young adult and old groups. No significant differences were observed in the elongation at failure or tangent modulus although the immature tendons seemed to have a lower modulus than the tendons from the young adult and old animals. The results from the tests are summarized in Table 3.17.

Discussion

The results from the elongation testing of human fetal and adult tendons by Rollhäuser (1950) and Blanton and Biggs (1970) are in direct disagreement with each other with the former indicating that the tensile strength is higher for adult than for infant tendinous tissue and the latter suggesting the opposite. Yamada's (1970) results instead indicate that the tensile strength of human tendinous tissue is approximately constant over the six, or maybe even seven, initial decades of life after which it diminishes. Considering instead ultimate strain, the studies also provide conflicting information, both in terms of the recorded magnitude and the potential effect of age.

Two of the three reviewed studies utilizing animal specimens reported higher values of ultimate strength and Young's modulus for specimens harvested from mature than immature subjects. These two studies, which utilized tissue from pig (Shadwick 1990) and rabbit (Nakagawa et al. 1996), also reported similar values of ultimate strength, Young's modulus, and ultimate strain when comparing the corresponding age groups of the two species. In contrast, Haut et al. (1992) reported substantially higher values of ultimate strength and strain and did not find any evidence of that these properties varied with age in the patellar tendon of dogs ranging in age from 6 months to 15 years.

While the majority of the reviewed human subject and animal studies indicate that the material properties of infant and adult tendinous tissue are different with the latter being stiffer and stronger, the conflicting information from the other studies suggests that this may not be the case, and that the effect of age on the material properties may be tendon dependent [as demonstrated in the study on pig extensor and flexor tendons by Shadwick (1990)]. The reviewed studies also do not provide any information on the possible effect of age on the material properties of tendinous tissue during maturation. Consequently, further research is needed to identify and quantify any potential effects of age on the material properties of tendinous tissue.

Finally, it is important to emphasize that the tensile strength values reported in the reviewed studies are determined as the ratio of the applied tensile force and the estimated initial cross-sectional area of the tendon specimens. Given the incompressible nature of biological soft tissues, the true cross-sectional area of the tendon at the time and location of tensile failure is therefore smaller than the initial cross-sectional area and the true ultimate tensile strength therefore higher than the reported nominal ultimate strength. This also applies to reported estimates of Young's modulus since this property is a function of the normal cross-sectional stress in the specimen. However, since Young's modulus is estimated at lower magnitudes of tensile elongation than the elongation associated with the ultimate strength, the difference between the true Young's modulus and the reported nominal value is smaller than the corresponding difference between true and reported nominal ultimate strength.

Ligament

Ligaments connect one bone to another and guide joint motion. They are flexible in order to permit joint movement but strong enough to resist abnormal movements at the joint. Ligaments have a structure similar to tendon but a more variable fiber composition and direction. They possess mechanoreceptors and nerve endings that help in joint stabilization by feeding information about joint position to the brain. The possible influence of age on the tensile properties of ligaments and bone–ligament–bone complexes has been investigated using specimens from both humans (Noyes and Grood 1976; Hollis et al. 1988; Rauch et al. 1988; Woo et al. 1991; Neumann et al. 1994; Lee et al. 1999; Fremerey et al. 2000) and animals (Booth and Tipton 1970; Tipton et al. 1978; Woo et al. 1986, 1990a, b; Lam et al. 1993). The majority of these studies have concluded that the tensile strength and tensile modulus of ligaments decrease from early adulthood to senescence. Unfortunately, few studies have addressed the influence of age on the tensile properties of ligaments during maturation. To our knowledge, there are no published results on the tensile properties of immature human ligaments.

Animal Studies

The by far most extensive studies on the tensile properties of immature ligaments have been conducted with specimens from rabbit (Woo et al. 1986, 1990a, b; Lam et al. 1993). In their 1986 article, Woo et al. reported on the structural properties of the rabbit femur–medial collateral ligament (MCL)–tibia complex and the material properties of the rabbit MCL. Specimens were removed from six animals aged 1.5 months (open epiphysis), ten animals aged 4–5 months (open epiphysis), nine animals aged 6–7 months (closed epiphysis), and eleven animals aged 12–15 months (closed epiphysis). Assuming a rectangular cross-section, Woo et al. estimated the cross-sectional area of the mid-ligament substance of each specimen using a vernier caliper to measure the width and thickness. The femur–MCL–tibia complex was mounted in an Instron machine such that a tensile load could be applied directly along the long axis of the ligament. Three parallel lines were stained on the ligament in a direction perpendicular to its long axis. The centerline was stained along the joint line directly above the tibial plateau, while the other two lines were stained 5 mm proximal and 5 mm distal to the midline. A preload of 0.5 N was applied to the specimen which then was subjected to ten cycles of loading-unloading with amplitude of 1.0 mm at an extension rate of 1.0 cm/min. Following cyclic testing, the specimen was stretched to failure at the same rate (which corresponded to a nominal strain rate of 0.003–0.004/s). By use of a video dimension analyzer, the change in distance between the outer two lines on the ligament was monitored during the test. The structural properties of the femur–MCL–tibia complex were determined from the load cell output and crosshead displacement. The load cell output divided by the initial cross-sectional area of the ligament gave the time history of ligament stress.

The stress–strain curves demonstrated that the tensile modulus of the rabbit MCL increases with age, but that the rate of increase levels off with age (no statistically significant differences in any of the structural or mechanical properties between specimens from the animals in the two oldest groups). The ultimate load, maximum deformation, and energy-absorption capability of the femur–MCL–tibia complex increased rapidly throughout maturation (from 1.5 to 6–7 months of age) and then leveled off. The tensile strength and ultimate strain of the ligaments from the animals in the two youngest groups could not be obtained since all the specimens from these animals failed by tibial avulsion rather than ligament rupture. In contrast, 67% of the specimens from the second oldest group and 56% of the specimens from the oldest group failed by ligament rupture.

Using the same methodology as in their 1986 study, Woo et al. (1990a) tested specimens of rabbit femur–MCL–tibia complexes removed from animals aged 3.5, 6, 12, 36, and 48 months. All except two of the 24 specimens from the 3.5 and 6-month-old animals failed by means of tibial avulsion, whereas all 38 specimens from the 12, 36, and 48-month-old animals failed by MCL rupture. The structural properties of the femur–MCL–tibia complexes and the material properties of the MCLs for the immature and mature specimens are shown in Table 3.18.

Table 3.18 Structural properties of the femur–MCL–tibia complex and material properties of the MCL as obtained in the tensile tests of rabbit femur–MCL–tibia complexes by Woo et al. (1990a)

Sex	Age (months)	n	Failure mode	Structural properties of the femur–MCL–tibia complex					Material properties of the MCL		
				k (N/mm)	F_f (N)	δ_f (mm)	J_f (mJ)	A (mm^2)	σ_f (MPa)	ε_f (%)	E (MPa)
Male	3.5	6	Tib avul	40.0±1.7	88.4±7.8	2.9±0.2	120±20	3.2±0.1	–	–	700±50
	6	6	Tib avul (n=4) MCL tear (n=2)	45.1±4.5	156.5±22.9	4.4±0.6	290±80	4.3±0.3	46.0±0.7[a]	11.2±2.1[a]	630±110
	12	6	MCL tear	64.2±5.7	313.2±18.8	5.9±0.3	860±70	3.6±0.2	84.4±5.7	10.6±0.8	1,180±90
	36	6	MCL tear	50.6±3.0	299.8±8.6	6.7±0.3	1,050±140	3.9±0.1	77.7±1.9	12.9±1.2	740±90
Female	3.5	6	Tib avul	31.6±2.0	87.8±4.4	3.5±0.3	130±20	2.6±0.1	–	–	750±70
	6	6	Tib avul	34.0±2.8	117.7±12.9	4.3±0.4	230±40	3.3±0.1	–	–	590±90
	12	14	MCL tear	48.2±3.1	290.4±20.2	6.5±0.5	770±150	4.0±0.1	75.8±4.8	9.4±0.7	950±80
	36	6	MCL tear	51.8±4.2	311.6±26.4	6.4±0.5	980±160	4.1±0.3	78.6±3.2	13.3±0.5	710±30
	48	6	MCL tear	50.3±2.9	267.0±26.7	5.8±0.6	770±150	3.9±0.3	68.9±4.9	11.9±1.3	520±120

The data are expressed as mean ± SEM
[a]Based on the two specimens that failed by MCL tear
Tib avul: tibial avulsion, n: number of specimens, k: linear stiffness, F_f: failure load, δ_f: failure elongation, J_f: failure energy, A: cross-sectional area, σ_f: failure stress, ε_f: failure strain, E: Young's modulus

Using the same type of specimens as in their 1986 and 1990a studies, Woo et al. (1990b) investigated the effects of extension rate on the structural properties of the femur–MCL–tibia complex and the material properties of the MCL in the immature and mature rabbit. Femur–MCL–tibia complexes from immature (3.5 months of age with open epiphysis) and mature rabbits (8.5 months of age with closed epiphysis) were stretched to failure at extension rates of 0.008, 0.10, 1.00, 10.0, and 113 mm/s (six specimens from each of the two age groups were tested at each extension rate). The corresponding strain rates for the MCLs were 0.00011 0.0015, 0.0159, 0.12, and 1.55/s for the immature specimens and 0.00011, 0.0015, 0.0166, 0.186, and 2.22/s for the mature specimens. All immature complexes failed by tibial avulsion, whereas all mature complexes failed by MCL rupture. Consequently, ultimate strength and ultimate strain of the MCL could only be obtained for the mature specimens. The modulus for the immature MCLs was found to increase with strain rate but not significantly. The stress–strain curves for the mature MCLs changed slightly with strain rate. The modulus was not significantly affected by strain rate ($p > 0.20$, two-way ANOVA). The tensile strength of the mature ligaments increased with strain rate ($p < 0.05$, one-way ANOVA), but the overall increase over the entire range of strain rates (four decades) was only 42%. The strain at failure also showed a significant increase with strain rate ($p < 0.05$, one-way ANOVA), but the increase over the entire range of strain rates (four decades) was only 37%. The structural properties of the femur–MCL–tibia complexes and the material properties of the MCLs for the immature and mature specimens are shown in Table 3.19.

Lam et al. (1993) investigated how the viscoelastic properties of bone–ligament–bone specimens change during maturation. Bone–MCL–bone specimens were obtained from the hind limbs of 24 New Zealand white rabbits aged 3, 6, 9, and 12 months. There were six animals in each of the four age groups. Using a materials testing machine, each specimen underwent a loading history consisting of 30 consecutive cycles of tensile distraction loading with an amplitude of 0.68 mm followed by 20 min of relaxation at 0.68 mm extension. The cross-head speed during the cyclic loading was 10 mm/min and translated to an average strain rate of 0.008/s. The 0.68 mm extension used both in the cyclic and subsequent static relaxation was chosen to produce an average tensile strain of 3% over the length of the ligament. The peak forces recorded in the first ten cycles of cyclic relaxation were normalized to the peak force in the first cycle. In order to allow for a comparison between the cyclic and static relaxations, the force signal recorded during static relaxation was normalized to the maximum force (at the start of the static relaxation) and retrieved for the same time points as for the peak forces in the cyclic testing.

The normalized peak load of the first ten cycles of cyclic relaxation decreased monotonically for all four age groups. However, while the results for the three older age groups were not significantly different from each other ($p \geq 0.05$), the relaxation for the 3-month-old specimens recorded at cycles five through ten was significantly higher than the corresponding levels recorded for the three older age groups. Similarly, the static relaxations recorded for the three oldest age groups were not significantly different from each other, whereas the corresponding relaxation recorded for 3-month-old specimens was significantly higher than the corresponding

Table 3.19 Structural properties of the femur–MCL–tibia complex and material properties of the MCL as determined by Woo et al. (1990b) in tensile tests with specimens from immature and mature rabbits. Six specimens from each of the two age groups were tested at each extension rate. The data are given as mean ± SEM

Age	Strain rate (per second)	Structural properties of the femur–MCL–tibia complex					Material properties of the MCL		
		k (N/mm)	F_f (N)	δ_f (mm)	J_f (J)	A (mm^2)	σ_f (MPa)	ε_f (%)	E (MPa)
Immature	0.00011	24.4 ± 1.2	54.3 ± 3.0	2.6 ± .1	0.07 ± .01	2.9 ± .1	–	–	610 ± 100
	0.0015	40.0 ± 2.4	84.5 ± 8.5	2.7 ± .1	0.09 ± .01	3.0 ± .1	–	–	620 ± 120
	0.0159	40.0 ± 1.9	88.4 ± 8.6	2.9 ± .2	0.12 ± .02	3.2 ± .1	–	–	700 ± 60
	0.12	41.2 ± 3.3	105.7 ± 8.9	3.2 ± .2	0.16 ± .02	3.1 ± .1	–	–	860 ± 70
	1.55	38.1 ± 1.4	123.7 ± 6.7	3.5 ± .2	0.20 ± .02	2.9 ± .1	–	–	970 ± 190
Mature	0.00011	52.8 ± 1.6	311.5 ± 12.1	6.4 ± .2	0.98 ± .08	4.1 ± .1	75.2 ± 2.4	9.5 ± 0.5	700 ± 70
	0.0015	54.8 ± 1.4	295.8 ± 28.1	5.6 ± .4	0.83 ± .14	3.6 ± .1	81.4 ± 7.4	10.0 ± 0.5	840 ± 150
	0.0166	54.3 ± 2.5	337.5 ± 23.7	6.6 ± .2	1.10 ± .11	4.0 ± .2	85.7 ± 7.7	11.0 ± 0.5	800 ± 60
	0.186	59.6 ± 2.8	347.1 ± 13.7	6.3 ± .2	1.10 ± .09	4.0 ± .2	87.2 ± 6.0	11.5 ± 1.0	910 ± 50
	2.22	60.7 ± 1.7	403.7 ± 7.5	6.6 ± .1	1.35 ± .06	3.8 ± .2	106.7 ± 6.8	13.0 ± 1.0	760 ± 160

k: linear stiffness, F_f: failure load, δ_f: failure elongation, J_f: failure energy, A: cross-sectional area, σ_f: failure stress, ε_f: failure strain, E: Young's modulus

relaxation recorded for the other age groups. As an example of the difference in relaxation behavior, Lam et al. compared the age group means of the static relaxation levels at the end of the 20 min hold time where the average load relaxation for the three older age groups ranged from 37 to 44%, and the corresponding relaxation for the youngest group was 68%. An additional effect of age on the viscoelastic properties became apparent when the static and cyclic relaxation results within each age group were compared. While the 3-month-old specimens behaved in accordance with the Boltzmann superposition principle for linear viscoelastic materials and thus demonstrated greater load relaxation during the static than during the cyclic testing, the results for the specimens from the older animals were the opposite, i.e., greater relaxation during the cyclic than during the static testing. Lam et al. proposed an explanation to their findings whereby the total relaxation was suggested to consist of two parts, a material relaxation and a water flux component, for which the individual contributions to the total relaxation varied with age.

Discussion

Several conclusions can be drawn from the studies of Woo et al. (1986, 1990a, b). The strength and stiffness of the rabbit femur–MCL–tibia complex increase with age during maturation. During maturation, the weakest link of the rabbit femur–MCL–tibia complex is not the ligament but instead the tibial insertion. The tensile modulus of the rabbit MCL is lower in the immature rabbit than in the young adult rabbit and seems to increase slightly with age during maturation. Since almost all of the immature complexes failed by tibial avulsion, it is not possible to conclude whether the ultimate strength and ultimate strain of the MCL in the rabbit is lower in the immature than in the adult rabbit and whether these properties increase with age during maturation. However, the results from the two immature male complexes that did fail by MCL rupture (see Table 3.18) indicate that the ultimate strength of the MCL is lower in the immature than in the mature rabbit, and that the ultimate strain of the MCL is approximately the same in the immature and mature rabbit. Strain rate seems to have a relatively low influence on the material properties of the MCL in the mature rabbit. Considering the tensile modulus only, strain rate seems to have a stronger influence on the immature than on the mature rabbit MCL. Further evidence of that age affects the viscoelastic properties (and thus the rate sensitivity) of the rabbit MCL is provided in the study by Lam et al. (1993).

While the studies by Woo et al. and Lam et al. provide strong indications that the material properties of ligamentous tissue in the rabbit vary, not only between immature and mature subjects, but also during maturation, the complete absence of corresponding human data prevents any attempts to determine whether the findings also apply to human ligamentous tissue. Finally, it should be emphasized that the values of stress and Young's modulus reported in the studies by Woo et al. are determined based on estimates of the initial cross-sectional area of the MCL specimens. Given the incompressible nature of biological soft tissues, the true cross-sectional area of the ligament at the time and location of tensile failure is therefore smaller than the initial cross-sectional area and the true stress and Young's modulus therefore higher than the corresponding reported ones determined using the initial cross-sectional area.

Articular Cartilage

The joints of the human skeleton include articular cartilage which enables bone to slip over one another smoothly. Articular cartilage can be divided into three distinct morphological zones that all have different properties. In the superficial zone (10–20% of the total thickness of the cartilage layer), the collagen fibers are oriented parallel to the articular surface; in the middle zone (40–60% of the total thickness of the cartilage layer), there is no preferred orientation for the collagen fibers; and in the deep zone (30% of the total thickness of the cartilage layer), the collagen fibers are approximately perpendicular to the articular surface. Several experimental studies have investigated the biomechanical properties of articular cartilage from different species in indentation, tension, unconfined and confined compression, stress relaxation, and creep and have reported very different results. To our knowledge, only two of those have reported results from testing of cartilage specimens from immature humans (Kempson 1982, 1991).

Human Subject Studies

Kempson (1982) conducted tensile tests with specimens of articular cartilage removed from the femoral condyles of 24 knee joints of PMHS aged 8–91 years. The dumbbell shaped test specimens (gauge length 3 mm, width 0.5 mm, and thickness 0.2 mm), which were oriented with their axes parallel to the predominant alignment of the collagen fibers in the superficial zone, were excised from the patellar groove and the central and posterior regions of the medial and lateral condyles and prepared from the superficial and deep articular layers. Using an Instron materials testing machine, the specimens were loaded at a constant rate of 5 mm/min in the plane parallel to the articular surface. The results for the specimens harvested from the five subjects aged 18 years or younger appear in Table 3.20.

The fracture stress of the superficial zone specimens increased with age up to the middle of the third decade and thereafter decreased with age. In contrast, the fracture stress of the deep zone specimens decreased monotonically with age over the entire age range investigated. The tensile stiffness for the superficial zone specimens at 5 and 10 MPa applied tensile stress increased with age up to the middle of the third decade and thereafter decreased with increasing age. Kempson suggested that the findings reflected changes in the organization of the collagen fiber mesh with age and possibly also changes in the collagen cross-links.

Kempson (1991) conducted tensile tests with specimens of articular cartilage removed from the femoral head and the talus of the ankle joint of human subjects aged 7–90 years. The specimens (gauge length 3 mm, width 1 mm, and thickness

Table 3.20 Mean values of fracture stress and tensile stiffness of human pediatric knee articular cartilage as reported by Kempson (1982)

Age (years)	Sex	$\sigma_{f\text{-}sup}$ (MPa)	$\sigma_{f\text{-}deep}$ (MPa)	$E_{5\text{-}sup}$ (MPa)	$E_{10\text{-}sup}$ (MPa)
8	M	25.5	27.0	139	67.1
9	F	31.3	31.7	101	79.9
16	M	33.6	–	122	77.0
17	M	29.1	19.8	122	82.9
18	M	30.1	17.5	127	89.5

$\sigma_{f\text{-}sup}$: superficial zone fracture stress, $\sigma_{f\text{-}deep}$: deep zone fracture stress, $E_{5\text{-}sup}$: superficial zone tensile stiffness at 5 MPa applied stress, $E_{10\text{-}sup}$: superficial zone tensile stiffness at 10 MPa applied stress

0.2 mm), which were oriented with their long axes parallel to the predominant alignment of the collagen fibers in the superficial zone, were prepared from the superficial and mid-depth cartilage layers. The specimens were loaded to failure in the direction of the long axis at a constant rate of 5 mm/min. The stiffness of each specimen at applied stress levels of 1 and 10 MPa was obtained from the gradient to the stress–strain curve. Fracture stress and stiffness at 1 and 10 MPa applied tensile stress were plotted as functions of age and regressed by exponential functions of the form $\exp[A + B \times age]$ in which A and B are constants determined from the regression and age the subject age in years.

The fracture stress recorded for the specimens from the superficial cartilage layer of the femoral head decreased from 33 MPa at 7 years of age to 10 MPa at age 90 years (Fig. 3.8), whereas the corresponding stress recorded for the specimens from the mid-depth cartilage layer of the femoral head decreased from 32 MPa at 7 years of age to 2 MPa at age 85 years. The fracture stress for the corresponding cartilage layers of the talus was less sensitive to age, decreasing from 24 MPa at age 8 years to 19 MPa at age 85 years (superficial layer) and from 17.5 MPa at age 18 years to 15 MPa at age 85 years (mid-depth level). The tensile stiffness at 10 MPa applied stress recorded for the specimens from the superficial cartilage layer of the femoral head decreased from 150 MPa at 7 years of age to 80 MPa at age 90 years (Fig. 3.8), whereas the corresponding stiffness recorded for the specimens from the mid-depth cartilage layer of the femoral head decreased from 60–100 MPa at 7 years of age to 10 MPa at age 60 years. The tensile stiffness at 10 MPa applied stress for the corresponding cartilage layers of the talus decreased from 125 MPa at age 10 years to 100 MPa at age 85 years (superficial layer) and increased from 35 MPa at 20 years to 40 MPa at age 85 years (mid-depth level). The tensile stiffness at 1 MPa applied stress recorded for the specimens from the superficial cartilage layer of the femoral head decreased from 70 MPa at 7 years of age to 50 MPa at age 90 years (Fig. 3.8), whereas the corresponding stiffness recorded for the specimens from the mid-depth cartilage layer of the femoral head decreased from 20–26 MPa at 15 years of age to 10–18 MPa at age 85 years. The tensile stiffness at 1 MPa applied stress for the corresponding cartilage layers of the talus decreased from 60 MPa at age 10 years to 55 MPa at age 85 years (superficial layer) and increased from 20 MPa at 18 years to 30 MPa at age 85 years (mid-depth level). Table 3.21 lists the values of the constants A and B of the fitted age dependent regression functions as well as the significance level of the constant B.

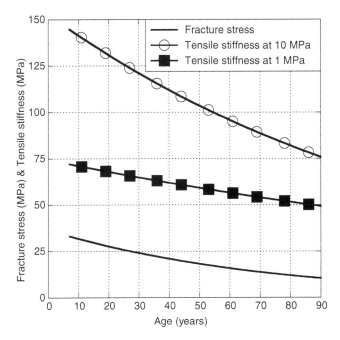

Fig. 3.8 Fracture stress and tensile stiffness at applied tensile stress levels of 1 and 10 MPa as a function of age for the human femoral head superficial cartilage layer as determined by Kempson (1991)

Discussion

The results from Kempson's (1991) testing demonstrate that the tensile fracture stress of the human femoral head superficial and mid-depth articular cartilage layers decrease with age of from childhood (age 7 years) through the ninth decade of life. The same holds true for the tensile stiffness of the superficial layer at applied tensile stress levels of 1 and 10 MPa and, for a smaller age range (7–60 years), for the tensile stiffness of the mid-depth layer at an applied tensile stress level of 10 MPa (a similar trend was also reported for the tensile stiffness of the mid-depth layer at an applied tensile stress level of 1 MPa but relied solely on data from specimens from subjects aged 15 years and older). While these findings provide strong indications that the tensile strength and stiffness of the superficial and mid-depth layer of human articular cartilage decrease monotonically with age from childhood, the corresponding results from Kempson's (1991) tensile testing of specimens of human talus superficial and mid-depth articular cartilage layers are much more limited (fewer data points overall and very few pediatric specimens) and therefore do not allow for verification of the findings from the femoral head cartilage testing. Consequently, it cannot be ruled out that the general findings presented for the superficial and mid-depth layers of human femoral head articular cartilage (monotonically decreasing

Table 3.21 Data from Kempson's (1991) tensile tests with specimens of superficial and mid-depth layer articular cartilage from the human femoral head and talus

Joint	Cartilage layer	Area	Number of joints	Number of specimens	Tensile property	A	B	p	Age range (years)
Fem head	Superficial	All	29	316	Fracture stress	3.6	−0.014	<0.001	7–90
Fem head	Superficial	All	25	271	Stiffness at 10 MPa	5.03	−0.0078	<0.001	7–90
Fem head	Superficial	All	25	280	Stiffness at 1 MPa	4.31	−0.0046	0.005	7–90
Fem head	Mid-depth	All	31	319	Fracture stress	3.66	−0.0341[a]	<0.001	7–85
Fem head	Mid-depth	Anterior	15	56	Stiffness at 10 MPa	4.42	−0.0287	0.05	7–60
Fem head	Mid-depth	Superior	15	60	Stiffness at 10 MPa	4.93	−0.0438	<0.001	7–60
Fem head	Mid-depth	Posterior	12	47	Stiffness at 10 MPa	3.91	−0.0073	<0.001	7–60
Fem head	Mid-depth	Anterior	21	85	Stiffness at 1 MPa	3.13	−0.016	0.01	15–85
Fem head	Mid-depth	Superior	29	138	Stiffness at 1 MPa	3.49	−0.016	0.1	15–85
Fem head	Mid-depth	Posterior	19	71	Stiffness at 1 MPa	3.20	−0.004	<0.1	15–85
Talus	Superficial	All	35	136	Fracture stress	3.19	−0.0034	0.7	8–85
Talus	Superficial	All	35	158	Stiffness at 10 MPa	4.84	−0.0027	<0.3	10–85
Talus	Superficial	All	40	159	Stiffness at 1 MPa	4.08	−0.011	<0.7	10–85
Talus	Mid-depth	All	25	97	Fracture stress	2.91	−0.0027	<0.2	18–85
Talus	Mid-depth	All	19	87	Stiffness at 10 MPa	3.34	0.0048	<0.4	20–85
Talus	Mid-depth	All	23	97	Stiffness at 1 MPa	2.96	0.0051	<0.2	18–85

The numbers in the columns "A" and "B" refer to the values of the constants A and B in the estimated age dependent regression functions of the form: tensile property in MPa=exp$[A+B \times age]$ in which age denotes the subject age in years. The numbers in the second last column refer to the statistical level of significance of the constant B, whereas the last column denotes the age range for which the regression expression is valid

[a]The value given in the original article is −0.334. However, a close examination of Fig. 3.2b in the original article indicates that the number should be −0.0341

tensile fracture stress and stiffness with age from childhood) are unique to femoral head articular cartilage. Extensive testing of articular cartilage from other joints is required to determine whether Kempson's findings also hold true for articular cartilage in general.

Furthermore, it needs to be pointed out that additional testing is needed for the purpose of characterizing the influence of age on the material properties of articular cartilage during maturation.

Structural Properties

Pelvis

The pelvis is a ring of bone inferiorly and superiorly bounded by the lower extremities and the spinal column, respectively. The two hemi-pelves and sacrum make up the pelvic ring which is connected at the pubic symphysis in the anterior of the body and with the sacrum in the posterior. The hemi-pelvis is formed from three bones: the ilium, ischium, and pubis. Until puberty, the ilium, ischium, and pubis are united by Y-shaped cartilage (triadiate cartilage) centered on the acetabulum. The triadiate cartilage is involved in the growth of the pediatric pelvis. The bone of the pelvis develops from three ossification centers in each of the bones of the hemi-pelves. The ischial and pubic rami fuse at approximately 7 years of age. At birth, the acetabulum is completely composed of cartilage. Between 12 and 17 years of age, this cartilage is replaced by bone. Peripheral strips of cartilage along the iliac crest, for example, will not ossify until approximately 25 years of age.

As with other pediatric bony structures, the pediatric pelvis has more cartilage and fewer brittle bones than the adult pelvis. This allows the pediatric pelvis to absorb more energy than the adult pelvis before fracture occurs. In addition, fractures often occur in the cartilaginous regions, so radiological diagnosis may be difficult (Torode and Zieg 1985). Adult pelvic fractures will usually demonstrate disruption of the ring in two places, whereas a single fracture of the ring is possible in children.

Demetriades et al. (2003) reviewed the 16,630 blunt trauma admissions [Glasgow Coma Scale score < 15; blood pressure < 90 mmHg; penetrating injuries of the head, neck, and trunk; spinal cord injury; diffuse abdominal tenderness; passenger space intrusion; fall > 15 ft (4.6 m); flail chest] at a Level I trauma center during an 8-year-period and reported that the incidence of pelvic fracture was more than twice as high (10.0% vs. 4.6%) in the adult (>16 years) as in the pediatric (≤16 years) patients ($p<0.0001$). The difference was greater when the population was limited to the 4,675 cases of falls from more than 15 ft (4.6 m) for which the adult incidence of pelvic fracture was 14.2% compared to only 2.0% for the pediatric patients ($p<0.0001$).

To the authors' knowledge, the only published results from experimental testing of the human pediatric pelvis are those from a study by Ouyang et al. (2003a) in which 12 intact pediatric PMHS aged 2–12 years were subjected to lateral pelvic impact. The subjects were positioned in a seated position on a test table with the right side of the pelvis (iliac wing and greater trochanter) facing a pneumatic flat plate impactor (mass 3.24 kg, height 18 cm, and width 14 cm) backed by a load cell. The face of the impactor was covered by Sorbothane of thickness 9.5 mm, whereas the support was covered by Neoprene of thickness 9.5 mm. The material properties of the Sorbothane and Neoprene were not provided. The subjects were arranged so that the buttocks were in full contact with the test table, the left side of the pelvis firmly positioned against the support, and the torso and head attached to a support boom using tape. The legs remained free, aligned at a right angle to the direction of impact. Each subject was impacted once by the nondriven impactor at a speed ranging from 7.0 to 9.1 m/s.

No pelvic injuries were generated in any of the tested subjects despite that pelvic compression levels exceeding 50% were recorded. Subject data and results are showed in Table 3.22. Figures 3.9 and 3.10 show the compensated impactor force as a function of pelvic deflection for the six younger (2–4 years) and six older (5–12 years) subjects, respectively. It is important to emphasize that the force is compensated for the mass of the impactor but not for the effective mass of the test subject, hence the force peak during the initial centimeters of pelvic deflection. The compensated force versus pelvic deflection curves shown in Figs. 3.9 and 3.10 appear to have been filtered given their smooth appearance. However, no information on filtering is provided in the original article.

Discussion

It is of interest to compare the levels of lateral pelvic compression recorded by Ouyang et al. (2003a) (21–55%) to corresponding levels reported in studies with adult PMHS. Viano (1989) subjected eight adult PMHS to lateral pelvic impact using a 23.4 kg pendulum that impacted the greater trochanter at speeds ranging from 4.2 to 10.3 m/s. Peak lateral pelvic compression was in the range of 6.3–30.6%, and only two of the 14 tests resulted in pelvic fracture (pubic ramus fracture). Logistic regression was used to derive a function that predicted that 27% of lateral pelvic compression corresponds to 25% risk of fracture. Cavanaugh et al. (1990) reported on side impact sled tests in which 17 adult PMHS were subjected to lateral impact. Peak values of lateral compression for the struck side half pelvis were in the range of 24–64%. Pelvic injuries were generated in seven of the 17 tests and included inferior and superior pubic rami fractures and sacroiliac joint separation. Using logistic regression, Cavanaugh et al. predicted that a lateral compression of the struck side half pelvis of 33% corresponds to 25% risk of pelvic fracture.

Given the absence of injury in the test series by Ouyang et al. (2003a), it is not possible to estimate any injury threshold or risk function for pelvic fracture. However, the high level of noninjurious peak lateral pelvic compression (21–55%) is a strong indication that the structural behavior of the pediatric pelvis is very

3 Experimental Injury Biomechanics of the Pediatric Extremities and Pelvis

Table 3.22 Test conditions and results from the lateral pelvic impact tests by Ouyang et al. (2003a)

Subject	Age (years)	Sex	Body weight[a] (kg)	Stature[a] (cm)	Cause of death[a]	Impact velocity[b] (m/s)	$(VC)_{max}$ (m/s)	F_{max} (N)	Lateral deflection at F_{max} (cm)	C_{max} (%)
1	2	Female	13 (73)	97 (>97)	Poisoning	7.6	1.1	2,390	2.8	42
2	2.5	Male	10.5 (<3)	87.5 (15)	Cerebral edema	7.5	1.2	2,900	1.7	48
3	3	Female	10.5 (<3)	85 (<3)	Heart disease	8.0	1.6	2,800	2.1	54
4	3	Male	13.5 (28)	93 (24)	Brain tumor	7.9	1.3	3,000	2.6	45
5	3	Male	10 (<3)	91 (11)	CHD	7.7	1.5	2,170	2.1	55
6	4	Male	14 (8)	109 (94)	CHD	7.3	1.1	2,790	2.1	41
7	5	Male	13 (<3)	101 (4)	Cerebritis	7.1	0.9	2,830	2.1	34
8	6	Male	20 (39)	109 (10)	MA	9.1	1.0	4,370	1.7	46
9	6	Male	16.5 (3)	108 (6)	Leukemia	7.1	0.8	3,180	1.6	34
10	7.5	Female	17 (<3)	117 (7)	Acute urinemia	7.7	0.9	3,210	2.6	34
11	12	Female	29 (<3)	142.5 (10)	Leukemia	7.8	0.3	3,520	2.6	28
12	12	Female	20 (<3)	140 (5)	CHD	7.0	0.6	3,400	2.1	21

Combinations of body weight and stature highlighted in gray are either as listed for the actual subject or as listed for the other subject of the same sex and age. Numbers in *brackets* in the body weight and stature columns denote the corresponding US percentiles according to the Center for Disease Control and Prevention (CDC)

CHD: congenital heart disease, *MA*: Mediterranean anemia, $(VC)_{max}$: maximum viscous criterion (the ratio of maximum lateral deflection and initial hip-to-hip width), F_{max}: compensated maximum impactor force, C_{max}: maximum percentage lateral compression of the pelvis

[a] Body weight, stature, and cause of death for the test subjects are not provided in Ouyang et al. (2003a) but are included in other publications by Ouyang et al. (2003b, 2005, 2006) that utilized the same PMHS for other testing

[b] The impact velocities used in the different tests are listed in multiple tables in the original article, but the information is not consistent. The values reported here are as listed in Table 3.1 in Ouyang et al. (2003a) and are believed to be the correct ones

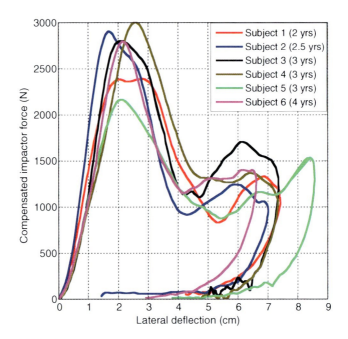

Fig. 3.9 Compensated impactor force as a function of lateral pelvic deflection for the six youngest pediatric PMHS tested by Ouyang et al. (2003a)

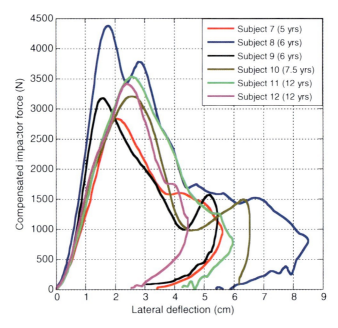

Fig. 3.10 Compensated impactor force as a function of lateral pelvic deflection for the six oldest padiatric PMHS tested by Ouyang et al. (2003a)

different to that of the adult pelvis. This is in agreement with Arbogast et al. (2002) who stated that the pelvis undergoes multidimensional structural change over the pediatric age range and is inherently a different mechanical structure in children than in adults.

While the results reported by Ouyang et al. (2003a) provide valuable information for the response of the pediatric pelvis to lateral impact, further testing is required for characterization of the fracture tolerance. Any further testing should also utilize a setup that allows for recording of the true force compressing the pelvis, i.e., the lateral impact force compensated for all inertial contributions. This would correspond to the contact force between the left side of the pelvis and the stationary support fixture attached to the test table in Ouyang et al.'s tests. Finally, it should be emphasized that most of the PMHS tested by Ouyang et al. were relatively small compared to the stature and weight of US children of corresponding ages, and that it is unknown to what extent the diseases suffered by the subjects prior to death affected the bone quality and thus the response and tolerance of the pelvis to lateral impact.[2]

The Lower and Upper Extremities

The lower extremities comprise all structures below the pelvis including the thigh, knee, leg, ankle, and foot, whereas the upper extremities comprise the arm, elbow, forearm, wrist, and hand. Data from human pediatric extremity testing are only available for the long bones of the thigh (femur), leg (tibia and fibula), arm (humerus), and forearm (radius and ulna).

The pediatric long bones are comprised of the cortical tubular shaft (diaphysis) that is inferiorly and superiorly bounded by the metaphysis, physis (growth plate), and epiphysis (Fig. 3.11). While transverse, spiral, and oblique long bone fractures occur both in adults and children, plastic bowing, torus, Greenstick, and growth plate fractures only occur in children. Plastic bowing occurs in response to a compression force and results in permanent deformation of the long bone. It occurs almost exclusively in children aged 2–5 years, and the most common bones affected are the radius and ulna. In children under 4 years, angulations not exceeding 20° will usually correct with growth. Torus (buckle) fractures can occur when there is a significant axial component to the force. They normally occur near the metaphysis of the bone where the cortex is less developed, and the bone literally buckles under the compressive force. As the metaphyseal cortex matures, the incidence of these fractures decreases. A Greenstick fracture occurs when the bone is incompletely broken with some of the cortex intact on the compression side of the fracture. Its name is derived from comparisons that can be drawn between this fracture pattern

---

[2] While a disease may not necessarily have a direct influence on bone quality, it may lead to prolonged periods of inactivity and bed rest and thus indirectly reduce skeletal strength.

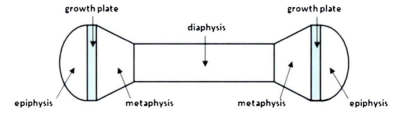

Fig. 3.11 Schematic showing the anatomical regions of a pediatric long bone

and the way that a living "green" stick will break. The fracture initiates on the tension surface of the bone but does not propagate through the whole thickness of the bone. The intact tissue often undergoes some plastic deformation leading to angulation at the fracture site. Growth plate properties and classification of growth plate injuries have been discussed in the material properties section of the current chapter and will not be further reviewed here.

Lower Extremity Studies

Asang et al. (1969) subjected intact tibiae from PMHS to quasi-static bending to failure. Prior to testing, the frozen tibiae were thawed for 6–10 h in physiological saline. The fixture used in the testing machine was designed to provide a loading situation similar to that experienced in skiing with a skiing boot supporting the distal section of the bone. The bending load was applied at one of three different points along the proximal section of the tibia. The recorded fracture moments for the three tibiae from subjects 18 years old or younger were 162 Nm (12-year-old male),[3] 345 Nm (17-year-old male), and 206 Nm (18-year-old female). Unfortunately, no information was provided as to the specific locations of the point of bending in those tests. Asang et al. also subjected intact tibiae to failure in torsion. The fracture torques recorded for the three pediatric tibiae (15–18 years) varied from approximately 40 to 117 Nm.

Jäger et al. (1973) conducted quasi-static three-point bending with tibiae harvested from PMHS ranging in age from 4 to 76 years. The recorded three-point bending force at fracture for the seven tibiae harvested from seven different subjects younger than 19 years were 903 N (4-year-old male), 7,505 N (14-year-old male), 6,818 N (14-year-old male), 7,652 N (14-year-old female), 5,494 N (14-year-old female), 8,240 N (18-year-old male), and 9,074 N (18-year-old male). The corresponding fracture moments or span lengths used in the tests were not reported.

Miltner and Kallieris (1989) conducted quasi-static three-point bending tests on 34 intact lower extremities removed from 18 pediatric PMHS ranging in age from 1 h to 6 years. The specimens were supported at the proximal end and just above the

---

[3] According to Stürtz (1980).

3 Experimental Injury Biomechanics of the Pediatric Extremities and Pelvis   137

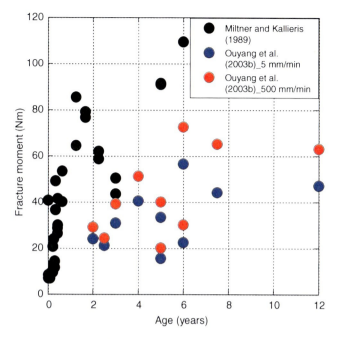

Fig. 3.12 Fracture moment as a function of subject age for the three-point bending tests with pediatric human femora by Miltner and Kallieris (1989) and Ouyang et al. (2003b)

knee and loaded from the lateral side with a blunt edge (radius 10 mm) applied to the middle of the thigh, bending it to the point of femur fracture at a loading rate of 50 mm/min. The recorded fracture moments ranged from 7.1 to 110 Nm and are plotted versus subject age in Fig. 3.12. Table 3.23 summarizes test conditions and results from the 34 tests.

In a second test series, Miltner and Kallieris subjected nine intact pediatric PMHS ranging in age from 2 to 12 months to dynamic lateral thigh loading by means of dropping the horizontally oriented subjects onto a sharp edge backed by a load cell. The drop height varied from 70 to 93 cm (impact velocity: 3.7–4.3 m/s). Each subject was dropped from the same height twice with impact onto the left and right lateral thigh, respectively. Peak forces were only presented for 11 of the 18 drop tests and ranged from 250 to 750 N. Only one of the 18 tests resulted in a complete femur fracture (a 2-month-old subject dropped from 90 cm). Another test was said to have caused a fissure. Subject anthropometry, drop height, and results for the tests are given in Table 3.24.

In a final test series, Miltner and Kallieris subjected one intact 27-month-old subject to lateral impact to the left and right thigh using a horizontal impactor with an edge like end contacting the subject. The mass of the impactor was not provided. The contact force recorded for the left and right thigh impact was 2,370 and 1,300 N, respectively. No fracture occurred in either test. Subject anthropometry, impact velocity, and results from the two tests appear in Table 3.24.

Table 3.23 Anthropometry, test conditions, and results from Miltner and Kallieris' (1989) quasi-static three-point bending tests on intact human pediatric lower limbs. The fracture deflection refers to the magnitude of vertical impactor translation from the point of thigh contact to femur fracture

Age (years)	Sex	Stature (cm)	Body weight (kg)	Aspect	Span length (cm)	Fracture force (N)	Fracture moment[a] (Nm)	Fracture deflection (mm)	Fracture type[b]	Femur width[c] (mm)
0.0001	Male	n/a	n/a	Left	6	2,720	40.8	16	Transverse	7
0.016	Male	51 (62)	2.8 (10)	Right	6	560	8.40	22	Transverse	5
0.016	Male	51 (62)	2.8 (10)	Left	6	470	7.05	22	Transverse	5
0.083	Male	53 (4)	3.7 (4)	Right	6	470	7.05	20	Transverse	7
0.083	Male	53 (4)	3.7 (4)	Left	6	595	8.93	21	Distal spongiosa	7
0.21	Female	n/a	n/a	Right	8	1,040	20.8	46	Distal spongiosa	8
0.21	Female	n/a	n/a	Left	8	480	9.60	21	Wedge	8
0.25	Female	63 (58)	5.3 (21)	Right	6	800	12.0	24	Distal spongiosa	8
0.25	Female	63 (58)	5.3 (21)	Left	6	815	12.2	23	Distal spongiosa	8
0.25	Male	58 (3)	4.9 (<3)	Right	8	660	13.2	22	Fissure	8
0.29	Female	62 (58)	5.1 (14)	Left	8	1,198	24.0	40	Distal spongiosa	8
0.29	Female	62 (58)	5.1 (14)	Right	6	780	11.7	21	Distal spongiosa	8
0.33	Male	67 (87)	7.0 (48)	Left	6	960	14.4	24	Transverse	8
0.33	Male	67 (87)	7.0 (48)	Right	6	2,440	36.6	39	Distal spongiosa	8
0.42	Male	65 (29)	6.5 (9)	Left	6	3,280	49.2	45	Distal spongiosa	8
0.42	Male	65 (29)	6.5 (9)	Right	6	1,920	28.8	35	Transverse	9
0.43	Female	64 (45)	6.6 (32)	Left	6	1,750	26.4	42	Distal spongiosa	9
0.43	Female	64 (45)	6.6 (32)	Right	6	2,750	41.5	32	Fissure	9
0.63	Female	69 (50)	7.0 (6)	Left	6	2,000	30.0	31	Fissure	9
0.63	Female	69 (50)	7.0 (6)	Right	6	3,560	53.4	45	Distal spongiosa	8
1.25	Male	78 (33)	11.7 (65)	Left	6	2,680	40.2	37	Transverse	8
1.25	Male	78 (33)	11.7 (65)	Right	6	5,700	85.5	45	Transverse	11
1.67	Male	92 (>97)	15.0 (97)	Left	6	4,300	64.5	40	Distal spongiosa	11
1.67	Male	92 (>97)	15.0 (97)	Right	12	2,640	79.2	60	Fissure	14
1.67	Male	92 (>97)	15.0 (97)	Left	12	2,560	76.8	60	Fissure	15

3 Experimental Injury Biomechanics of the Pediatric Extremities and Pelvis

2.25	Male	92 (71)	14.0 (71)	Right	n/a	1,300	n/a	n/a	n/a
				Left	n/a	2,370	n/a	n/a	n/a
2.25	Female	95 (96)	24.0 (>97)	Right	10	2,352	58.8	Fissure	13
				Left	10	2,480	62.0	Transverse	13
3	Female	93 (28)	13.5 (41)	Right	10	1,744	43.6	Wedge	15
				Left	10	2,016	50.4	Transverse	15
5	Male	109 (50)	18.0 (42)	Right	15	2,440	91.5	Oblique	17
				Left	13	2,800	91.0	Transverse	18
6	Female	117 (63)	26.0 (92)	Left	15	2,920	109.5	Oblique	16

Numbers in brackets in the stature and body weight columns denote the corresponding US percentiles according to the CDC

[a] The fracture moments given in the original article were calculated using the formula "Bending moment = $0.5 \times$ (fracture force) × (distance between supports)" which is incorrect since the factor in front of the fracture force should be 0.25 rather than 0.5. The fracture moments reported here are calculated as the ones given in the original article divided by 2 and should consequently be the correct ones

[b] It is not clear from the original article what fracture types that correspond to "Distal spongiosa" and "Fissure"

[c] The original article does not provide a definition for "femur width" but it is assumed to be the lateral diameter at the midshaft location

Table 3.24 Anthropometry, test conditions, and results from Miltner and Kallieris' (1989) drop tests and horizontal impactor test with intact pediatric PMHS

Age (months)	Sex	Stature (cm)	Body weight (kg)	Drop height (cm)	Impact velocity (m/s)	Aspect	Peak force (N)	Fracture type[a]	Femur width[b] (mm)
2	Female	60 (74)	5.60 (71)	90	4.2	Right	440	No fracture	9
						Left	600	No fracture	9
2	Male	58 (22)	5.25 (28)	90	4.2	Right	n/a	No fracture	8
						Left	n/a	Transverse	8
2.5	Male	60 (37)	5.20 (17)	93	4.3	Right	420	No fracture	n/a
						Left	450	No fracture	n/a
3	Male	63 (62)	6.20 (40)	90	4.2	Right	410	No fracture	9
						Left	320	No fracture	9
4.5	Male	64 (21)	6.40 (7)	70	3.7	Right	250	No fracture	8
						Left	250	No fracture	8
7	Female	72 (93)	8.10 (59)	80	4.0	Right	500	No fracture	n/a
						Left	450	No fracture	n/a
9	Male	62 (<3)	n/a	93	4.3	Right	n/a	No fracture	8
						Left	n/a	Fissure	10
9.5	Male	73 (41)	8.80 (15)	86	4.1	Right	n/a	No fracture	11
						Left	n/a	No fracture	11
12	Female	71 (13)	7.40 (<3)	90	4.2	Right	n/a	No fracture	9
						Left	750	No fracture	9
27	Male	92 (71)	14.00 (71)	n/a	3.8[c]	Right	1,300	No fracture	n/a
					4.7	Left	2,370	No fracture	n/a

The two tests in which the horizontal impactor was used are highlighted in gray. Numbers in brackets in the stature and body weight columns denote the corresponding US percentiles according to the CDC

[a] It is not clear from the original article what fracture types that correspond to "Distal spongiosa" and "Fissure"

[b] The original article does not provide a definition for "femur width" but it is assumed to be the lateral diameter at the midshaft location

[c] The impact velocity in this test is also reported as 4.4 m/s. Given the large difference in peak force between the impact to the right and left thigh, the lower of the two velocities reported for the right thigh (3.8 m/s) is assumed to be the correct one

Table 3.25 Data for the pediatric PMHS used in the long bone tests conducted by Ouyang et al. (2003b)[a]

Age (years)	Sex[b]	Weight (kg)	Stature (cm)	Cause of death
2	Female	13 (73)	97 (>97)	Poisoning
2.5	Male	10.5 (<3)	87.5 (15)	Cerebral edema
3	Male	13.5 (28)	93 (24)	Brain tumor
3	Male	10 (<3)	91 (11)	Congenital heart disease
4	Male	14 (8)	109 (94)	Congenital heart disease
5	Male	13 (<3)	101 (4)	Cerebritis
5	n/a	13 (<3)[c]	91.3 (<3)[c]	n/a
6	Male	16.5 (3)	108 (6)	Leukemia
6	Male	20 (39)	109 (10)	Mediterranean anemia
7.5	Female	17 (<3)	117 (7)	Acute urinemia
12	Female	20 (<3)	140 (5)	Congenital heart disease

Numbers in brackets denote the corresponding US percentiles according to the CDC
[a]The long bones were harvested from the same pediatric subjects that were used in the pelvic impact testing conducted by Ouyang et al. (2003a). However, the PMHS data for the long bone testing provided in Table 3.1 in Ouyang et al. (2003b) exclude two of the subjects used in the pelvic testing (a 3- and a 12-year-old female) but include an additional 5-year-old subject (sex and cause of death not provided) that was not utilized in the pelvic impact testing or any of the other test series that utilized the same subjects for other testing (Ouyang et al. 2005, 2006). It is therefore possible that this additional 5-year-old subject in fact was one of the two "excluded" subjects for which the age, body weight, and stature provided in Table 3.1 in Ouyang et al. (2003b) thus are incorrect
[b]The sex of the test subjects are not provided in Ouyang et al. (2003b) but are included in other publications by Ouyang et al. (2003a, 2005, 2006) that utilized the same subjects for other testing
[c]The weight and stature percentiles given for this subject apply regardless of whether the subject was male or female

Ouyang et al. (2003b) reported on symmetric three-point bending tests on femora, tibiae, and fibulae harvested from 11 unembalmed pediatric PMHS ranging in age from 2 to 12 years. Anthropometry and cause of death for the subjects are listed in Table 3.25. The ends of each bone were embedded in a steel box with polymethylmethacrylate (PMMA). The boxes were shaped such that rotation could take place at both ends. In addition, one end could move horizontally. The long bones from one of the two extremities of each subject were loaded at a rate of 5 mm/min, whereas the corresponding bones from the contralateral limb were loaded at 500 mm/min. The femora and tibiae were all subjected to sagittal plane bending by an anteroposterior directed force, whereas the fibulae were subjected to lateral plane bending. The maximum bending moments recorded for the femora, tibiae, and fibulae were in the ranges 16–73, 11–51, and 2.4–21 Nm, respectively, and are plotted versus subject age in Figs. 3.12 (femora) and 3.13 (tibiae and fibulae). The complete results from the tests are listed in Table 3.26.

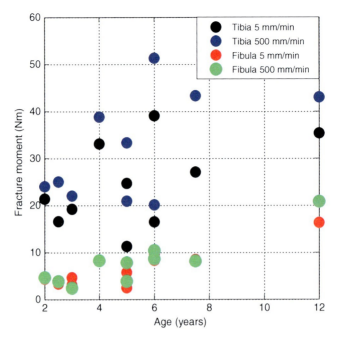

Fig. 3.13 Fracture moment as a function of subject age for the three-point bending tests with pediatric human tibiae and fibulae by Ouyang et al. (2003b)

Discussion

The femoral fracture moments recorded in the three-point bending tests with intact thighs by Miltner and Kallieris are higher than the corresponding moments recorded by Ouyang et al. (2003b) (Fig. 3.12). One likely contributor to this discrepancy is the difference in size between the subjects utilized in the two studies. Figures 3.14 and 3.15 show body weight as a function of age for the male and female subjects in the two studies in comparison to the 5, 50, and 95 body weight percentiles for US children. As shown, the subjects utilized by Miltner and Kallieris were in general heavier than Ouyang et al.'s subjects at any given age and closer to the US median. The body weight percentiles for the subjects utilized in Miltner and Kallieris' three-point bending tests ranged from 4 to greater than 97 with a median of 36.5, whereas the subjects utilized by Ouyang et al. for three-point bending had a body weight percentile ranging from less than 3 to 73 with seven of the 11 subjects less than or equal to the third percentile. Figures 3.16 and 3.17 show stature as a function of age for the male and female subjects in the two studies in comparison to the 5th, 50th, and 95th stature percentiles for US children. Similar to the case of body weight, the subjects utilized by Miltner and Kallieris were in general taller and closer to the US median than the subjects utilized by Ouyang et al. The stature percentiles for the subjects utilized in Miltner and Kallieris' three-point bending tests ranged from 3 to greater than 97 with a median of 54. The corresponding range for the subjects tested

Table 3.26 Results from Ouyang et al.'s (2003b) symmetric three-point bending tests on femora, tibiae, and fibulae harvested from unembalmed pediatric human subjects

Loading rate (mm/min)	Bone	Measure	Subject age (years)										
			2	2.5	3	3	4	5	5	6	6	7.5	12
5	Femur	F_{max} (N)	559	621	869	—	810	489	715	535	1,137	753	935
		SP (mm)	172	137	142	—	200	128	187	168	199	235	202
		M_{max} (Nm)	24.1	21.3	30.9	—	40.5	15.7	33.4	22.5	56.6	44.2	47.2
		D_{max} (mm)	8.1	8.6	8.1	—	13.7	5.3	8.6	9.4	6.9	16.0	13.3
		k (N/mm)	113	145	216	—	102	152	111	79	220	114	162
		E_{max} (J)	3.2	3.9	5.1	—	7.9	1.8	4.1	3.3	5.1	9.6	9.8
	Tibia	F_{max} (N)	570	545	595	—	748	387	677	490	925	596	818
		SP (mm)	150	122	129	—	177	117	146	135	169	182	173
		M_{max} (Nm)	21.4	16.6	19.2	—	33.1	11.3	24.7	16.5	39.1	27.1	35.4
		D_{max} (mm)	8.2	8.2	6.4	—	8.5	4.5	8.1	5.9	6.4	8.6	10.2
		k (N/mm)	137	134	131	—	136	129	180	104	207	132	168
		E_{max} (J)	3.5	3.1	2.5	—	4.4	1.1	4.0	1.8	4.0	3.8	6.2
	Fibula	F_{max} (N)	128	114	137	104	195	87	161	237	209	191	277
		SP (mm)	136	115	135	119	172	112	142	161	159	177	235
		M_{max} (Nm)	4.35	3.28	4.62	3.09	8.39	2.44	5.72	9.54	8.31	8.45	16.3
		D_{max} (mm)	14.4	11.1	16.0	12.4	18.6	9.4	9.6	15.6	10.6	18.7	17.0
		k (N/mm)	13.6	16.6	14.2	10.8	17.0	18.5	22.3	15.2	28.5	10.3	21.9
		E_{max} (J)	1.3	0.89	1.6	0.81	2.5	0.60	0.99	2.0	1.5	2.1	3.0

(continued)

Table 3.26 (continued)

Loading rate (mm/min)	Bone	Measure	Subject age (years)									
			2	2.5	3	4	5	5	6	6	7.5	12
500	Femur	F_{max} (N)	675	707	1,108	1,026	595	891	719	1,459	1,109	1,249
		SP (mm)	173	138	142	200	136	180	168	199	235	202
		M_{max} (Nm)	29.2	24.4	39.3	51.3	20.2	40.1	30.2	72.6	65.2	63.1
		D_{max} (mm)	9.6	9.2	12.4	17.9	10.1	15.1	10.1	11.0	14.7	15.2
		k (N/mm)	109	134	167	89	137	115	99	210	119	149
		E_{max} (J)	4.1	4.0	9.6	12.2	4.2	9.7	4.6	10.4	11.2	13.2
	Tibia	F_{max} (N)	649	782	681	876	662	913	600	1,214	951	995
		SP (mm)	148	128	129	177	126	146	134	169	182	173
		M_{max} (Nm)	24.0	25.0	22.0	38.8	20.9	33.3	20.1	51.3	43.3	43.0
		D_{max} (mm)	9.6	10.9	7.3	12.7	7.8	10.6	6.4	8.8	11.4	9.2
		k (N/mm)	108	126	144	133	134	175	110	219	150	168
		E_{max} (J)	3.9	5.8	3.2	8.0	3.3	6.8	2.0	6.6	7.6	6.2
	Fibula	F_{max} (N)	127	120	–	203	131	214	242	201	171	325
		SP (mm)	146	127	135	162	117	146	172	172	190	256
		M_{max} (Nm)	4.64	3.81	2.40	8.22	3.83	7.81	10.4	8.64	8.12	20.8
		D_{max} (mm)	19.2	21.1	11.0	22.9	17.0	17.8	25.6	15.6	26.1	25.0
		k (N/mm)	12.0	10.7	8.3	15.0	13.7	20.5	16.7	18.9	9.0	16.2
		E_{max} (J)	1.8	1.8	0.57	3.3	1.6	2.7	4.4	2.1	2.9	4.9

F_{max}: maximum force, SP: span (distance between supports), M_{max}: maximum bending moment, D_{max}: maximum deflection, k: bending stiffness, E_{max}: maximum energy absorbed

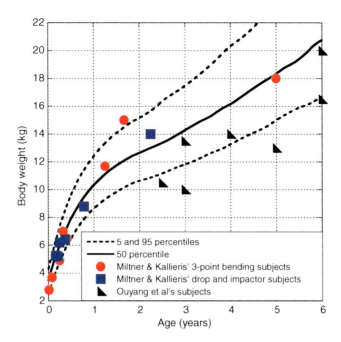

Fig. 3.14 Body weight as a function of age for the male PMHS tested by Miltner and Kallieris (1989) and Ouyang et al (2003a, b) compared to 5, 50, and 95-percentile body weight for male US children. The weight for the 5-year-old subject of unknown gender utilized by Ouyang et al. (2003b) is not plotted

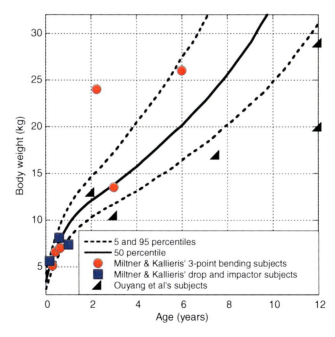

Fig. 3.15 Body weight as a function of age for the female PMHS tested by Miltner and Kallieris (1989) and Ouyang et al (2003a, b) compared to 5, 50, and 95-percentile body weight for female US children. The weight for the 5-year-old subject of unknown gender utilized by Ouyang et al. (2003b) is not plotted

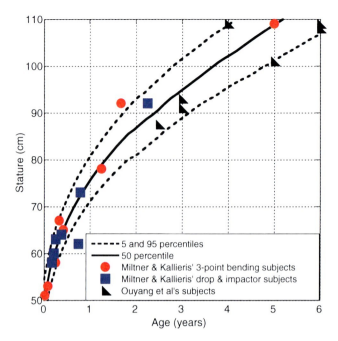

Fig. 3.16 Stature as a function of age for the male PMHS tested by Miltner and Kallieris (1989) and Ouyang et al (2003a, b) compared to 5, 50, and 95-percentile stature for male US children. The stature for the 5-year-old subject of unknown gender utilized by Ouyang et al. (2003b) is not plotted

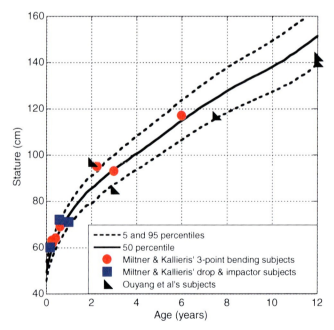

Fig. 3.17 Stature as a function of age for the female PMHS tested by Miltner and Kallieris (1989) and Ouyang et al (2003a, b) compared to 5, 50, and 95-percentile stature for female US children. The stature for the 5-year-old subject of unknown gender utilized by Ouyang et al. (2003b) is not plotted

by Ouyang et al. was approximately the same (<3–97) but with a distribution skewed to smaller stature (median 10).

Another potential contributor to the difference in fracture moments between the two studies is that Miltner and Kallieris tested intact thighs, whereas Ouyang et al.'s data are from femora stripped of all soft tissues. It is possible that the soft tissues of the thighs tested by Miltner and Kallieris had a slightly reinforcing effect and thus increased the fracture moment as compared to if the specimens had been stripped of all soft tissues prior to testing. It should also be emphasized that Miltner and Kallieris tested their specimens in lateral plane bending, whereas Ouyang et al. utilized anteroposterior bending. Results from bending tests on adult human femora do, however, indicate that the response and tolerance of the human femur is relatively insensitive to bending direction (Funk et al. 2004), a finding that is consistent with the approximately rotationally symmetric cross section of the femur. Consequently, the difference in bending direction most likely had minor, if any, influence on the fracture moments. Finally, there are of course several other potential factors, such as differences in subject bone quality and experimental conditions between the two studies, that may have contributed to the difference in fracture moment.

The results from Ouyang et al.'s (2003b) tibiae and fibulae testing (Fig. 3.13) are, as previously pointed out for the corresponding results from their pelvis (Ouyang et al. 2003a) and femur testing, possibly not representative of the bending tolerance of the tibia and fibula in healthy US children of the same ages. This is for two reasons: the subjects being smaller than US children of the same ages and the potential direct and indirect effects of the diseases suffered by the subjects prior to death on bone quality. The fact that Miltner and Kallieris (1989) reported higher femur fracture moments than did Ouyang et al. for pediatric subjects of similar ages suggests that the corresponding tibia and fibula fracture moments reported by Ouyang et al. also are lower than what would be expected for US children of the same ages.

Upper Extremity Studies

The only published experimental results from mechanical testing of pediatric human upper extremities are those of Ouyang et al. (2003b). They reported on symmetric three-point bending tests with humeri, radii, and ulnae harvested from 11 unembalmed PMHS aged 2–12 years. Anthropometry and cause of death for the subjects are listed in Table 3.25. The ends of each bone were embedded in a steel box with PMMA. The boxes were shaped such that rotation could take place at both ends. In addition, one end could move horizontally. The long bones from one of the two extremities of each subject were loaded at a rate of 5 mm/min, whereas the corresponding bones from the contralateral limb were loaded at 500 mm/min. All specimens were subjected to sagittal plane bending by an anteroposterior directed bending force. The tests were stopped after the bending force had reached its maximum. The maximum bending moment recorded for the humeri, radii, and ulnae were in the ranges 7.6–48.6, 2.3–17.0, and 3.1–21.2 Nm, respectively, and are plotted versus age in Fig. 3.18. The complete results are listed in Table 3.27.

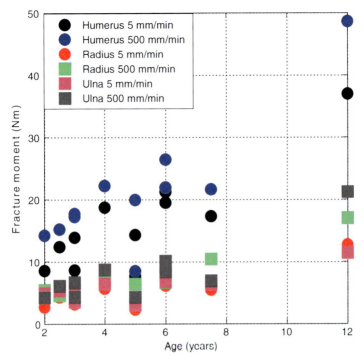

Fig. 3.18 Fracture moment as a function of subject age for the three-point bending tests with pediatric human humeri, radii, and ulnae by Ouyang et al. (2003b)

Discussion

The results from Ouyang et al.'s (2003b) humerus, radius, and ulna testing (Fig. 3.18) are, as previously pointed out for the corresponding results from their pelvis (Ouyang et al. 2003a), femur, tibia, and fibula testing, possibly not representative of the bending tolerance of the humerus, radius, and ulna in healthy US children of the same ages. This is for two reasons: the subjects being smaller than US children of the same ages and the potential direct and indirect effects of the diseases suffered by the subjects prior to death on bone quality. The fact that Miltner and Kallieris (1989) reported higher femur fracture moments than did Ouyang et al. for pediatric subjects of similar ages suggests that the corresponding humerus, radius, and ulna fracture moments reported by Ouyang et al. also are lower than what would be expected for US children of the same ages. Further testing is required to verify the results of Ouyang et al. (2003b).

Table 3.27 Results from Ouyang et al.'s (2003b) symmetric three-point bending tests on humeri, radii, and ulnae harvested from unembalmed pediatric PMHS

Loading rate (mm/min)	Bone	Measure	Subject age (years)										
			2	2.5	3	3	4	5	5	6	6	7.5	12
5	Humerus	F_{max} (N)	280	488	417	290	516	288	432	580	554	418	673
		SP (mm)	122	102	133	119	145	106	132	146	141	166	220
		M_{max} (Nm)	8.54	12.4	13.9	8.63	18.7	7.63	14.3	21.2	19.5	17.3	37.0
		D_{max} (mm)	3.6	4.5	8.5	7.6	9.3	4.3	8.2	10.0	4.8	8.9	10.7
		k (N/mm)	101	121	101	59	110	106	100	103	164	65	88
		E_{max} (J)	0.64	1.4	2.5	1.5	3.3	0.83	2.5	4.1	1.8	2.4	4.7
	Radius	F_{max} (N)	120	208	134	141	181	113	190	241	233	190	308
		SP (mm)	89	82	111	90	125	81	94	116	105	115	165
		M_{max} (Nm)	2.67	4.26	3.72	3.17	5.66	2.29	4.47	6.99	6.12	5.46	12.7
		D_{max} (mm)	3.1	8.9	11.7	5.3	9.7	3.2	5.2	8.6	4.5	9.7	14.6
		k (N/mm)	48	64	20	29	37	75	61	47	67	35	36
		E_{max} (J)	0.23	1.5	1.1	0.42	1.3	0.28	0.70	1.5	0.63	1.3	3.1
	Ulna	F_{max} (N)	211	249	171	149	221	–	133	213	243	212	260
		SP (mm)	93	86	120	94	116	–	92	128	120	120	175
		M_{max} (Nm)	4.91	5.35	5.13	3.50	6.41	–	3.06	6.82	7.29	6.36	11.4
		D_{max} (mm)	3.9	3.8	7.4	7.6	17.2	–	5.6	10.4	6.6	–	22.9
		k (N/mm)	75	101	41	26	35	–	35	39	56	–	22
		E_{max} (J)	0.54	0.65	0.89	0.74	3.0	–	0.54	1.6	1.1	–	4.4

(continued)

Table 3.27 (continued)

Loading rate (mm/min)	Bone	Measure	Subject age (years)										
			2	2.5	3	3	4	5	5	6	6	7.5	12
500	Humerus	F_{max} (N)	429	543	517	355	588	331	603	619	681	505	864
		SP (mm)	132	112	133	119	151	102	132	141	155	171	225
		M_{max} (Nm)	14.2	15.2	17.2	17.7	22.2	8.44	19.9	21.9	26.4	21.6	48.6
		D_{max} (mm)	6.8	9.7	11.9	7.2	10.5	4.6	8.3	10.1	6.8	17.4	16.0
		k (N/mm)	102	129	104	63	104	116	117	121	143	55	82
		E_{max} (J)	1.8	3.6	4.4	1.5	4.1	0.79	3.2	4.4	2.6	6.0	8.7
	Radius	F_{max} (N)	194	198	190	153	283	—	272	250	238	354	411
		SP (mm)	112	92	100	94	107	—	94	106	119	118	165
		M_{max} (Nm)	5.43	4.55	4.75	3.60	7.57	—	6.39	6.63	7.08	10.4	17.0
		D_{max} (mm)	5.6	8.7	10.5	7.3	10.5	—	6.8	10.5	6.8	14.6	16.4
		k (N/mm)	35	45	34	27	55	—	72	47	47	48	38
		E_{max} (J)	2.3	1.2	1.5	0.68	1.9	—	1.3	1.7	1.0	3.6	4.5
	Ulna	F_{max} (N)	143	266	241	181	296	—	183	305	331	203	484
		SP (mm)	119	92	111	96	118	—	94	110	122	136	175
		M_{max} (Nm)	4.25	6.12	6.69	4.34	8.73	—	4.30	8.39	10.1	6.90	21.2
		D_{max} (mm)	6.8	7.3	9.2	8.2	18.8	—	8.3	10.5	14.2	7.7	24.3
		k (N/mm)	25	61	47	33	36	—	41	56	53	34	42
		E_{max} (J)	0.60	1.4	1.4	0.95	4.3	—	1.1	2.3	3.5	0.94	8.7

F_{max}: maximum force, SP: span (distance between supports), M_{max}: maximum bending moment, D_{max}: maximum deflection, k: bending stiffness, E_{max}: maximum energy absorbed

Conclusions

A summary of the available literature on material and structural properties of the pediatric pelvis and extremities reveals data collected in experiments with a range of different biological models. While the data are limited, and even nonexistent for some tissues and structures, some aspects of the properties of the pediatric pelvis and extremities can be stated with reasonable certainty. For example, the stiffness and strength of cortical bone are greater in the adult than the child and increase with age during maturation. Other examples are growth plate cartilage, for which PMHS and animal model studies strongly indicate that the strength increases with age during maturation, and articular cartilage, for which PMHS testing provides significant evidence of that the tensile fracture stress decrease with age from childhood. On the structural level, there are existing data for pelvic response in lateral impact that are useful for ATD development and computational model validation, but further testing is necessary for characterizing the tolerance in this loading mode. Structural data are also available for the extremity long bones in the form of fracture moments in quasi-static bending. However, additional data from controlled dynamic tests are required for characterization of the loading response as well as for the development of age dependent injury thresholds and risk functions for the pediatric arm, forearm, thigh, and leg.

References

Aerssens J, Boonen S, Lowet G, Dequeker J (1998) Interspecies differences in bone composition, density, and quality: potential implications for *in vivo* bone research. Endocrinology 139(2):663–670

Amamilo SC, Bader DL, Houghton GR (1985) The periosteum in growth plate failure. Clin Orthop Relat Res 194:293–305

Arbogast KB, Mari-Gowda S, Kallan MJ et al (2002) Pediatric pelvic fractures in side impact collisions. Stapp Car Crash J 46:285–296

Asang E, Posch P, Engelbrecht R (1969) Experimentelle Untersuchungen über die Bruchfestigkeit des Menschlichen Schienbeins. Monatsschr Unfallheilkd Versicher Versorg Verkehrsmed 72:8336–8344

Blanton PL, Biggs NL (1970) Ultimate tensile strength of fetal and adult human tendons. J Biomech 3:181–189

Booth FW, Tipton CM (1970) Ligamentous strength measurements in pre-pubescent and pubescent rats. Growth 34:177–185

Bright RW, Burstein AH, Elmore SM (1974) Epiphyseal-plate cartilage: a biomechanical and histological analysis of failure modes. J Bone Joint Surg Am 56:688–703

Burstein AH, Currey JD, Frankel VH et al (1972) The ultimate properties of bone tissue: the effects of yielding. J Biomech 5:35–44

Burstein AH, Reilly DT, Martens M (1976) Aging of bone tissue: mechanical properties. J Bone Joint Surg Am 58:82–86

Cavanaugh JM, Walilko TJ, Malthora A et al (1990) Biomechanical response and injury of the pelvis in twelve sled side impacts. Stapp Car Crash Conf 34:1–12, SAE Tech Paper #902305

Ching R, Nuckley D, Hertsted S et al (2001) Tensile mechanics of the developing cervical spine. Stapp Car Crash J 45:329–336
Chung SMK, Batterman SC, Brighton CT (1976) Shear strength of the human femoral capital epiphyseal plate. J Bone Joint Surg Am 58:94–103
Cohen B, Chorney GS, Phillips DP et al (1992) The microstructural tensile properties and biochemical composition of the bovine distal femoral growth plate. J Orthop Res 10:263–275
Cowin SC, van Buskirk WC, Ashman RB (1987) Properties of bone. In Skalak R and Chien S (eds) Handbook of Bioengineering. McGraw-Hill Book Company
Currey JD (1979) Changes in the impact energy absorption of bone with age. J Biomech 12:459–469
Currey JD (2001) Ontogenetic changes in compact bone material properties. In: Cowin SC (ed) Bone mechanics handbook, 2nd edn. CRC Press LLC, Boca Raton, FL
Currey JD, Butler G (1975) The mechanical properties of bone tissue in children. J Bone Joint Surg Am 57:810–814
Currey JD, Brear K, Zioupos P (1996) The effects of aging and changes in mineral content in degrading the toughness of human femora. J Biomech 29:257–260
Demetriades D, Karaiskakis M, Velmahos GC et al (2003) Pelvic fractures in pediatric and adult trauma patients: are they different injuries? J Trauma 54:1146–1151
Ding M (2000) Age variations in the properties of human tibial trabecular bone and cartilage. Acta Orthop Scand Suppl 71:1–45
Ding M, Dalstra M, Danielsen CC et al (1997) Age variations in the properties of human tibial trabecular bone. J Bone Joint Surg Br 79:995–1002
Elliot DH (1965) Structure and function of mammalian tendon. Biol Rev Camb Philos Soc 40:392–421
Evans FG, Lebow M (1951) Regional differences in some of the physical properties of the human femur. J Appl Phys 3:563–572
Flahiff CM, Brooks AT, Hollis JM et al (1995) Biomechanical analysis of patellar tendon allografts as a function of donor age. Am J Sports Med 23:354–358
Fremerey R, Bastian L, Siebert WE (2000) The coracoacromial ligament: anatomical and biomechanical properties with respect to age and rotator cuff disease. Knee Surg Sports Traumatol Arthrosc 8:309–313
Fujii T, Takai S, Arai Y et al (2000) Microstructural properties of the distal growth plate of the rabbit radius and ulna: biomechanical, biochemical, and morphological studies. J Orthop Res 18:87–93
Funk JR, Kerrigan JR, Crandall JR (2004) Dynamic bending tolerance and elastic–plastic material properties of the human femur. Annu Proc Assoc Adv Automot Med 48:215–233
Guse RJ, Connolly JF, Alberts R et al (1989) Effect of aging on tensile mechanical properties of the rabbit distal femoral growth plate. J Orthop Res 7:667–673
Haut RC, Lancaster RL, DeCamp CE (1992) Mechanical properties of the canine patellar tendon: some correlations with age and the content of collagen. J Biomech 25:163–173
Hirsch C, Evans FG (1965) Studies on some physical properties of infant compact bone. Acta Orthop Scand 35:300–313
Hollis JM, Lyon RM, Marcin S et al (1988) Effect of age and loading axis on the failure properties of the human ACL. Trans Orthop Res Soc 13:81
Hubbard RP, Soutas-Little RW (1984) Mechanical properties of human tendon and their age dependence. J Biomech Eng 106:144–150
Irwin A, Mertz HJ (1997) Biomechanical basis for the CRABI and Hybrid III child dummies. Stapp Car Crash Conf 41:261–272, SAE Tech Paper #973317
Ivarsson BJ, Crandall J, Longhitano D et al (2004) Lateral injury criteria for the 6-year-old pedestrian – part I. Criteria for the head, neck, thorax, abdomen and pelvis. SAE Tech Paper #2004-01-0323
Ivarsson BJ, Crandall J, Longhitano D et al (2004) Lateral injury criteria for the 6-year-old pedestrian – part II. Criteria for the upper and lower extremities. SAE Tech Paper #2004-01-1755

Jäger M, Dietschi C, Ungethüm M (1973) Über das Verhalten der unversehrten und durchborten Tibia bei Biegebeanspruchung. Arch Orthop UnfallChir 76:188–194

Johnson GA, Tramaglini DM, Levine RE et al (1994) Tensile and viscoelastic properties of human patellar tendon. J Orthop Res 12:796–803

Kempson GE (1982) Relationship between the tensile properties of articular cartilage from the human knee and age. Ann Rheum Dis 41:508–511

Kempson GE (1991) Age-related changes in the tensile properties of human articular cartilage: a comparative study between the femoral head of the hip joint and the talus of the ankle joint. Biochim Biophys Acta 1075:223–230

Kent R, Stacey S, Kindig M et al (2006) Biomechanical response of the pediatric abdomen, part 1: development of an experimental model and quantification of structural response to dynamic belt loading. Stapp Car Crash J 50:1–26

Ko R (1953) The tension test upon the compact substance of the long bones of human extremities. J Kyoto Prefect Univ Med 53:503–525

Kubo K, Kanehisa H, Kawakami Y et al (2001) Growth changes in the elastic properties of human tendon structures. Int J Sports Med 22:138–143

Lam TC, Frank CB, Shrive NG (1993) Changes in the cyclic and static relaxations of the rabbit medial collateral ligament complex during maturation. J Biomech 26:9–17

Lee TQ, Dettling J, Sandusky MD et al (1999) Age related biomechanical properties of the glenoid-anterior band of the inferior glenohumeral ligament–humerus complex. Clin Biomech 14:471–476

Lindahl O (1976) Mechanical properties of dried defatted spongy bone. Acta Orthop Scand 47:11–19

McCalden RW, McGeough JA, Barker MB et al (1993) Age-related changes in the tensile properties of cortical bone. The relative importance of changes in porosity, mineralization, and microstructure. J Bone Joint Surg Am 75:1193–1205

Mertz HJ, Jarrett K, Moss S et al (2001) The Hybrid III 10-year-old dummy. Stapp Car Crash J 45:316–326

Miltner E, Kallieris D (1989) Quasistatische und dynamische Biegebelastung des kindlichen Oberschenkels zur Erzeugung einer Femurfraktur. Z Rechtsmed 102:535–544

Morscher E (1968) Strength and morphology of growth cartilage under hormonal influence of puberty. Animal experiments and clinical study on the etiology of local growth disorders during puberty. Reconstr Surg Traumatol 10:3–104

Morscher E, Desaulles PA, Schenk R (1965) Experimental studies on tensile strength and morphology of the epiphyseal cartilage at puberty. Ann Paediatr 205:112–130

Mosekilde L, Mosekilde L, Danielsen CC (1987) Biomechanical competence of vertebral trabecular bone in relation to ash density and age in normal individuals. Bone 8:79–85

Nafei A, Danielsen CC, Linde F et al (2000) Properties of growing trabecular bone. Part I: mechanical and physical properties. J Bone Joint Surg Br 82:910–920

Nakagawa Y, Hayashi K, Yamamoto N et al (1996) Age-related changes in biomechanical properties of the Achilles tendon in rabbits. Eur J Appl Physiol Occup Physiol 73:7–10

Neumann P, Ekström LA, Keller TS et al (1994) Aging, vertebral density, and disc degeneration alter the tensile stress–strain characteristics of the human anterior longitudinal ligament. J Orthop Res 12:103–112

Noyes FR, Grood ES (1976) The strength of the anterior cruciate ligament in humans and rhesus monkeys. J Bone Joint Surg Am 58:1074–1082

Ouyang J, Zhu QA, Zhao WD et al (2003a) Experimental cadaveric study of lateral impact of the pelvis in children. Di Yi Jun Yi Da Xue Xue Bao 23:397–401

Ouyang J, Zhu Q, Zhao W et al (2003b) Biomechanical character of extremity long bones in children. Chin J Clin Anat 21:620–623

Ouyang J, Zhu Q, Zhao W et al (2005) Biomechanical assessment of the pediatric cervical spine under bending and tensile loading. Spine 30:716–723

Ouyang J, Zhao W, Xu Y et al (2006) Thoracic impact testing of pediatric cadaveric subjects. J Trauma 61:1492–1500

Rauch G, Allzeit B, Gotzen L (1988) Biomechanische Untersuchungen zur Zugfestigkeit des vorderen Kreuzbandes unter besonderer Berücksichtigung der Altersabhängigkeit. Unfallchirurg 91:437–443

Reilly DT, Burstein AH, Frankel VH (1974) The elastic modulus for bone. J Biomech 7:271–275

Rollhäuser H (1950) Konstruktions- und Altersunterschiede in Festigkeit kollagener Fibrillen. Gegenbaurs Morphol Jahrb 90:157–179

Rudicel S, Pelker RR, Lee KE et al (1985) Shear fractures through the capital femoral physis of the skeletally immature rabbit. J Pediatr Orthop 5:27–31

Salter RB, Harris WR (1963) Injuries involving the epiphyseal plate. J Bone Joint Surg Am 45:587–622

Sedlin ED, Hirsch C (1966) Factors affecting the determination of the physical properties of femoral cortical bone. Acta Orthop Scand 37:29–48

Shadwick RE (1990) Elastic energy storage in tendons: mechanical differences related to function and age. J Appl Physiol 68:1033–1040

Stucke K (1950) Über das elastische Verhalten der Achillessehne im Belastungsversuch. Langenbecks Arch Klin Chir 265:579–599

Stürtz G (1980) Biomechanical data of children. Stapp Car Crash Conf 24:513–559, SAE Tech Paper #801313

Tipton CM, Matthes RD, Martin RK (1978) Influence of age and sex on the strength of bone–ligament junctions in knee joints of rats. J Bone Joint Surg Am 60:230–234

Torode I, Zieg D (1985) Pelvic fractures in children. J Pediatr Orthop 5:76–84

Torzilli PA, Takebe K, Burstein AH (1982) The material properties of immature bone. J Biomech Eng 104:12–20

Tsuda K (1957) Studies on the bending test and impulsive bending test on human compact bone. J Kyoto Prefect Univ Med 61:1001–1025

van Ratingen MR, Twisk D, Schrooten M et al (1997) Biomechanically based design and performance targets for a 3-year old child crash dummy for frontal and side impact. Stapp Car Crash Conf 41:243–260, SAE Tech Paper #973316

Viano DC (1989) Biomechanical responses and injuries in blunt lateral impact. Stapp Car Crash Conf 33:113–142, SAE Tech Paper #892432

Vinz H (1969) Die festigkeitsmechanischen Grundlagen der typischen Frakturformen des Kindesalters. Zentralblatt Chir 94:1509–1515

Vinz H (1970) Die Änderung der Festikeitseigenschaften des kompakten Knochengewebes im Laufe der Altersentwicklung. Gegenbaurs Morphol Jahrb 115:257–272

Vinz H (1972) Die Festigheit der reinen Knochensubstanz. Näherungsverfahren zur Bestimmung der auf den hohlraumfreien Querschnitt bezogenen Festigkeit von Knochengewebe. Gegenbaurs Morphol Jahrb 117:453–460

Wall JC, Hutton WC, Cyron BM (1974) The effects of age and strain rate on the mechanical properties of bone. Proc International Meeting on Biomechanics of Trauma in Children, Lyon, France, Sep 17–19, 1974. International Research Committee on the Biokinetics of Impacts: 185–193

Wall JC, Chatterji SK, Jeffery JW (1979) Age-related changes in the density and tensile strength of human femoral cortical bone. Calcif Tissue Int 27:105–108

Weaver JK, Chalmers J (1966) Cancellous bone: its strength and changes with aging and an evaluation of some methods for measuring its mineral content. I. Age changes in cancellous bone. J Bone Joint Surg Am 48:289–299

Williams JL, Do PD, Eick JD et al (2001) Tensile properties of the physis vary with anatomic location, thickness, strain rate and age. J Orthop Res 19:1043–1048

Woo SL, Orlando CA, Gomez MA et al (1986) Tensile properties of the medial collateral ligament as a function of age. J Orthop Res 4:133–141

Woo SL, Ohland KJ, Weiss JA (1990a) Aging and sex-related changes in the biomechanical properties of the rabbit medial collateral ligament. Mech Ageing Dev 56:129–142

Woo SL, Peterson RH, Ohland KJ et al (1990b) The effects of strain rate on the properties of the medial collateral ligament in skeletally immature and mature rabbits: a biomechanical and histological study. J Orthop Res 8:712–721

Woo SL, Hollis JM, Adams DJ et al (1991) Tensile properties of the human femur–anterior cruciate ligament–tibia complex. The effects of specimen age and orientation. Am J Sports Med 19:217–225

Yamada H (1970) Strength of biological materials. In: Evans FG (ed). Williams and Wilkins Co., Baltimore, MD

Chapter 4
Experimental Injury Biomechanics of the Pediatric Head and Brain

Susan Margulies and Brittany Coats

Introduction

Traumatic brain injury (TBI) is a leading cause of death and disability among children and young adults in the United States and results in over 2,500 childhood deaths, 37,000 hospitalizations, and 435,000 emergency department visits *each year* (Langlois et al. 2004). Computational models of the head have proven to be powerful tools to help us understand mechanisms of adult TBI and to determine load thresholds for injuries specific to adult TBI. Similar models need to be developed for children and young adults to identify age-specific mechanisms and injury tolerances appropriate for children and young adults. The reliability of these tools, however, depends heavily on the availability of pediatric tissue material property data. To date the majority of material and structural properties used in pediatric computer models have been scaled from adult human data. Studies have shown significant age-related differences in brain and skull properties (Prange and Margulies 2002; Coats and Margulies 2006a), indicating that the pediatric head cannot be modeled as a miniature adult head, and pediatric computer models incorporating age-specific data are necessary to accurately mimic the pediatric head response to impact or rotation. This chapter details the developmental changes of the pediatric head and summarizes human pediatric properties currently available in the literature. Because there is a paucity of human pediatric data, material properties derived from animal tissue are also presented to demonstrate possible age-related differences in the heterogeneity and rate dependence of tissue properties. The chapter is divided into three main sections: (1) brain, meninges, and cerebral spinal fluid (CSF); (2) skull; and (3) scalp.

S. Margulies, Ph.D. (✉)
Department of Bioengineering, University of Pennsylvania, 240 Skirkanich Hall, 210 South 33rd Street, Philadelphia, PA 19104, USA

B. Coats, Ph.D.
Department of Mechanical Engineering, University of Utah, 50 South Central Campus Drive, Salt Lake City, UT 84112, USA

Brain, Meninges, and CSF

Anatomy and Development

At birth, the human brain is typically between 350 and 450 g and continues to grow rapidly until about 2 years old, after which the growth process slows dramatically (Dobbing and Sands 1973; Dekaban 1978). A series of studies from researchers at the National Institute of Mental Health (Giedd and Snell 1996; Lange et al. 1997; Lenroot et al. 2007) used magnetic resonance imaging (MRI) to measure brain volumes in children and adolescents (4–20 years old). While total brain volume slightly increases in children older than 4 years old, the brain substructures still undergo significant alterations. For example, lateral ventricles and white matter tracks significantly increase past the age of 4 years old, while gray matter significantly decreases (Giedd and Snell 1996; Wilke et al. 2006; Lenroot et al. 2007). These developmental changes are consistent for both male and female genders, but there is a distinct difference in the timing of the changes. Peak brain volume for females and males occurs at 10.5 and 14.5 years old, respectively (Fig. 4.1). Females also reach peak gray and white matter volumes earlier than males. These trajectory differences may be related to the overall larger volume of the male cerebrum and cerebellum to that of females (8–11 % difference; Giedd and Snell 1996; Goldstein and Seidman 2001).

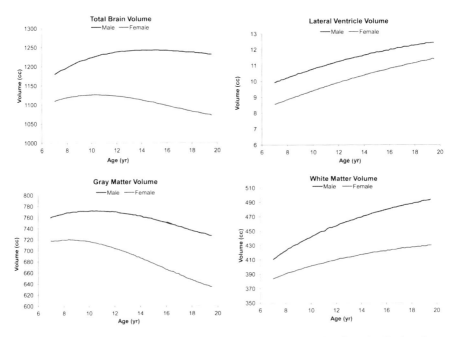

Fig. 4.1 Male and female brain, ventricle, gray and white matter growth trajectories. Brain volume is smaller in females and reaches peak volumes developmentally earlier than males. Data digitally extracted from figures in Lenroot et al. (2007)

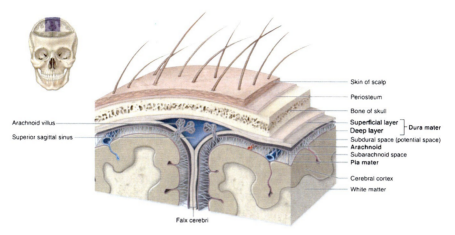

Fig. 4.2 Sketch of the meninges in the midsagittal vertex (image courtesy of McGraw Hill)

The three major layers that surround and protect the brain from injury are the scalp, skull, and meninges. The anatomy and development of scalp and skull will be discussed in detail later in the chapter. The meninges consists of three membrane layers called the dura, arachnoid, and pia mater (Fig. 4.2). The dura mater has a superficial and deep layer that firmly attaches to the inner surface of the skull and provides support to the neurocranium during skull development. The two dural layers run tightly together for the majority of the skull surface except near the midsagittal vertex and occiput of the brain where the deep dural layer splits from the superficial layer and reflects into the brain. The falx cerebri dural reflection lies between the right and left hemispheres of the brain, extending nearly to the corpus callosum. The space created by the separation of the superficial and deep dural layers at this location is called the superior sagittal sinus. The deep dural layer protrusion near the occiput, called the tentorium cerebelli, separates the cerebellum and the occipital lobe of the cerebrum. The arachnoid mater attaches to the inner surface of the deep dural layer and the pia mater attaches to the outer surface of the brain. The space between the two layers is called the subarachnoid space. Cerebrospinal fluid (CSF) flows within this space and interacts with the transversing blood vessels, providing nutrients to the brain while removing wastes.

Myelin is a phospholipid layer that surrounds the axons (white matter) of neurons. In rats it has been reported that the lipid layer provides protection to axons, and that unmyelinated axons are more vulnerable to injury than myelinated axons (Reeves et al. 2005). The human infant brain is only 34 % myelinated at birth (Dobbing and Sands 1973), and myelination generally progresses inferior to superior, central to peripheral, and posterior to anterior of the brain (Barkovich et al. 1988). There is a rapid increase in myelin from birth to approximately 18 months, after which myelination slows down and is nearly completed by 4–5 years old (Dobbing and Sands 1973). Lipid deposition during myelination occurs at the expense of water content; thus the pediatric brain is 88 % water at birth and rapidly *decreases* until 18 months old where it levels off to approximately 80 %, and eventually decreases to adult values of approximately 78 %. (These anatomical attributes are important for understanding material property changes of the brain with age.)

Material Properties

Brain

Age dependence. To date, there is only a single published material property measurement ($n=1$) of pediatric human brain tissue (Prange 2002). Fresh human temporal cortex gray matter was obtained and tested within 3 h of excision from a 5-year-old patient during a temporal lobectomy procedure. For comparison, temporal cortex specimens from similar lobectomy procedures on adults ($n=5$) were also collected. All specimens were tested using shear stress–relaxation procedures in a humidified parallel-plate testing device (Arbogast and Thibault 1997). The specimens were deformed sequentially to 2.5, 5, 10, 20, 30, 40, and 50 % shear strains with each strain anteceded by a 60-s hold period and a 60-s rest interval. At each strain level, the stress–relaxation test was repeated a total of three times, but only data from the third test was recorded. At the completion of tests from all strain levels, the specimens were retested at 5 % strain to ensure that no material damage had occurred at the higher deformations. Resulting strain rates for the material property testing ranged from 0.42 to 8.33 s^{-1}. The time-dependent shear modulus for adult and pediatric tissue was calculated by combining all the measured load-displacement data from each age group (i.e., adult or pediatric) and fitting it to a first-order Ogden hyperelastic material model modified to include energy dissipation (Prange and Margulies 2002).

$$\mu(t) = \frac{\alpha T_{12}(\lambda + \lambda^{-1})}{2(\lambda^{\alpha} - \lambda^{-\alpha})}, \tag{4.1}$$

where $\mu(t)$ is the time-dependent shear modulus of the tissue, T_{12} is the shear stress at time t, α is a material constant, and λ is the principal stretch ratio at time t. The derivation of this constitutive equation assumed that each specimen was incompressible, isotropic, and homogeneous. The shear modulus was calculated at five time points (isochrones) during the test (100, 300, 800, 1,800, and 60,000 ms). Instantaneous shear modulus (μ_0) was determined by fitting the time-dependent shear moduli to a single second-order Prony series (4.2).

$$\mu(t) = \mu_0 \left(1 - \sum_{i=1}^{2} c_i \left(1 - e^{\frac{-t}{\tau_i}}\right)\right) \tag{4.2}$$

When undergoing large deformations of the brain, the instantaneous shear modulus of a single brain tissue specimen from a 5-year-old (380 Pa) was *stiffer* than that of the average of the adult specimens (296 Pa, Table 4.1). The shear moduli at later

4 Experimental Injury Biomechanics of the Pediatric Head and Brain

Table 4.1 Summary of published properties for human pediatric brain, meninges, and cerebral vasculature

Material	Testing method	Property	Age	Value	Source
Brain (temporal cortex)	Shear stress–relaxation	Instantaneous shear modulus	5 years	304 Pa	Prange (2002)
Dura	Axisymmetric tension		30–42 weeks gest	$3,377 \pm 1,052$ N/m	Kriewall et al. (1983)
	Biaxial tension	Stiffness		$1,814 \pm 590$ N/m	Bylski et al. (1986)
Bridging veins	Uniaxial tension	Stiffness	3–9 years	235 ± 199 N/m	Meaney (1991)
		Ultimate stress		12.02 ± 5.9 MPa	
		Stretch ratio		1.67 ± 0.27	

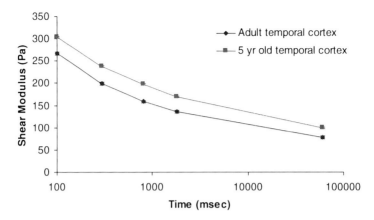

Fig. 4.3 Shear moduli of human cortex gray matter from adults ($n=5$) and a 5-year-old child ($n=1$). The pediatric specimen was stiffer than the adult specimens at all isochrones (Prange and Margulies 2002)

isochrones were also consistently stiffer in the 5-year-old compared to the adult (Fig. 4.3).

In the same study (Prange and Margulies 2002), adult and immature porcine cerebrum specimens were tested in a similar manner to the human specimens (Table 4.2). The adult porcine tissues tested were thalamic gray matter and white matter in the corpus callosum and corona radiata. The small size of the young piglet brain hampered isolation of pure white and gray matter, so mixed gray/white matter specimens were tested. Similar to the human data, the instantaneous shear modulus of the 3–5-day-old piglets' (maturational similarity to an infant; Duhaime et al. 2000) mixed gray/white matter specimens (527 Pa) was stiffer than that of the adult porcine specimens (258 Pa, thalamic and corona radiata stiffness were averaged together to create a mixed gray/white stiffness comparable to the piglet). The instantaneous shear modulus for a 4-week-old (maturational similarity to a toddler; Duhaime et al. 2000) mixed gray/white matter cerebrum (216 Pa) was not significantly different than that of the adult porcine brain cerebrum. Immature cerebrum has also been reported to be significantly stiffer than mature cerebrum in the rat in unconfined compression (Gefen et al 2003).

All of these studies focus on measuring the material property response to large brain deformation. Thibault and Margulies (1998) investigated the linear viscoelastic complex shear modulus (4.3a)–(4.3c) of mixed gray and white matter specimens from juvenile (2–3-day-old) and adult porcine cerebrum using a similar testing protocol as that of Prange and Margulies (2002), but measured the brain response to small deformations over a wide range of frequencies (20–200 Hz).

Table 4.2 Summary of immature animal brain material property studies reported in the literature

Species	Model	Test method	Property	Age	Region	Value		Source
Pig	Linear viscoelastic	Harmonic shear (2.5 % strain)	Dynamic modulus $G'(\omega) + iG''(\omega)$	2–3 days old (infant)	Mixed cortical gray/ white matter	20 Hz: 50: 100: 150: 200:	$771 + i222$ Pa $798 + i371$ $925 + i413$ $1,092 + i484$ $1,257 + i678$	Thibault and Margulies (1998)
				Adult	Cortex	20 Hz: 50: 100: 150: 200:	$1202 + i341$ Pa $1,186 + i735$ $1,424 + i1,256$ $1,576 + i1,694$ $1,756 + i2,139$	
		Harmonic shear (5.0 % strain)		2–3 days old (infant)	Mixed cortical gray/ white matter	20 Hz: 50: 100: 150: 200:	$731 + i217$ Pa $782 + i433$ $916 + i587$ $1025 + i763$ $1,340 + i973$	
				Adult	Cortex	20 Hz: 50: 100: 150: 200:	$550 + i309$ Pa $629 + i504$ $816 + i720$ $1,050 + i1,052$ $1,215 + i1,232$	

(continued)

Table 4.2 (continued)

Species	Model	Test method	Property	Age	Region	Value	Source
Pig	Ogden hyperelastic	Shear stress–relaxation (2.5–50 % strain)	Instantaneous shear modulus (see Figs. 4.2 and 4.3 for later time points)	3–5 days old (infant)	Mixed cortical gray/white matter	527 Pa	Prange and Margulies (2002)
				4 weeks old (toddler)	Mixed cortical gray/white matter	217 Pa	
				Adult	Corona radiata	254 Pa	
					Corpus callosum	182 Pa	
					Thalamus	264 Pa	
					Cortex	220 Pa	Coats and Margulies (2006b)
Pig	Transversely isotropic viscoelastic	Shear stress–relaxation (2.5–50 % strain)	Instantaneous shear modulus	4 weeks old (toddler)	Brainstem pons	Parallel/perpendicular: 13 Pa Cross-sectional: 121 Pa	Ning et al. (2006)
Rat	Linear viscoelastic	Cortical indentation (1 mm @ 1 mm/s)	Instantaneous (G_i) and long term (G_∞) modulus	13–17 days old (toddler–teen)	Cortical gray matter	$G_i = 1{,}754 \pm 462$ Pa $G_\infty = 626 \pm 184$ Pa	Gefen et al. (2003)
				43–90 days old (adult)		$G_i = 1{,}232 \pm 343$ Pa $G_\infty = 398 \pm 119$ Pa	
Rat	Ogden hyperelastic	Atomic force microscope indentation (30 % strain)	Elastic modulus	10 days old (infant)	Average over cortex and hippocampus	451 Pa	Elkin et al. (2010)
				17 days old (teen)		775 Pa	
				Adult		1,048 Pa	
		Unconfined compression (30 % strain)	Long-term modulus	10 days old (infant)	Cortex	645 Pa	
					Hippocampus	601 Pa	
				17 days old (teen)	Cortex	901 Pa	
					Hippocampus	667 Pa	
				Adult	Cortex	1,773 Pa	
					Hippocampus	1,172 Pa	

$$G^* = G'(\omega) + iG''(\omega). \qquad (4.3a)$$

$$G'(\omega) = \left(\frac{\tau_0}{\gamma_0}\right)\cos\delta. \qquad (4.3b)$$

$$G''(\omega) = \left(\frac{\tau_0}{\gamma_0}\right)\sin\delta. \qquad (4.3c)$$

G' is the elastic and G'' is the viscous component of the complex modulus at frequency ω, shear stress amplitude τ_0, shear strain amplitude γ_0, and with phase angle δ between the sinusoidal varying responses. At 2.5 % strain, the juvenile cerebrum had a significantly ($p<0.05$) lower range of shear storage moduli ($G' = 771$–1,257 Pa) than the adult cerebrum ($G' = 1{,}202$–1,756 Pa) across all frequencies tested (Table 4.2). At 5 % strain, however, the juvenile moduli were generally stiffer than the adult, but these differences were not statistically significant ($p=0.21$). To compare results between the two data sets, Prange and Margulies fit their 5 % strain data to the same linear viscoelastic model reported in Thibault and Margulies (4.3a)–(4.3c) and compared the storage modulus to that reported in Thibault and Margulies at 20 Hz (the most similar frequency between the two studies). The juvenile and adult storage moduli reported in Prange and Margulies were not significantly different than those reported in Thibault and Margulies ($p=0.07$ and 0.97, respectively), showing good agreement between the two studies. Furthermore, Prange and Margulies report that the juvenile shear moduli from 5 % strain were stiffer than the adult ($p<0.001$), similar to what was reported in Thibault and Margulies. Combined, these studies suggest that the immature cerebrum is significantly stiffer than the adult cerebrum when undergoing large deformations, but age-related differences in brain tissue response to small deformations may be minimal or reversed.

Because of the paucity of data on pediatric human brain tissue, computational models of the pediatric head typically use ratios of adult to juvenile porcine brain material properties to extrapolate pediatric material properties from human adult material properties. From the Prange and Margulies study, the ratio of 3–5-day-old porcine (maturational similarity to an infant) instantaneous shear modulus to adult porcine shear modulus in mixed gray/white matter cerebrum is 2.04. Applying this ratio to the instantaneous shear modulus of adult human brain tissue (296 Pa) results in an estimated instantaneous shear modulus of 604 Pa in the infant human brain. The stiffness of a 5-year-old brain would be expected to lie somewhere between the adult and infant moduli. In accordance with this estimate, the instantaneous shear modulus of a single sample of brain tissue from a 5-year-old (380 Pa) is between the adult and infant shear moduli, but its stiffness is closer to that of the adult human brain tissue. Biomechanical models have reported that axons, rather than the surrounding matrix of astrocytes and oligodendrocytes, contribute more to the stiffness of brain tissue (Arbogast and Margulies 1999; Ning et al. 2006). Axons in the pediatric brain are undergoing rapid myelination during the first year of life. Lipids have a low shear modulus (100 Pa; Yamada 1970), suggesting that the increasing volume

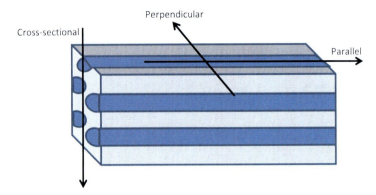

Fig. 4.4 Ning et al. (2006) tested 4-week-old porcine brainstem in finite shear deformations that were parallel, perpendicular, or cross-sectional to axon orientation

of myelin in the brain will decrease the effective shear modulus of brain tissue. The pediatric brain reaches adult levels of myelination at approximately 18 months old, making the composition of the 5-year-old brain more similar to an adult brain than an infant brain. These data suggest that the age-dependent material properties of brain tissue are not linear, but rather change rapidly during the first couple of years and more gradually in early childhood. Therefore, linear scaling ratios derived from animal data may have limited utility in estimating pediatric brain stiffness, and therefore more human pediatric brain material property data is needed to accurately determine the nonlinear age-dependent relationship of the human brain.

Anisotropy. The only study published investigating the directional dependence of immature brain tissue is in the pons from brainstem of 4-week-old piglets (maturational similarity to toddlers). Following a similar protocol to that of Prange and Margulies (2002), Ning et al. (2006) excised rectangular specimens for shear stress–relaxation testing. The orientation of the specimens allowed for shear to be tested parallel, perpendicular, or cross-sectional to the axonal directions in the brainstem (Fig. 4.4). All specimens were tested in vitro and deformed sequentially to 50, 40, 30, 20, 10, 5, 2.5 %, and back to 50 % to verify reproducibility. Each deformation was anteceded by a 60-s hold period and a 60-s rest interval. At each strain level, the deformation was repeated a total of three times, but only data from the third test was recorded. A finite element model of the shear stress–relaxation tests and a genetic optimization algorithm were used to determine the material parameters of the tissue based on a transversely isotropic, viscoelastic model. Because the tissue was assumed to be transversely isotropic, cross-sectional material parameters were compared to combined parallel and perpendicular material parameters. The cross-sectional shear modulus (121.2 Pa) was approximately ten times stiffer than the combined parallel and perpendicular shear modulus (12.7 Pa). This is in contrast to an earlier study on adult porcine brainstem (Arbogast 1997) which reported no significant differences among specimens parallel, perpendicular, or cross-sectional to the axons in the brainstem. However, the adult porcine investigation imposed small deformations (2.5–5 %) whereas the Ning et al. study focused on large deformations.

In large deformations, Prange and Margulies (2002) reported a regional dependence of anisotropy in adult porcine cerebrum, with the shear modulus of white matter being more influenced by test direction than gray matter. The authors attribute this to the linear-oriented axons in the white matter tracks compared to the homogeneous dense network of nuclei in cerebral gray matter. Because the structural organization of the axons and nuclei in the white and gray matter is not thought to change during maturation, it is likely that pediatric cerebrum will also have similar differences. Further studies are needed to evaluate the anisotropy of pediatric brain tissue properties.

Regional dependence. It may be assumed that due to the structural and compositional differences between white and gray matter (throughout the brain) there will be a significant difference in the regional material properties. There are no published studies of regional differences in human pediatric brain tissue, but a rheology study of human adult brain tissue (Shuck and Advani 1972) reported small directional dependent differences between the corona radiata (white matter) and thalamus (gray matter) at small strains and strain rates. The authors felt these differences were minor, however, and averaging across all directions resulted in no significant difference between the two regions. Prange and Margulies (2002) further explored differences between white and gray matter, and within regions of white matter, in porcine adult brain tissue using shear stress–relaxation tests at large deformations. They report that the instantaneous shear modulus (μ_0) of thalamic gray matter (264 Pa) was stiffer than that of corona radiata (254 Pa), which was stiffer than that of corpus callosum (182 Pa). To determine material property differences within regions of gray matter, Coats and Margulies (2006b) repeated the Prange and Margulies testing protocol on cortical gray matter specimens from adult pigs. They reported the instantaneous shear modulus of cortical tissue (220 Pa) to be less stiff than that of thalamic gray matter and corona radiata, but stiffer than that of corpus callosum (Fig. 4.5). As the pediatric brain develops, the volume of gray and white matter alters significantly. If regional differences exist in pediatric brain tissue as they do in adult porcine brain tissue, then computational models of the pediatric brain should include the volumetric and material property changes of the regions of the brain to better mimic the age-dependent response of the brain to loading.

Rate dependence. Using porcine brain tissue, Thibault and Margulies (1998) have provided the only study investigating the rate dependence of material properties in the immature brain. In the same shear testing protocol described above, they reported a significant increase in the complex shear modulus of cortical gray matter from 2- to 3-day-old piglets when frequency is increased from 20 to 200 Hz, encompassing strain rates between 2 and 40 s^{-1}. They reported a similar increase in stiffness with increase in strain rate in the adult porcine brain, a finding that has been corroborated in many other material property investigations in adult porcine, bovine, and caprine brain specimens (Koeneman 1966; Shuck and Advani 1972; Peters et al. 1997; Arbogast and Margulies 1998; Bilston et al. 2001; Lippert et al. 2004).

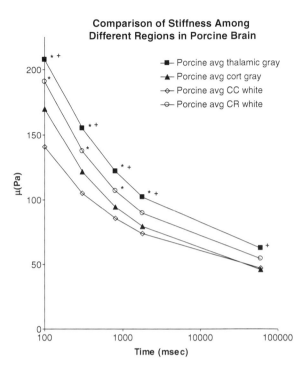

Fig. 4.5 Shear moduli of porcine gray and white matter regions from Prange and Margulies (2002) compared to shear moduli of porcine cortical gray matter measured in Coats and Margulies (2006b). Average corpus callosum was significantly different from both thalamus ($p < 0.002$) and corona radiata ($p < 0.005$) at the isochrones indicated (*asterisk*). Cortex was significantly different than the thalamus ($p < 0.006$) at all isochrones (*plus*). (SD bars not shown for clarity)

Meninges

The meninges provide support and protection to the brain during head rotation and impact. A recent finite element model parametric study reported that the inclusion of the pia mater significantly reduced cortical contact strains during repetitive rotational load to the infant head (Couper and Albermani 2008). In computational models of adult head trauma, the inclusion of the dura, tentorium, and falx has also been reported to significantly affect the frequency response and intracranial pressure from side head impact (Ruan et al. 1991; Kumaresan and Radhakrishnan 1996). As more studies underscore the importance of the meninges on the response of the pediatric head, the material properties of the dura, pia, and arachnoid become important to accurate traumatic brain injury simulation.

To date, two studies by the same group (Kriewall et al. 1983; Bylski et al. 1986) have published the only characterization of human pediatric dural tissue (Table 4.1). Axisymmetric biaxial tension tests were performed on fetal (30–42 weeks gestation) dura mater at 0.02 mm/s. In the biaxial testing experiments (Bylski et al. 1986), each specimen experienced increasingly larger deformations six times before pulling to failure. The final failure load was always greater than the six cyclic loads, so only the failure load was analyzed. The material properties were reported to be best characterized by a nonlinear elastic constitutive model developed by Skalak et al (1973).

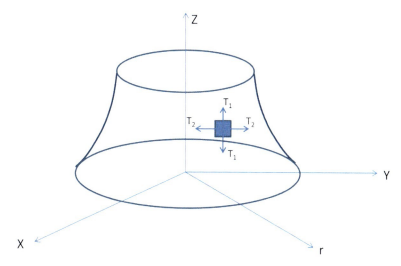

Fig. 4.6 Membrane configuration for the biaxial tension tests performed on pediatric dura by Bylski et al. (1986). The STZC constitutive equations calculated the meridional (T_1) and circumferential (T_2) forces on the membrane during biaxial loading

$$T_1 = \left(\frac{C\lambda_1}{2\lambda_2}\right)\left[\Gamma\left(\lambda_1^2 - 1\right) + \lambda_2^2\left(\lambda_1^2\lambda_2^2 - 1\right)\right], \qquad (4.4a)$$

$$T_2 = \left(\frac{C\lambda_2}{2\lambda_1}\right)\left[\Gamma\left(\lambda_2^2 - 1\right) + \lambda_1^2\left(\lambda_1^2\lambda_2^2 - 1\right)\right], \qquad (4.4b)$$

where T_1 and T_2 are the forces acting in the meridional and circumferential axes on the membrane (Fig. 4.6); Γ is a ratio of two material constants (B and C), each with units of force per unit length; and λ_1 and λ_2 are stretch ratios along the meridional and circumferential axes of the membrane.

No significant difference was found between specimens removed from the parietal left, parietal right, or frontal regions of the brain. The average constant, C, resulting from fitting the data to (4.4a) and (4.4b) for all specimens, was $1{,}850 \pm 602$ gf/cm ($1{,}814 \pm 590$ N/m). No dural property data on older children and adolescents has been published, but the dynamic complex modulus of adult human dura in tension is reported as 31.7 MPa (Galford and McElhaney 1970). Given the differences in test rate and methodology between the pediatric and adult studies, no direct comparison can be made. There are no pediatric human or animal studies on the pia or arachnoid, but investigators have begun publishing data on the adult animal pia and pia–arachnoid complex. In bovine adult pia mater pulled in quasistatic tension, the pia tissue specimens (width = 35 mm, no thickness reported) exhibited stiff, elastic behavior with a large 0.024 N/mm toe region and a subsequent linear elastic region with an approximate stiffness of 0.19 N/mm (Aimedieu and Grebe 2004). The entire

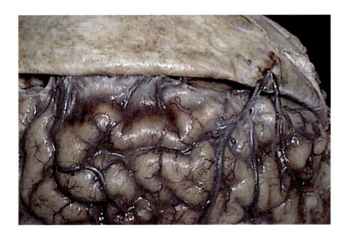

Fig. 4.7 Image of bridging veins exiting from the dura (*folded back* in picture) and entering the brain. (Image from http://library.med.utah.edu/WebPath/CNSHTML/CNS315.html)

adult bovine pia–arachnoid complex has been tested in tension and shear (Jin et al. 2006; Jin et al. 2007), and the complex is reported to be stiffer than the brain tissue, to have a significant material dependence on rate, and to be characteristic of an isotropic and linearly elastic material until failure. It is unknown if these material characteristics are applicable to the pediatric brain, and should be used with caution in pediatric head injury models until age-dependent studies on the meninges can be completed.

Cerebral Vasculature

The cerebral vasculature transports oxygen, nutrients, and other important substances to the brain to ensure its proper functioning. In studies of TBI and subdural hematoma, the focus is primarily on bridging veins spanning the subdural space. These vessels exit from the dura, span the subdural space, and pass through the pia and arachnoid into the brain (Fig. 4.7). Meaney (1991) measured the material properties of bridging veins ($n=59$) from a wide age range of unembalmed cadavers (3–62 years old) to determine the age dependence of the material properties. Specimens were tested in tension within 24 h postmortem at dynamic strain rates ranging from 0.4 to 241 s^{-1}. The ultimate stretch ratio (1.67 ± 0.27), ultimate tension (235 ± 199 N/m), and ultimate stress (12.02 ± 5.9 MPa) from 3- to 9-year-old cadaver specimens (Table 4.1) were reported not to be significantly different than those of the adult (1.51 ± 0.17, 326 ± 196 N/m, and 18.57 ± 14.2 MPa, respectively). Meaney also reported no significant effect of strain rate on any of the material properties. This lack of rate dependence is in agreement with an earlier study (Lee and Haut 1989) of 65 unperfused bridging veins from elderly adult (62–85 years old) cadavers tested at either low (0.1–2.5 s^{-1}) or high (100–250 s^{-1}) strain rates, and more recent tensile studies measuring the material properties of cadaver bridging vein-superior sagittal sinus at strain rates spanning 0.1–3.8 s^{-1} (Delye et al. 2006) and cortical arteries and veins at strain rates spanning 0.01–500 s^{-1} (Monson et al. 2003). The only study that disagrees with the lack of rate dependence of bridging veins is by Lowenhielm (1974), who reported a significant decrease in elongation with an increase in strain rate (5 vs. 200 s^{-1}) in bridging veins from 22 specimens from 11 cadavers (13–87 years old). Meaney postulated that the difference in the reported rate dependence may be due to slippage of specimens in the grips in the Lowenhielm study.

Monson et al. (2003, 2005, 2008) have published several studies that characterize the mechanical properties of cerebral vasculature in adults. Their key findings from these studies are that (1) cerebral arteries are significantly ($p<0.05$) stiffer (21.4 MPa) than cerebral veins (3.41 MPa) in longitudinal tension and will fail with significantly ($p<0.05$) less deformation; (2) the longitudinal modulus (1.8–5.5 MPa) of cerebral arteries is significantly ($p<0.025$) stiffer than the circumferential modulus (0.8–2.6 MPa); (3) cadaveric cortical arteries are stiffer (31.1 ± 12.9 MPa) than fresh cortical arteries (19.3 ± 6.9 MPa) and stretch 35 % less before failure. No significant difference is found for modulus and stretch to failure between cortical veins from autopsy and surgery; and (4) small arteries (cortical) tend to be stiffer (31.1 ± 12.9 MPa) than large arteries (middle cerebral artery, 18.2 ± 9 MPa), but small veins (cortical) are significantly *less stiff* (3.5 ± 2.6 MPa) than the larger bridging veins (6.4 ± 3.4 MPa) and stretch 12 % less before failure. In summary, the mechanical properties of cerebral vessels are dependent on test direction, vessel type, size, and freshness of the tissue. The donor age encompassing specimens in all of these studies ranged 14–62 years old, so it is unknown if these cerebral vessel characteristics extend to the younger pediatric age group.

From the published data it appears that the material properties of adult bridging veins may be used to mimic pediatric bridging veins; however, due to the absence of data this finding has not been supported in the young infant and toddler (<3 years old). In addition, studies on characteristics of other cerebral vasculature in this age group are needed to verify the vessel characteristics reported in adults. Computational models of adult traumatic head injury (Zhou et al. 1995; Huang et al. 1999; Zhang et al. 2002; Kleiven 2003) have begun incorporating bridging veins to determine the likelihood of subdural hematoma from TBI, and this investigation should be extended to models of pediatric head injury. Adult computational models investigating the importance of including the more detailed cerebral vasculature are contradictory in their findings (Zhang et al. 2002; Ho and Kleiven 2007) and further evaluation is warranted before conclusions can be extended to the pediatric population.

Cerebral Spinal Fluid

CSF is formed in the ventricles and distributed throughout the central nervous system via the subarachnoid space. In adult computational models, CSF has typically been modeled with material properties similar to water (Hosey and Liu 1980; Ruan et al. 1994; Zhou et al. 1995; Kumaresan and Radhakrishnan 1996). Bloomfield

Table 4.3 Cerebrospinal fluid (CSF) viscosity compared to the viscosity of distilled water. Relative CSF viscosity is the ratio between CSF and distilled water viscosity. (Data extracted from Bloomfield et al. 1998)

Patient age (years)	Patient gender	Average water viscosity (mPa·s)	Average CSF viscosity (mPa·s)	CSF relative viscosity
1/12	M	0.705	0.727	0.97
14	M	0.700	0.760	1.10
15	F	0.750	0.745	0.99
17	F	0.705	0.748	0.94
18	F	0.750	0.650	0.87
19	F	0.930	0.800	0.86

et al. (1998) used rheometry to measure the viscosity of CSF at high (360–1,460 s^{-1}) and low (25–120 s^{-1}) shear rates to determine if it met the standards of a Newtonian fluid. Samples of CSF were collected clinically from patients ranging from 1 month to 69 years old and tested using a rheometer. Distilled water samples were tested for comparison. Only specimens from patients 19 years old and younger ($n=6$) are reported here. The viscosities of all specimens were similar to the measured viscosities of distilled water (Table 4.3) and the authors reported a significant increase in shear stress with an increase in strain rate, validating the modeling of CSF as a Newtonian fluid similar to water. It should be noted that there was only a single specimen from an infant and none from a toddler or young adolescent (<12 years old). The infant CSF had the lowest measured viscosity (0.727 mPa·s), but it was not much lower than CSF from a 73-year-old male (0.735 mPa·s) or a 17-year-old female (0.748 mPa·s). Additionally, the ratio of infant CSF viscosity to distilled water viscosity (0.97) was well within the range of CSF-to-water ratios for older patients (0.86–1.24). Bloomfield reported that changes in the biochemical composition of CSF had little effect on the viscosity of the fluid, suggesting that any biochemical changes that may occur during the developmental period of a child will be negligible to the overall viscosity of the fluid.

Structural Testing and Load Tolerances

In adult head injury, load tolerances for concussion have been suggested based on measurements made from surrogate recreations of concussive impacts in football players (Pellmen et al. 2003; Zhang et al. 2004) and instrumented boxers (Breton et al. 1991), and load tolerances specific for subdural hemorrhage (SDH) and diffuse axonal injury (DAI) have been suggested based on scaled rotational accelerations from inertial head injury studies in primates (Gennarelli et al. 1979; Margulies and Thibault 1992). In pediatric head injury, there is no human data on load tolerances causing concussion, SDH, or DAI, but several animal injury models have investigated the immature brain response to injury. Fluid percussion (Armstead and Kurth 1994; Prins et al. 1996; Giza et al. 2005) and weight-drop models (Grundl et al. 1994; Adelson et al. 1996; Bittigau et al. 1999) in pigs and rats have provided valuable insight into the age-dependent brain response to injury (i.e., behavior, recovery time, and cellular changes), but do not attempt to establish the minimum load necessary to cause injury in the immature brain. The applied load is often adjusted in weight-drop models to maintain a constant force to animal brain weight ratio across age groups, but this is rarely the case in fluid percussion injury. This may be the reason why some fluid percussion studies report the immature brain to be more vulnerable to injury (Armstead and Kurth 1994; Prins et al. 1996), while other weight-drop studies report no significant differences between the adult and immature animals (Adelson et al. 1996). The weight-drop models are mostly carried out using a small animal rodent model. Rodents have lissencephalic brains with little white matter and have a maturational sequence that is very different from humans, making translation from these studies to humans difficult. In contrast, piglets have a distinct gyral pattern, differentiation of white and gray matter, and the growth pattern in the postnatal pig brain is similar to infants. Therefore, pigs are often selected as a large animal model to investigate the age-dependent response to traumatic head injury.

In a controlled cortical impact study, Duhaime et al. (2000) produced cortical lesions across three age groups of piglets (5-day "infant," 1-month "toddler," and 4-month "adolescent"). The impacted area and vertical displacement were scaled to brain dimensions such that the percentage of total brain volume displaced was equal for each age group. Velocity of indentation was held constant among age groups. They report that the infant piglets had the smallest cortical lesion and the adolescent group had the largest at 1 week post injury, indicating that the younger infant brain was either less vulnerable to injury or was able to have a quicker recovery than the older pigs.

In nonimpact, inertial studies, anesthetized 3–5-day-old piglets ($n=7$) underwent a single rapid, horizontal head rotation ($\omega = 250 \pm 10$ rad/s, $\alpha = 117 \pm 21$ krad/s^2, $\Delta t = 11.4 \pm 0.8$ ms; Raghupathi and Margulies 2002). Gross examination revealed that all piglets had subdural and subarachnoid hemorrhage over the frontal lobes, and histology detected diffuse traumatic axonal injury. When comparing the axonal damage in these piglets to axonal damage in adult pigs undergoing the same rotational load ($\omega = 249 \pm 31$ rad/s, $\alpha = 173 \pm 56$ krad/s^2; Smith et al. 2000), the piglets had 3.4 times more injured axons/mm^2 on cerebral histology slices compared to adults. Additionally, after scaling the angular velocity according to brain mass, the peak angular velocity experienced by the piglet was 22 % lower than the velocity experienced by the adult, suggesting that the immature brain is more vulnerable to rotational injury than the adult. In a subsequent study, the effect of repeating the rapid horizontal head rotation 15 min following the initial rotation was investigated (Raghupathi et al. 2004). The velocities were lower than those in the previous studies (172 ± 17 rad/s for animals receiving a single moderate rotation, $n=5$; and 136 ± 8 followed by 140 ± 6 rad/s for animals experiencing two mild rotations, $n=6$). On gross examination, small amounts of blood were located on the frontal lobe and brainstem in the single moderate rotation group (3/5 animals) and the double mild rotation group (6/6 animals). No subarachnoid hemorrhage was found during histological analysis. For animals experiencing a single moderate rotation, there were a few regions with one or two injured axons. In animals experiencing two mild rotations, there were many more regions with one or two injured axons and several with three or more injured axons. These data suggest that there exists a cumulative effect of injury and two insults to the head result in greater trauma in the immature brain than a single insult—even if the repeated insult has a slightly lower load. To further validate this finding, future studies investigating the severity of pediatric brain injury from cyclic low-velocity rotational loading of the head over time should be performed and compared to brain injury from a range of single, higher velocity loads similar to those experienced following accidental falls or during motor vehicle crashes.

The above studies focused only on rotation in the horizontal direction. However, recent data indicates that there is a directional dependence on pediatric head injury (Eucker, Smith et al. 2011). Anesthetized 3–5-day-old piglets underwent rapid head rotations in the horizontal (187 ± 16 rad/s), sagittal (165 ± 3 rad/s), and coronal (212 ± 11 rad/s) directions. The animals undergoing sagittal rotations had the worst clinical findings with the greatest loss of consciousness, 100 % apnea, and lowest cerebral blood flow. Animals undergoing horizontal rotations had mixed outcomes with variable durations of unconsciousness and less cerebral infarction than the animals with sagittal head rotation, but animals with coronal head rotations had the least severe findings and were not significantly different than control animals. Comparisons of coronal and horizontal head rotation in the adult pig have had similar findings. Smith et al. (2000) reported significantly longer unconscious times and more subdural and/or subarachnoid blood in animals undergoing horizontal rotation compared to those undergoing coronal rotation. Interestingly, the animals with horizontal head rotation had significantly more white matter damage in the brainstem while the animals with coronal rotation had significantly more white matter damage in the cerebrum. Similar head injury studies in the adult primate also report significant effects of rotational direction on brain trauma (Gennarelli et al. 1982, 1987), but differences between head and neck orientation in the biped compared to the quadruped make direct comparison difficult.

The above-mentioned studies have been focused on understanding mechanisms of pediatric TBI, but not determining minimum load tolerances necessary to produce concussion, SDH, or DAI. Further studies that span both injurious and non-injurious scenarios are needed to develop reasonable load tolerances for pediatric head injury. What has been learned, however, from the current animal models is that load tolerances will be dependent on injury modality, developmental age of the child, rotational direction and/or impact location, number of insults to the brain, and time between those insults.

Skull

Anatomy and Development

Skull development is rapid in utero and continues throughout childhood. In utero, the superior portion of the human skull (parietal, frontal, and the upper portion of the occipital and temporal bones) is formed through intramembranous ossification, while

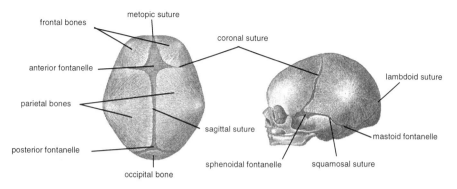

Fig. 4.8 Illustration of the major pediatric cranial bones, sutures, and fontanelles in the infant. The striations from the center of each bone plate are the trabeculae radiating from the ossification center (*lighter regions*). Image adapted from Wikimedia Commons

Table 4.4 Approximate ages for closure of sutures and fontanelles

Sutures and fontanelles	Approximate age at closure
Posterior fontanelle	Birth to 6 months[a], 8 weeks[b]
Anterolateral (sphenoid) fontanelle	Birth to 6 months[a]
Posterolateral (mastoid) fontanelle	1 year[a]
Anterior fontanelle	1–2 years[a,b,c]
Most sutures in Basilcranium	3–5 years[a]
Metopic suture	2–4 years[c], 5–6 years[a]
Basisphenoid suture	25 years[a] (for complete fusion)
Coronal suture	30–40 years[a,b] (for complete fusion)
Sagittal suture	30–40 years[a,b] (for complete fusion)
Lambdoid suture	30–40 years[a,b] (for complete fusion)

[a]Meschan (1959)
[b]Gooding (1971)
[c]Scheuer and Black (2004)

the basal portion of the skull (lower segment of occipital and temporal bones, ephemoid, and sphenoid) is formed through endochondral ossification. A single layer of mesenchymel cells (mesenchyme) surrounds the developing fetal brain. Ossification centers appear on the membrane and begin to form bone tissue on the membrane (intramembranous ossification) or induce mesenchymel cells to differentiate into cartilage prior to ossification (endochondral ossification). Once ossification begins, bone grows outward radially from the ossification centers, forming bony islands on the mesenchyme. The mesenchyme membrane marks the path for future bone growth, and allows the bone plates to move relative to each other and cause head deformation during childbirth. Cranial bone formation is nearly complete at birth, leaving only narrow regions of membrane between the bones, called suture and fontanelles (Fig. 4.8). The continued presence of the membrane after birth allows additional brain growth throughout infancy and childhood until the sutures and fontanelles are obliterated by new bone. While the edges of many bone plates join together in early childhood, they do not become completely fused until adulthood (Table 4.4).

Table 4.5 Suture width for 0- to 5-month-old infants

Age (months)	Coronal (mm)	Sagittal (mm)	Lambdoid (mm)
0.5	1.17±0.26 ($n=4$)	1.09±0.35 ($n=2$)	1.25±0.21 ($n=5$)
1.0	1.88±0.35 ($n=2$)	1.14 ($n=1$)	1.00±0.29 ($n=2$)
1.5	0.99±0.32 ($n=13$)	0.88±0.22 ($n=9$)	1.05±0.49 ($n=14$)
2.0	1.02±0.31 ($n=12$)	1.07±0.44 ($n=8$)	0.89±0.48 ($n=14$)
2.5	0.79±0.25 ($n=16$)	0.98±0.27 ($n=12$)	1.00±0.34 ($n=16$)
3.0	1.13±0.66 ($n=12$)	1.04±0.41 ($n=10$)	1.17±0.54 ($n=13$)
3.5	1.07±0.34 ($n=4$)	0.89±0.07 ($n=2$)	0.65±0.23 ($n=4$)
4.0	0.92±0.26 ($n=4$)	0.83±0.20 ($n=3$)	0.72±0.24 ($n=5$)
4.5	0.69±0.14 ($n=3$)	0.98±0.07 ($n=2$)	0.91±0.26 ($n=3$)
5.0	1.33±0.42 ($n=2$)	0.84 ($n=1$)	0.37 ($n=1$)
Avg. 0–5 months	0.89±0.35	0.93±0.28	0.96±0.39

Data extracted by digitizing graphs in Soboleski et al. (1997)

X-ray was the first technique used to document the change in suture width from childbirth to adulthood. X-rays of the skull found a large variation in coronal, sagittal, and lambdoid suture width in neonates (from 1.5 to 17 mm), but this data set may include children born prematurely (Gooding 1971). In older children, suture width is more consistently defined and easier to measure. Coronal suture is <3 mm in children between 2 and 12 months and <2 mm in children greater than 3 years old (Gooding 1971). More recently, ultrasound has been used to obtain precise measurements of suture in the infant. Soboleski et al. (1997) measured the coronal, sagittal, and lambdoid suture thickness and width in fifty 0.5–5-month-old infants (developmental age based on a full term of 40 weeks). The authors report the average thickness of coronal, sagittal, and lambdoid suture across all ages to be 1.97±0.54, 1.88±0.56, and 2.49±0.86 mm, respectively. The average width of infant coronal, sagittal, and lambdoid suture was reported as 0.89±0.35, 0.93±0.28, and 0.96±0.39 mm, respectively (Table 4.5). No significant difference in suture width was found between the medial and lateral regions of the coronal and lambdoid sutures and both measurements were used to determine the average thickness and width of these sutures.

Cranial bone is a classic lamellar structure. At birth, the cranial skull begins as a single cortical bone layer with a fine network of trabeculae that can be macroscopically seen emanating from the ossification centers (Fig. 4.8). As the bone continues to develop, the single layer of cortical bone differentiates into three layers: an outer table of cortical bone, a middle spongy diploe layer, and an inner table of cortical bone. The trabeculae that are visible in pediatric skull tissue are not visible in the adult skull. At birth the frontal bone is the thickest cranial bone (4.12±0.43 mm), but the rapid rate of growth of the occipital bone causes it to overtake the thickness of the frontal bone at 8 years of age. The parietal bone is the thinnest bone at birth and grows to be only slightly thicker than frontal bone by adulthood (Table 4.6).

4 Experimental Injury Biomechanics of the Pediatric Head and Brain

Table 4.6 Cranial thickness increases with age

Age	Frontal (mm)	Parietal Anterior (mm)	Parietal Posterior (mm)	Occipital (mm)
20–40 weeks gestation			0.78 ± 0.20 (19)[a,b]	1.16 ± 0.40 (7)[b]
Term (>40 weeks) to 6 months	4.12 ± 0.43 (32)[c]	0.96 ± 0.50 (39)[b,c]	1.14 ± 0.64 (36)[c,d]	1.32 ± 0.35 (37)[b,c]
7–11 months	4.95 ± 0.35 (32)[c]	0.76 ± 0.33 (35)[b,c]	1.83 ± 1.02 (37)[c,d]	1.54 ± 0.88 (34)[b,c]
1–2 years	5.9 ± 0.51 (32)[c]	1.84 ± 0.38 (32)[c]	2.37 ± 0.88 (50)[c,d]	2.88 ± 0.34 (32)[c]
3–5 years	6.48 ± 0.81 (32)[c]	3.16 ± 0.39 (32)[c]	3.34 ± 0.87 (42)[c,d]	4.22 ± 0.50 (32)[c]
6–7 years	5.70 ± 1.43 (32)[c]	3.68 ± 0.5 (32)[c]	3.50 ± 1.10 (42)[c,d]	5.00 ± 0.53 (32)[c]
8–11 years	4.76 ± 1.13 (32)[c]		4.31 ± 0.95 (53)[c,d]	5.63 ± 0.64 (32)[c]
12–15 years	4.20 ± 0.76 (32)[c]		4.89 ± 0.95 (53)[c,d]	6.14 ± 0.79 (32)[c]
16–18 years	4.15 ± 0.8 (32)[c]		5.29 ± 0.83 (34)[c,d]	6.40 ± 0.88 (32)[c]

Values are mean ± SD followed by the number of calvaria in *parentheses*
[a]Contributing source to average: Kriewall et al. (1981)
[b]Contributing source to average: Coats and Margulies (2006a)
[c]Contributing source to average: Roche (1953)
[d]Contributing source to average: Loder (1996)

Material Properties

Cranial Bone

Anisotropy. At birth, pediatric cranial bone has a visible fiber orientation due to the trabeculae in the bone, suggestive of anisotropic material properties found in other fiber-oriented tissues (e.g., tendon, muscle). In a series of studies, McPherson and Kriewall (1980; Kriewall 1982) tested 554 specimens from 16 fetal calvaria (20–42 weeks gestation) in quasistatic (0.5 mm/min) three-point bending with the trabeculae of the bone running either parallel or perpendicular to the long axis of the specimen. They report the elastic modulus of parietal bone across all ages to be nearly three times larger when the trabeculae are oriented parallel ($3{,}686 \pm 1{,}440$ MPa) to the long axis of the specimen than when the trabeculae are oriented perpendicular to the long axis of the specimen ($1{,}309 \pm 881$ MPa, Table 4.7), confirming significant anisotropy in the immature cranial bone. In contrast, adult cranial bone has no trabeculae (Dempster 1967) and no significant difference between moduli when tested in two orthogonal directions tangential to the surface of the skull (McElhaney et al. 1970).

Age dependence. In addition to directional dependence, McPherson and Kriewall report a significant increase in cranial bone elastic modulus with increased fetal development (20–40 weeks gestation). In their initial study (1980), they report that the quasistatic elastic modulus of preterm infants (24–30 weeks gestation) was lower than the elastic modulus of term infants (36–40 weeks gestation) for both tests parallel and perpendicular to the long axis of the specimen. This finding was confirmed in their later studies with additional calvaria (1982). More recent

Table 4.7 Summary of material properties reported in the literature for parietal bone

Age	McPherson and Kriewall (0.05 mm/min)		Margulies and Thibault (2.54–2,540 mm/min)			Coats and Margulies (1.6–2.8 m/s)		
	$E_\parallel$ (MPa)	$E_\perp$ (MPa)	$E_\perp$ (MPa)	$\sigma_\perp$ (MPa)	$U_\perp$ (N·mm/mm^3)	$E_\perp$ (MPa)	$\sigma_\perp$ (MPa)	$\varepsilon_\perp$ (mm/mm)
20–40 weeks gest	3,441 ± 1,093 (n=15)	1,309 ± 881.1	305.3 ± 230.1	9.1 ± 5.5 (n=2)	0.055 ± 0.016	371.5 ± 329.9	26.3 ± 28.8 (n=5)	0.01 ± 0.02
Term (>40 weeks) to 3 months	7,360 (n=1)	–	820.9	10.6 (n=1)	0.061	531.3 ± 298.4	31.8 ± 19.4 (n=6)	0.07 ± 0.03
4–6 months	–	–	2,641 ± 673.8	52.8 ± 16.4 (n=1)	0.253 ± 0.160	552.3	23.7 (n=1)	0.05
7–9 months	–	–	–	–	–	–	–	–
10–12 months	–	–	–	–	–	531.0 ± 242.2	42.9 ± 15.8 (n=2)	0.06 ± 0.06
1–2 years	–	–	–	–	–	–	–	–
3–5 years	–	–	–	–	–	–	–	–
6–7 years	7,380 ± 840	5,860 ± 690 (n=1)	–	–	–	–	–	–

Values reported in mean ± SD. Number in *parenthesis* indicates the number of calveria tested

Table 4.8 Summary of material properties reported in the literature for frontal and occipital bone

Age	Frontal bone: McPherson and Kriewall (0.05 mm/min)		Occipital bone: Coats and Margulies (1.6–2.8 m/s)		
	$E_\parallel$ (MPa)	$E_\perp$ (MPa)	$E_\perp$ (MPa)	$\sigma_\perp$ (MPa)	$\varepsilon_\perp$ (mm/mm)
20–40 weeks gest	3,057 ± 230 ($n=2$)	1,700 ± 790 ($n=1$)	161.1 ± 128.3	11.54 ± 7.3 ($n=7$)	0.030 ± 0.027
Term (>40 weeks) to 3 months	–	–	511.4 ± 354.0	15.3 ± 14.0 ($n=4$)	0.030 ± 0.014
4–6 months	–	–	355.3 ± 53	17.9 ± 2.2 ($n=1$)	0.050 ± 0.001
7–9 months	–	–	–	–	–
10–12 months	–	–	412.7 ± 247.2	20.6 ± 13.0 ($n=2$)	0.060 ± 0.075

Values reported in mean ± SD. Number in *parenthesis* indicates the number of calveria tested

material property studies on pediatric cranial bone have verified this finding and reported the same to be true in older infants. Margulies and Thibault (2000) tested parietal bone from four human cadaver infants (25 weeks gestation to 6 months) in three-point bending at 2.54 and 2,540 mm/min. Small sample size prevented statistical analysis of material property age dependence, but the elastic modulus and ultimate stress of parietal bone from the 6-month-old infant were at least five times larger than those in the preterm age group (25–40 weeks gestation, Table 4.7). More recently, Coats and Margulies (2006a) tested 46 parietal and occipital bones from 21 infant cadavers (36 weeks gestation to 1 year) in three-point bending at 1.58 and 2.81 m/s (94,800 and 168,600 mm/min, respectively), rates similar to the impact velocity from an unimpeded fall from 1 and 3 ft (0.3 and 0.9 m, respectively). They also report a significant increase of elastic modulus and ultimate stress with infant donor age (Tables 4.7 and 4.8).

The oldest donor age of cranial bone tested to date is from a 6-year-old child (McPherson and Kriewall 1980). The parietal bone specimens from the 6-year-old had a quasistatic average elastic modulus nearly 18 times larger than the dynamic average modulus reported for a 1-year-old infant (Coats and Margulies 2006a), indicating that the material properties continue to be age dependent past the infant stage of development (>1 year). The quasistatic modulus from the 6-year-old child is still 1.5 times less stiff than that reported for adult (Hubbard 1971).

No material property data is available in the literature on children older than 6 years old. To estimate the elastic moduli in older children, Irwin and Mertz (1997) used a cubic spline curve fit to connect the pediatric elastic moduli from McPherson and Kriewall's studies to the average adult elastic modulus (Hubbard 1971). From the curve fit, the authors estimated the elastic modulus for older infants and children (Table 4.9). Since the publication of this interpolated relationship, the average elastic modulus for a 6- and 12-month-old infant has been measured as 2.6 GPa

Table 4.9 Comparison of measured and predicted elastic modulus for older children

Age	Predicted[a] elastic modulus (GPa)	Average measured elastic modulus (GPa)
6 months	2.8	2.6[b]
12 months	3.2	3.2[c]
18 months	3.6	N/A
3 years	4.7	N/A
Adult	–	9.9[d]

[a]Predicted values interpolated from cubic spline fit of neonate and adult data (Irwin and Mertz 1997)
[b]Average value from Margulies and Thibault (2000)
[c]Average value from Coats and Margulies (2006a)
[d]Average value from Hubbard (1971)

(Margulies and Thibault 2000) and 3.2 GPa (Coats and Margulies 2006a), respectively. The Irwin and Mertz spline curve fit predictions for these young ages (2.8 GPa for 6 months and 3.2 GPa for 12 months) are similar to these limited data sets, suggesting that the curve fit may provide suitable estimations until actual material property data can be measured.

Rate dependence. At the high rates tested, Coats and Margulies (2006a) reported a significant rate dependence of ultimate strain, but no significant rate dependence on elastic modulus or ultimate stress of infant cranial bone. A small sample size prevented statistical analyses between the two test rates (2.54 and 2,540 mm/min) in Margulies and Thibault (2000), but graphical representation of the data shows overlapping values of elastic modulus and ultimate stress for both rates, suggesting that the data was also not rate dependent. When the elastic moduli of preterm (<42 weeks gestation) parietal bone measured perpendicular to the trabeculae from all three studies are plotted on the same graph (Fig. 4.9), no rate dependence is evident over eight decades of loading rate. This finding is in contrast to adult cranial bone which has been shown to be highly rate dependent (Wood 1971). There have been no published studies investigating rate dependence in pediatric cranial bone tested with the trabeculae oriented parallel to the long axis of the specimen, so it is unknown if pediatric cranial bone is rate dependent in this direction. Further investigations into the anisotropy of pediatric cranial bone need to be performed to elucidate this possibility.

Regional differences. Coats and Margulies (2006a) investigated the regional material property differences of pediatric cranial bone. They found parietal bone to be significantly stiffer (461.1±63.8 MPa) and have a higher ultimate stress (30.2±4.8 MPa) than occipital cranial bone (329.0±55.3 and 14.7±2.1 MPa, respectively). McPherson and Kriewall (1980) included four frontal bone specimens from two calvaria in their study, but the small sample size prevented statistical comparison to the parietal specimens (Tables 4.7 and 4.8).

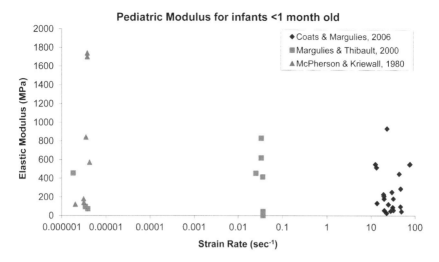

Fig. 4.9 Compilation of elastic moduli reported in the literature for preterm (20–42 weeks gestation) infant parietal bone. No significant rate dependence is found across eight magnitudes of strain rates when testing with the trabeculae perpendicular to the long axis of the bone. *Note*: Logarithmic scale used for *x*-axis. Data from McPherson and Kriewall (1980), Margulies and Thibault (2000), and Coats and Margulies (2006a)

Cranial Suture

Very few studies on material properties of pediatric cranial suture exist in the literature. To date the only one that investigates *human* pediatric cranial suture is from Coats and Margulies (2006a). They tested 14 coronal suture specimens from 11 infant (21 weeks gestation to 1 year old) calvaria in tension at 1.20 and 2.38 m/s, rates similar to those that might occur at impact from a fall of 1 and 3 ft (0.3 and 0.9 m, respectively). Using a two-way ANOVA (age, rate), they found no significant independent effect of age or strain rate on the elastic modulus, ultimate stress, or ultimate strain of coronal suture. The mean and standard deviation across all age groups for each of these properties was 8.1 ± 4.7 MPa, 4.7 ± 1.6 MPa, and 1.5 ± 1.3 mm/mm, respectively. Elastic modulus was significantly affected by the interaction between donor age and strain rate, such that the modulus of specimens from older donors tended to increase with strain rate while the modulus of specimens from younger donors tended to decrease with strain rate. The lack of a donor age–strain rate interaction for ultimate stress and ultimate strain may be due to testing a bone–suture–bone segment instead of suture alone. Failure of the bone–suture–bone specimens consistently occurred at the bone–suture junction. Consequently, it is reasonable to conclude that the reported ultimate stress and strain more accurately correspond to the stress and strain of the suture material at the time when the junction between suture and bone fails, and it is likely that the ultimate stress and ultimate strain of suture are higher than the values reported.

Structural Testing and Load Tolerances

Kriewall (1982) performed 0.05 mm/min bending load-displacement tests on 20 *whole* parietal bones from neonates (20–42 weeks gestation) and reported a significant increase in stiffness with gestational age. This finding is not surprising given the increase in structural thickness of the parietal bone with age (Table 4.5). The average stiffness of parietal bone across all ages in the study was 1.02 ± 0.9 kg/mm (10 ± 8.8 N/mm) and was found to significantly decrease with increasing dry density, but not be affected by bulk density.

Prange et al. (2004) performed quasistatic and dynamic 5 % compression tests on the whole head of three neonate cadavers (1, 3, and 11 days after birth). These nondestructive tests were performed by compressing the head in the anterior–posterior (frontal bone to occipital bone) and lateral (right parietal bone to left parietal bone) directions at rates of 0.05 mm/s (quasistatic) and 1, 10, and 50 mm/s (dynamic). They reported the overall stiffness of the head to be significantly larger ($p<0.01$) when tested dynamically (27.3 ± 9.8 N/mm) than when tested quasistatically (7.4 ± 1.9 N/mm). Additionally, there was no significant effect of test direction (anterior–posterior and lateral) across all test velocities ($p>0.21$). These findings are in contrast to those reported for adult skull with closed sutures (Hodgson et al. 1967), which report adult cranial stiffness to be 50 % greater during anterior–posterior compression compared with lateral compression.

In addition to the compression tests, Prange et al. (2004) dropped the same three infant cadaver heads from 15 and 30 cm onto a hard steel surface outfitted with a uniaxial force plate. One drop from each height was performed onto the vertex, occipital, left parietal, right parietal, and frontal bones of each cadaver head. The cranial cavity was filled with water and sealed to prevent the intracranial contents from coming out during testing. After completion of the tests, dissection and visual inspection found no skull fractures or permanent deformation of the head. The average impact force onto the vertex of the skull (335 ± 107 N) was generally higher than the other impact locations (246 ± 61 N); however it was not statistically significant ($p=0.18$). These tests suggest that impact loads able to cause skull fracture in newborn infants must be greater than 335 N for impacts onto the vertex and 246 N for impact onto the occipital, frontal, or parietal bones. The calculated peak acceleration and HIC in the study were not significantly different across all impact locations ($p=0.18$ and 0.78, respectively), suggesting that the average peak acceleration (55 g) and HIC calculation (84) are also below fracture tolerances in newborn infants.

In a case study investigation, Holck (2005) positioned a single infant cadaver head so that the left and right parietal bones were compressed between a force plate and a 15-mm-thick rubber pad. A 9-kg iron cylinder was dropped from increasingly higher heights onto the rubber plate and the impact force measured on the other side of the skull. Two drops with measured forces of 200 and 600 N did not appear to damage the skull (based on inspection of the smooth force–time curve measured). However, at 800 N a visible discontinuity in the force–time curve indicated that some damage had occurred to the skull. An external inspection of the skull (the scalp was still intact) was unable to reveal the location of damage. After performing subsequent drops resulting in 1,000 and 600 N, the skull was autopsied. The autopsy revealed that the skull had a bilateral fracture across the left and right parietal bones and that the coronal suture was completely ruptured throughout its entire length. Conclusions for this single specimen are not directly applicable to a real-world scenario because the infant cadaver was a stillborn and the developmental age of the infant is unknown (i.e., a premature infant is likely to have much wider sutures than a term infant). Additionally, the infant cadaver had been sitting in alcohol for 50 years. A study on the effects of storage media on bone (Lucksanasomboool et al. 2001) reports that storing bovine femur and tibia in alcohol for 1 week significantly increases the fracture load of bone compared to storage in saline. After 50 years of storage in alcohol, it is highly likely that the fracture loads reported by Holck are appreciably greater than those of fresh cadaveric tissue.

In another set of cadaver studies, Weber (1984, 1985) dropped 40 non-perfused infant cadavers (1–9 months) onto the occipital–parietal region of the skull from 82 cm onto stone tile, carpet, foam-padded linoleum, a folded camel hair blank, and a 2-cm-thick foam mat. All 40 (100 %) of the cadavers dropped onto the stone tile, carpet, and foam-padded linoleum had skull fracture. Two out of 10 cadavers (20%) dropped onto the foam mat and 4 out of 25 cadavers (16%) dropped onto the folded blanket had skull fracture. The height in Weber's study is more than double the height tested in Prange's study. The impact forces from this higher height are likely greater, but the difference in cadaver ages and lack of impact surface material property descriptions make it difficult to make a reasonable estimate of the loads experienced by the cadavers in Weber's study.

Scalp

Anatomy and Development

The extracalvarial soft tissue (i.e., scalp) consists of three muscles and five tissue layers (skin, dense connective tissue, galea, loose connective tissue, and periosteum). The muscles are responsible for the motion of the scalp and forehead. The skin and galea are tightly bound to each other by the dense fibrous and fatty connective tissue layer that houses the blood and nerves of the scalp. The galea is not tightly bound to the periosteum and there is potential for fluid or blood to collect within this space. Any movement of the scalp relative to the skull occurs between these layers (Ruff et al. 1985).

The thickness of scalp increases with age and is generally thicker in males than in females (Young 1959). Specifically, there is an increase in scalp thickness along the mid-sagittal line between 1 and 9 months of age. This thickness remains consistent until about 3 or 4 years old when the scalp begins to thicken again until adulthood is reached (Table 4.10). Not only is the overall thickness of the scalp changing with age, but the thickest regions of scalp tissue are shifting from one location to another

Table 4.10 Thickness of scalp (in mm) with age

Age (years)	N	Lambda	Bregma	Nasion
1 month	11–16	3.2	2.4	7.2
3 months	8–13	3.5	2.9	7.2
6 months	11–15	3.5	3.1	7.6
9 months	10–15	4.1	3.4	8.2
1	15–18	3.7	3.6	8
2	16–19	3.6	3.3	7.5
3	17–20	3.6	3.3	5.8
4	17–19	3.7	3.5	5.6
6	18–20	3.8	3.5	5.4
8	17–18	4	3.7	5.4
10	19–20	4.4	4.1	5.5
12	20	4.9	4.9	5.6
14	19–20	5.5	5	5.8
16	12	6.2	5	6.1
Adult (female)	48–50	7	5.6	5.7
Adult (male)	49–50	6.4	5.5	5.7

Measurements are taken along the mid-sagittal line near the lambda (posterior parietal), bregma (anterior parietal), and nasion (forehead). Values extracted from Young (1959)

(Young 1959). For example, in children <8 years old the space between the eyebrows just above the nose is the thickest part of the scalp tissue. However, between 8 and 9 years old this predominance diminishes, and by 10 years old, the lambda (a region behind the vertex of the skull) becomes the location with the thickest scalp.

Material and Structural Properties

There exist no published studies on the material or mechanical properties of pediatric scalp tissue (human or animal). The only mechanical property study on human adult tissue is a tensiometric test in adult males (Raposio and Nordstrom 1998). The authors reported an initial linear increase with load up to 500 g, after which the scalp exhibits an exponential growth of stiffness with load (up to 5,000 g). It is unknown if the pediatric scalp would respond in a similar manner. Regardless, pediatric head injury models incorporating scalp tissue are generally more interested in the effect of scalp on mitigating skull and brain injury in impact scenarios (Klinich et al. 2002; Coats et al. 2007). This leads to the need for rate-dependent compressive studies of pediatric scalp. Galford and McElhaney (1970) tested scalp from adult rhesus monkeys in tension (stress–relaxation and free vibration) and compression (creep). The authors reported a higher modulus of scalp (1.54 MPa) from the dynamic tests compared to the range of instantaneous moduli in the quasistatic relaxation tests (1.03–1.38 MPa), confirming the rate dependence of the tissue. To fully characterize the pediatric scalp response to impact, rate- and age-dependent compression studies measuring the compressive modulus of the material are needed.

Summary

To date, the majority of pediatric brain material property testing has been in rodents and pigs, and reveals a two-fold increase in stiffness in the infant relative to the adult. However, it is premature to use these data to extrapolate the adult data set to children until property relationships across age are validated with human tissue samples. The material properties of CSF have been reported to be independent of age, and adult human properties may be suitable in a pediatric model. Similarly, properties of bridging veins have been reported to be independent of age from 3 years old to an adult, but it is important to determine properties in younger children (<3 years old). Importantly, membrane structures of the brain play a critical role in computational models evaluating the adult brain response to load, but little is known about the properties of these structures in children or young adults.

Cranial bone has the largest body of published material property data of all structures in the pediatric head. However, in order to model the tissue properly in infant computational simulations, further investigation is warranted to characterize the anisotropy and rate dependence of cranial bone, and to identify differences between regions of the cranium. Additionally, the material properties for the ages of 1–18 years are absent as only a single human calvarium (6 years old) has been reported. Therefore, further material property testing needs to be performed to verify the current practice of estimating modulus by interpolating between infant and adult data sets.

As technology advances, increasingly complex computational models of the adult human head will be developed. The paucity of age-specific material properties limits the development of accurate pediatric head models. Further, the human pediatric data that are available demonstrate that linear scaling of human adult material properties to children is not accurate and immature animal material properties are not always representative of pediatric human values. However, until more data from human pediatric tissue become available, published immature animal data will continue to provide insight into the roles of age, region, direction, and test rate on the material properties of pediatric tissue.

References

Adelson P, Robichaud P et al (1996) A model of diffuse traumatic brain injury in the immature rat. J Neurosurg 85:877–884

Aimedieu P, Grebe R (2004) Tensile strength of cranial pia mater: preliminary results. J Neurosurg 100:111–114

Arbogast K (1997) A characterization of the anisotropic mechanical properties of the brainstem. Department of Bioengineering, University of Pennsylvania, Pennsylvania, PA

Arbogast K, Margulies S (1998) Material characterization of the brainstem from oscillatory shear tests. J Biomech 31:801–807

Arbogast K, Margulies S (1999) A fiber-reinforced composite model of the viscoelastic behaviour of the brainstem in shear. J Biomech 32:865–870

Arbogast K, Thibault K (1997) A high frequency shear device for testing soft biological tissues. J Biomech 30:757–759

Armstead W, Kurth C (1994) Different cerebral hemodynamic responses following fluid percussion brain injury in the newborn and juvenile pig. J Neurotrauma 11:487–497

Barkovich A, Kjos B et al (1988) Normal maturation of the neonatal and infant brain: MR imaging at 1.5 T. Neuroradiology 166:173–180

Bilston L, Liu Z et al (2001) Large strain behaviour of brain tissue in shear: some experimental data and differential constitutive model. Biorheology 38:335–345

Bittigau P, Sifringer M et al (1999) Apoptotic neurodegeneration following trauma is markedly enhanced in the immature brain. Ann Neurol 45:724–735

Bloomfield I, Johnston I et al (1998) Effects of proteins, blood cells, and glucose on the viscosity of cerebrospinal fluid. Pediatr Neurosurg 28:246–251

Breton F, Pincemaile Y et al (1991) Event-related potential assessment of attention and the orienting reaction in boxers before and after a fight. Biol Psychol 31:57–71

Bylski DI, Kriewall TJ et al (1986) Mechanical behavior of fetal dura mater under large deformation biaxial tension. J Biomech 19:19–26

Coats B, Margulies SS (2006a) Material properties of human infant skull and suture at high rates. J Neurotrauma 23:1222–1232

Coats B, Margulies SS (2006b) Material properties of porcine parietal cortex. J Biomech 39:2521–2525

Coats B, Ji S et al (2007) Parametric study of head impact in the infant. Stapp Car Crash J 51:1–15

Couper Z, Albermani F (2008) Infant brain subjected to oscillatory loading: material differentiation, properties, and interface conditions. Biomech model Mechanobiol 7:105–125

Dekaban AS (1978) Changes in brain weights during the span of human life: relation of brain weights to body heights and body weights. Ann Neurol 4:345–356

Delye H, Goffin J et al (2006) Biomechanical properties of the superior sagittal sinus-bridging vein complex. Stapp Car Crash J 50:625–636

Dempster WT (1967) Correlation of types of cortical grain structure with architectural features of the human skull. Am J Anat 120:7–32

Dobbing J, Sands J (1973) Quantitative growth and development of human brain. Arch Dis Child 48:757–767

Duhaime AC, Margulies SS et al (2000) Maturation-dependent response of the piglet brain to scaled cortical impact. J Neurosurg 93:455–462

Elkin BS, Ilankovan et al (2010) Age-dependent regional mechanical properties of the rat hippocampus and cortex. J Biomech Eng 132:011010

Eucker S, Smith C et al (2011) Physiological and histopathological responses following closed rotational head injury depend on direction of head motion. Exp Neuro 227:79–88

Galford J, McElhaney J (1970) A viscoelastic study of scalp, brain, and dura. J Biomech 3:211 221

Gefen A, Gefen N et al (2003) Age-dependent changes in material properties of the brain and braincase of the rat. J Neurotrauma 20:1163–1177

Gennarelli T, Abel J et al (1979) Differential tolerance of frontal and temporal lobes to contusion induced by angular acceleration. Presented at 23rd Stapp car crash conference, Oct. 17–19. Coronado, CA

Gennarelli T, Thibault L et al (1982) Diffuse axonal injury and traumatic coma in the primate. Ann Neurol 12:564–574

Gennarelli TA, Thibault LE et al (1987) Directional dependence of axonal brain injury due to centroidal and non-centroidal acceleration. 31st Stapp car crash conference, San Diego, CA

Giedd J, Snell J (1996) Quantitative magnetic resonance imaging of human brain development: ages 4–18. Cereb Cortex 6:551–560

Giza C, Griesbach G et al (2005) Experience-dependent behavioral plasticity is disturbed following traumatic injury to the immature brain. Behav Brain Res 157:11–22

Goldstein J, Seidman L (2001) Normal sexual dimorphism of the adult human brain assessed by in vivo magnetic resonance imaging. Cereb Cortex 11:490–497

Gooding C (1971) Cranial sutures and fontanelles. The C.V. Mosby Company, St. Louis, MO

Grundl P, Biagas K et al (1994) Early cerebrovascular response to head injury in immature and mature rats. J Neurotrauma 11:135–148

Ho J, Kleiven S (2007) Dynamic response of the brain with vasculature: a three-dimensional computational study. J Biomech 40:3006–3012

Hodgson V, Gurdjian E et al (1967) Development of a model for the study of head injury. 11th Stapp car crash conference, Warrendale, PA

Holck P (2005) What can a baby's skull withstand? Testing the skull's resistance on an anatomical preparation. Forensic Sci Int 151:187–191

Hosey R, Liu Y (1980) A homeomorphic finite-element model of impact head and neck injury. Presented at Finite Elements in Biomechanics, Tucson, AZ

Huang H, Lee M et al (1999) Three-dimensional finite element analysis of subdural hematoma. J Trauma Inj Infect Crit Care 47:538–544

Hubbard R (1971) Flexure of layered cranial bone. J Biomech 4:251–263

Irwin A, Mertz H (1997) Biomechanical bases for the crabi and hybrid III child dummies. SAE Tech Paper #973317

Jin X, Lee J et al (2006) Biomechanical response of the bovine pia-arachnoid complex to tensile loading at varying strain-rates. Stapp Car Crash J 50:637–649

Jin X, Ma C et al (2007) Biomechanical response of the bovine pia-arachnoid complex to normal traction loading at varying strain rates. Stapp Car Crash J 51:115–125

Kleiven S (2003) Influence of impact direction on the human head in prediction of subdural hematoma. J Neurotrauma 20:365–379

Klinich K, Hulbert G et al (2002) Estimating infant head injury criteria and impact response using crash reconstruction and finite element modeling. Stapp Car Crash J 46:165–194

Koeneman J (1966) Viscoelastic properites of brain tissue. Case Institute of Technology, Cleveland, OH

Kriewall T (1982) Structural, mechanical, and material properties of fetal cranial bone. Am J Obstet Gynecol 142:707–714

Kriewall T, McPherson F et al (1981) Bending properties and ash content of fetal cranial bone. J Biomech 14:73–79

Kriewall TJ, Akkas N et al (1983) Mechanical behavior of fetal dura mater under large axisymmetric inflation. J Biomech Eng 105:71–76

Kumaresan S, Radhakrishnan S (1996) Importance of partitioning membranes of the brain and the influence of the neck in head injury modelling. Med Biol Eng Comput 34:27–32

Lange N, Giedd J et al (1997) Variability of human brain structures size: ages 4–20 years. Psych Res Neuroimag 74:1–12

Langlois J, Rutland-Brown W et al (2004) Traumatic brain injury in the United States: emergency department visits, hospitalizations, and deaths. CDC, NCIPC, Atlanta, GA

Lee M, Haut R (1989) Insensitivity of tensile failure properties of human bridging veins to strain rate: implications in biomechanics of subdural hematoma. J Biomech 22:537–542

Lenroot R, Gogtay N et al (2007) Sexual dimorphism of brain developmental trajectories during childhood and adolescence. Neuroimage 36:1065–1073

Lippert SA, Rang EM et al (2004) The high frequency properties of brain tissue. Biorheology 41:681–691

Loder R (1996) Skull thickness and halo-pin placement in children: the effects of race, gender, and laterality. J Pediatr Orthop 16:340–343

Lowenhielm P (1974) Dynamic properties of the parasagittal bridging veins. Z Rechtsmed 74:55–62

Lucksanasombool P, Higgs W et al (2001) Fracture toughness of bovine bone: influence of orientation and storage media. Biomaterials 22:3127–3132

Margulies S, Thibault L (1992) A proposed tolerance criterion for diffuse axonal injury in man. J Biomech 25:917–923

Margulies SS, Thibault KL (2000) Infant skull and suture properties: measurements and implications for mechanisms of pediatric brain injury. J Biomech Eng 122:364–371

McElhaney JH, Fogle JL et al (1970) Mechanical properties of cranial bone. J Biomech 3:495–511

McPherson G, Kriewall T (1980) The elastic modulus of fetal cranial bone: a first step toward understanding of the biomechanics of fetal head molding. J Biomech 13:9–16

Meaney D (1991) Biomechanics of acute subdural hematoma in the subhuman primate and man. Department of Bioengineering, University of Pennsylvania, Pennsylvania, PA

Monson KL, Goldsmith W et al (2003) Axial mechanical properties of fresh human cerebral blood vessels. J Biomech Eng 125:288–294

Monson KL, Goldsmith W et al (2005) Significance of source and size in the mechanical response of human cerebral blood vessels. J Biomech 35:737–744

Monson KL, Barbaro NM et al (2008) Biaxial response of passive human cerebral arteries. Ann Biomed Eng 36:2028–2041

Ning X, Zhu Q et al (2006) A transversely isotropic viscoelastic constitutive equation for brainstem undergoing finite deformation. J Biomech Eng 128:925–933

Pellmen EJ, Viano DC et al (2003) Concussion in professional football: reconstruction of game impacts and injuries. Neurosurgery 53:799–814

Peters G, Meulman J et al (1997) The applicability of the time/temperature superposition principle to brain tissue. Biorheology 34:127–138

Prange M (2002) Biomechanics of traumatic brain injury in the infant. PhD thesis, Department of Bioengineering. University of Pennsylvania, PA. p 186

Prange M, Margulies S (2002) Regional, directional, and age-dependent properties of brain undergoing large deformation. J Biomech Eng 124:244–252

Prange M, Luck J et al (2004) Mechanical properties and anthropometry of the human infant head. Stapp Car Crash J 48:279–299

Prins M, Lee S et al (1996) Fluid percussion brain injury in the developing and adult rat: a comparative study of mortality, morphology, intracranial pressure and mean arterial blood pressure. Dev Brain Res 95:272–282

Raghupathi R, Margulies S (2002) Traumatic axonal injury after closed head injury in the neonatal pig. J Neurotrauma 19:843–853

Raghupathi R, Mehr M et al (2004) Traumatic axonal injury is exacerbated following repetitive close head injury in the neonatal pig. J Neurotrauma 21:307–316

Raposio E, Nordstrom R (1998) Biomechanical properties of scalp flaps and their correlations to reconstructive and aesthetic surgery procedures. Skin Res Technol 4:94–98

Reeves T, Phillips L et al (2005) Myelinated and unmyelinated axons of the corpus callosum differ in vulnerability and functional recovery following traumatic brain injury. Exp Neurol 196:126–137

Roche A (1953) Increase in cranial thickness during growth. Hum Biol 25:81–92

Ruan J, Khalil T et al (1991) Human head dynamic response to side impact by finite element modeling. J Biomech Eng 113:276–283

Ruan JS, Khalil T et al (1994) Dynamic response of the human head to impact by three-dimensional finite element analysis. J Biomech Eng 116:44–50

Ruff R, Osborn A et al (1985) Extracalvarial soft tissues in cranial computed tomography: normal anatomy and pathology. Invest Radiol 20:374–380

Scheuer L, Black S (2004) The juvenile skeleton. Elsevier Academic Press, San Diego, CA

Shuck L, Advani S (1972) Rheological response of human brain tissue in shear. J Basic Eng 94:905–911

Skalak R, Tozeren A et al (1973) Strain energy function of red blood cell membranes. Biophys J 13:245–264

Smith DH, Nonaka M et al (2000) Immediate coma following inertial brain injury dependent on axonal damage in the brainstem. J Neurosurg 93:315–322

Soboleski D, McCloskey D et al (1997) Sonography of normal cranial suture. AJR Am J Roentgenol 168:819–821

Thibault K, Margulies S (1998) Age-dependent material properties of the porcine cerebrum: effect on pediatric inertial head injury criteria. J Biomech 31:1119–1126

Weber W (1984) Experimental studies of skull fractures in infants. Z Rechtsmed 92:87–94

Weber W (1985) Biomechanical fragility of the infant skull. Z Rechtsmed 94:93–101

Wilke M, Krageloh-Mann I et al (2006) Global and local development of gray and white matter volume in normal children and adolescents. Exp Brain Res 178:296–307

Wood JL (1971) Dynamic response of human cranial bone. J Biomech 4:1–12

Yamada H (1970) Strength of biological materials. Williams and Wilkins Co, Baltimore, MD

Young R (1959) Age changes in thickness of scalp in white males. Hum Biol 31:74–79

Zhang L, Bae J et al (2002) Computational study of the contribution of the vasculature on the dynamic response of the brain. Stapp Car Crash J 46:145–163

Zhang L, Yang KH et al (2004) A proposed injury threshold for mild traumatic brain injury. J Biomech Eng 126:226–236

Zhou C, Khalil T et al (1995) A new model comparing impact responses of the homogeneous and inhomogeneous human brain. Presented at 39th Stapp car crash conference, San Diego, CA

Chapter 5
Experimental Injury Biomechanics of the Pediatric Neck

Roger W. Nightingale and Jason F. Luck

Injury Incidence and Significance

Motor vehicle related crashes rank as the most common cause of spinal related injuries in the pediatric population (Platzer et al. 2007; Brown et al. 2001; Kokoska et al. 2001; Eleraky et al. 2000; Hamilton and Myles 1992a; Bonadio 1993; Babcock 1975). Pediatric spinal related trauma accounts for between 1 and 12 % of all spinal related injuries (Hamilton and Myles 1992a; Hadley et al. 1988; Aufdermaur 1974). Cervical spine trauma in children accounts for approximately 2 % of all cervical spinal injuries (Henrys et al. 1977). Approximately 1–2 % of all children admitted for traumatic injury are related to injuries to the cervical spine (Platzer et al. 2007; Brown et al. 2001; Kokoska et al. 2001; Orenstein et al. 1994; Rachesky et al. 1987). Overall, pediatric neck injury rates are significantly lower than adult rates; however, the neck injury rate in children between the ages of 11 and 15 years approaches the adult rate of 18.8 per 100,000 (McGrory et al 1993; Myers and Winkelstein 1995). For children less than 11 years of age, neck injuries are relatively rare (1.2 per 100,000), but have particularly devastating consequences (McGrory et al. 1993). The overall mortality rate amongst victims of pediatric spinal trauma is approximately 16–41 % but considerably higher for the youngest ages (Platzer et al. 2007; Brown et al. 2001; Kokoska et al. 2001; Eleraky et al. 2000; Givens et al. 1996; Orenstein et al. 1994; Hamilton and Myles 1992b).

With the advances in occupant protection devices that have occurred over the last three decades, there has been a dramatic reduction in serious injuries and fatalities for all motor vehicle occupants (NHTSA SCI 2009; Viano 1995; Huelke et al. 1981; Hartemann et al. 1977). However, with the ability to survive large decelerations comes an increased potential for neck injuries from inertial loading

R.W. Nightingale, Ph.D. (✉) • J.F. Luck
Department of Biomedical Engineering, Pratt School of Engineering, Duke University, Room 136 Hudson Hall, Durham, NC 27708-0281, USA
e-mail: rwn@duke.edu

(Huelke et al. 1978, 1992, 1993, 1995). Although inertial neck injuries occur throughout the population, the anatomy of developing children may make them more susceptible. Greater head mass in relation to overall body proportions, ligamentous laxity of the cervical spine, decreased facet angle, and a relatively slender cervical spine are all potential factors that may heighten the risk of cervical spine injury in young children (Kasai et al. 1996; Swischuk 1977; Cattell and Filtzer 1965; Bailey 1952).

The importance of understanding the effects of tension and bending in the cervical spine as a mechanism of injury was highlighted by crash data during the 1990s indicating that airbags were causing severe injuries and deaths to front seat occupants in minor to moderate crashes where only minor injuries were expected (Dalmotas et al. 1995; Brown et al. 1995; Giguere et al. 1998; NHTSA SCI 2009). In the period prior to multiple stage and less aggressive airbags, Graham et al. (1998) observed that airbags increased the mortality risk for children younger than 12 years. A Special Crash Investigations (SCI) Program was established by NHTSA to investigate injuries and deaths associated with airbag deployment in minor to moderate severity motor vehicle accidents. Data from the SCI showed that as of January of 2009 there were a total of 296 confirmed fatalities since 1990. An additional three unconfirmed fatalities were under investigation, with all occurring since 2008. Out of the 296 fatalities, 105 were either an adult driver or passenger, whereas 191 were children. In 1997 and 1998 the highest total number of fatalities reported was 52 and 48, respectively. Since the late 1990s, new regulatory influences, advances in airbag systems, warnings, and education campaigns have dramatically reduced child fatalities in out-of-position airbag deployments; however, it is clear that children are particularly vulnerable.

Pediatric cervical spine injuries involve the full spectrum of the bony elements, connective tissues, and the spinal cord. Approximately 60–72 % of all spinal injuries in the pediatric population occur in the cervical spine (Eleraky et al. 2000; Hadley et al. 1988). The epidemiological data indicate that children under the age of 8–9 years typically present with injuries to the upper cervical spine as opposed to the lower cervical spine (Platzer et al. 2007; Eleraky et al. 2000; Orenstein et al. 1994; Nitecki and Moir 1994; Hamilton and Myles 1992a; Hadley et al. 1988). These injuries are usually associated with the cartilaginous growth regions and occur in both the atlas (C1) (Thakar et al. 2005; Mikawa et al. 1987) and axis (C2) (Fuchs et al. 1989; Schippers et al. 1996; Shaw and Murphy 1999; Sanderson and Houten 2002; Blauth et al. 1996; Odent et al. 1999; Lui et al. 1996; Heilman and Riesenburger 2001). Atlantooccipital (O–C1) dislocations are of particular concern because of their high mortality rate (Roche and Carty 2001; Giguere et al. 1998; Angel and Ehlers 2001; Saveika and Thorogood 2006). These dislocations often result in large tractions on the upper spinal cord and brainstem. Dislocations of the atlantoaxial (C1–C2) joint due to ligament disruption have also been reported (Wigren and Amici 1973; Lui et al. 1996). Fractures through the neural arches of C2, typically referred to as "hangman's fractures," have been reported in children; however, they are quite rare (Swischuk 1998; Bodenham et al. 1992; Weiss and Kaufman 1973). In general, bony fracture is uncommon in children who are skeletally

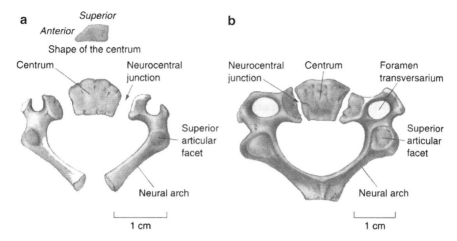

Fig. 5.1 Cervical vertebrae of a perinate (**a**) and a 3-year-old (**b**) showing the synchondroses and other relevant vertebral anatomy (Scheuer and Black 2004). The opening in the posterior arch of the vertebra in (**a**) is the posterior synchondrosis

immature (Aufdermaur 1974). Most lower cervical spine fractures occur in the growth zone at the inferior or superior end-plate of the vertebral body. This was demonstrated by Aufdermaur (1974), in a necropsy study of 32 juvenile spines. The author examined 12 injured spines and manually loaded an additional 20 to failure. He was unable to produce anything other than cartilaginous end-plate fractures.

Anatomy and Development

The anatomy of the pediatric cervical spine is different in many ways from the adult cervical spine and changes throughout development. It is clear from the epidemiology of pediatric cervical spine injury that injury patterns in children are quite different than those in adults and the reason for those differences are related to differences in both exposures, and in morphology and anatomy.

Developmental Anatomy

Perhaps the most significant difference between the adult and pediatric cervical spine is the synchondroses—the cartilaginous regions that are necessary for growth. The vertebrae of the lower cervical spine (C3 through C7) have three synchondroses: the posterior synchondrosis and two neurocentral synchondroses (Fig. 5.1). The posterior synchondrosis is located at the midpoint of the posterior arch where the future spinous process will form. The neurocentral synchondroses connect the vertebral centrum to the lateral masses. In the neonate, the vertebrae are composed of three parts—a centrum and two lateral arches. The posterior synchondrosis of

Table 5.1 Appearance of ossification centers (Scheuer and Black 2004)

	Vertebral centrum	Neural arch	Anterior arch	Dens	Ossiculum terminale[a]
C1	–		Year 2	–	–
C2	Month 4–5 (prenatal)		–	Month 4–5 (prenatal)	Year 2 (postnatal)
C3					
C4		Month 2+ (prenatal); prior to centra			
C5	Month 3–4 (prenatal)			–	–
C6					
C7					

– not applicable to vertebra, *grayed areas* grouped vertebrae
[a]Cartilaginous precursor to ossiculum terminale denoted the chondrum terminale

Table 5.2 Fusion of synchondroses (Scheuer and Black 2004)

	Neurocentral	Posterior	Dentoneural	Dentocentral[a]	Intradental	Ossiculum terminale to dens
C1	Year 5–6 (postnatal)	Year 4–5 (postnatal)	–	–	–	–
C2	Year 4–6 (postnatal)	Year 3–4 (postnatal)	Year 3–4 (postnatal)	Year 4–6 (postnatal)	Month 7–8 (prenatal)	Year 12[b]
C3						
C4						
C5	Year 3–4 (postnatal)	Year 2 (postnatal)	–	–	–	–
C6						
C7						

– not applicable to vertebra, *grayed areas* grouped vertebrae
[a]Also referred to as subdental
[b]Lack of fusion known as "ossiculum terminale persistens Bergmen"

the lower cervical spine typically ossifies in the second year, and the neurocentral synchondroses follow by age three or four (Tables 5.1 and 5.2).

The upper cervical spine of the child is especially important in the study of pediatric spine biomechanics due to the high rate of injury associated with these vertebrae (Klinich et al. 1996). The upper cervical spine contains two vertebrae, the first cervical vertebra typically referred to as the atlas and the second cervical vertebra referred to as the axis. At birth, the atlas contains two cartilaginous regions, one anterior and one posterior connecting the two lateral masses (Bailey 1952; Ogden 1984a). Eventual ossification in the future location of the anterior tubercle leads to a third ossification region and a clearer delineation of cartilaginous growth regions posteriorly between the two lateral masses and anteriorly between the newly developed anterior ossification location and the lateral masses bilaterally (Fig. 5.2). The axis exhibits an even more complicated growth structure with a total of four postnatal synchondroses (Ogden 1984b; Fesmire and Luten 1989). Similar to the

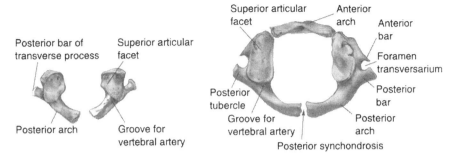

Fig. 5.2 The atlas (C1) of a perinate (*left*) and a 2–3-year-old (*right*) (Scheuer and Black 2004)

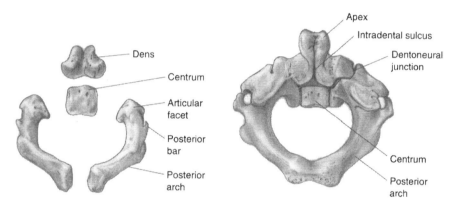

Fig. 5.3 Development of the axis (C2) showing the perinate (*left*) and the 3-year-old (*right*) (Scheuer and Black 2004)

lower cervical spine, the axis has a posterior synchondrosis and two dentoneural/neurocentral synchondroses (Fig. 5.3). Additionally, the axis has a dentocentral synchondrosis that runs superior to the centrum and inferior to the dens. These cartilaginous growth regions are frequently the site of injuries in the pediatric population (Adams 1992; Clasper and Pailthorpe 1995; Keller and Mosdal 1990; Seimon 1977; Bhattacharyya 1974; Ewald 1971; Smith et al. 1993).

There are some biomechanically important trends in the development of the intervertebral discs (Kumaresan et al. 1998; Peacock 1952; Taylor 1975). At full term, there is a relatively large amount of nucleus pulposus (NP) occupying approximately one half of the diameter of the disc. Initially, there is not much differentiation between the NP and the annulus. By age three, the NP still occupies a relatively large percentage of the disc volume as compared to the adult; however, the fibers of the annulus are better organized and there is a clearer demarcation between the two. The collagen fibers in the annulus become coarser and clearly defined by age 6. Disc geometry also changes with age. At full term, the disc is biconvex. The biconvexity remains through at least age 10, although it becomes progressively less pronounced until skeletal maturity is reached (Kumaresan et al. 1998; Peacock 1952; Taylor 1975).

Table 5.3 Facet angle as a function of age and level (Kasai et al. 1996)

| | Total facet angle (°) | | | | |
Age (year)	C3	C4	C5	C6	C7
1	66.1±4.4	63.3±3.5	56.6±3.6	52.5±3.2	41.7±2.4
2	58.3±3.9	55.3±5.3	52.8±3.2	48.9±3.4	37.8±3.1
3	51.5±3.8	50.0±2.4	49.6±3.8	44.0±3.7	34.5±2.4
4	48.4±4.6	47.3±3.6	47.7±3.2	42.0±2.5	32.6±2.9
5	45.7±2.2	44.8±3.3	46.3±3.3	39.2±2.4	29.7±2.6
6	44.9±2.6	43.5±3.2	44.0±3.2	38.5±2.8	29.3±2.1
7	43.6±3.2	47.8±3.9	42.6±4.0	37.8±2.7	29.0±3.0
8	42.2±2.9	42.5±3.4	42.4±2.7	37.4±2.6	28.6±3.0
9	41.0±3.2	41.2±3.5	42.2±3.0	37.0±3.0	27.9±2.5
10	39.9±3.2	40.8±3.2	41.7±3.4	36.7±3.7	27.3±3.3
11	40.0±3.2	40.6±3.2	41.6±2.8	35.8±4.0	27.0±2.6
12	39.5±3.4	39.7±2.9	40.5±3.2	36.0±3.5	26.8±3.3
13	39.3±3.0	40.2±2.3	41.6±2.8	36.6±3.3	27.4±3.3
14	40.2±4.1	40.4±3.5	41.3±4.0	35.6±3.6	27.0±3.7
15	40.0±3.9	40.8±3.3	40.7±3.2	34.9±3.0	27.3±3.9
16	40.0±3.6	40.3±3.7	41.0±3.1	35.9±2.9	28.1±3.9
17	38.1±3.1	40.7±2.7	41.1±3.0	35.5±4.2	27.3±4.1
18	39.0±3.9	40.4±3.5	42.0±2.7	35.8±3.6	26.3±3.5

Quantitative Anatomy

One of the most thorough studies of pediatric anatomy was done by Kasai et al. (1996). Radiographs of 180 boys and 180 girls from Japan were obtained from clinical scans to quantify the diameters and central heights of the cervical vertebrae, the anterior to posterior vertebral height ratio, body height index, facet joint angles, tilting and sliding motions, cervical length from C3 to C7, and the cervical lordosis angle from C3 to C7. This manuscript provides a wealth of quantitative information. The measurements were examined to correlate growth in the vertebral body and the facet joint with cervical lordosis and intervertebral motion in the pediatric cervical spine. Vertebral body diameter (anteroposterior) and cervical length (the total height of the C3 through C7 vertebral bodies) increased with age. A decrease in the angle of cervical lordosis and body height index until 9 years of age was observed followed by an increase in angle and body height index in the older ages. Significant correlations between lordosis angle, cervical length, and body height index were identified along with strong correlations to age. Facet angle also changes with age (Table 5.3). The facet angle starts out relatively flat and becomes more vertical, until approximately 10 years of age at which point the orientation stays constant. Facet angle was correlated to the "sliding" motion of adjacent vertebrae. This sliding motion was measured as the relative horizontal displacement of adjacent vertebrae as a percent of disc diameter during flexion and extension. As the facet angle became more vertical the total sliding of adjacent vertebrae decreased (Table 5.4).

Table 5.4 Changes in horizontal motion of adjacent vertebrae with age and level (Kasai et al. 1996)

	Total sliding (%)				
Age (year)	C2–C3	C3–C4	C4–C5	C5–C6	C6–C7
1	34.4±6.1	29.8±5.9	24.6±4.3	18.9±5.2	11.8±3.7
2	31.9±5.3	28.6±6.0	23.8±5.0	18.1±4.5	11.9±3.2
3	30.6±6.9	26.6±4.4	23.5±4.8	16.8±3.6	12.3±3.2
4	29.1±7.1	24.5±3.5	21.2±5.0	15.7±2.8	11.1±3.2
5	26.4±3.4	23.2±4.5	17.1±4.3	14.9±2.4	9.9±3.3
6	24.1±5.6	20.3±4.0	16.8±4.2	12.8±2.7	9.9±3.1
7	22.1±4.4	17.6±3.5	15.6±3.4	13.6±1.9	9.1±2.8
8	16.8±3.4	14.5±3.3	17.4±3.6	13.6±2.9	8.6±2.3
9	17.5±3.1	14.4±2.4	14.6±3.3	14.0±2.9	8.5±2.3
10	15.1±3.2	13.9±3.1	15.9±3.6	13.0±2.2	8.2±1.9
11	16.2±2.8	13.9±3.2	14.8±2.3	13.3±2.3	7.7±2.8
12	14.3±2.6	12.4±3.0	13.3±3.1	11.9±2.8	8.4±2.0
13	15.1±3.3	12.6±3.6	12.7±3.7	12.6±3.8	6.7±1.8
14	13.4±2.7	11.9±4.4	12.6±3.8	12.5±3.2	7.0±2.2
15	12.8±3.5	12.1±4.2	11.7±3.0	12.1±2.9	7.1±2.3
16	13.2±4.4	12.0±3.7	12.9±3.2	11.8±3.3	6.4±2.4
17	13.0±3.3	11.8±3.9	13.4±3.3	12.3±3.9	7.0±2.8
18	12.9±4.1	11.0±4.1	12.0±3.5	11.6±3.4	6.6±3.0

Additional resources for quantitative and developmental anatomy are Kuhns (1998) and Yoganandan et al. (2002).

Structural Properties of the Human Pediatric Neck

As with all the body regions, there are very few biomechanical studies of the human pediatric cervical spine. The few studies in the current literature have differing research objectives and small sample sizes. As a result, it is difficult to take the information gained from one study and compare or combine it with the other studies in order to form a more complete understanding of the structural responses of the pediatric cervical spine. Because there are so few peer reviewed studies, they will be individually summarized and discussed in the following sections.

Duncan (1874)

One of the few, and certainly the earliest, published studies on pediatric neck mechanics comes from Duncan (1874). Duncan's investigation was based on the desire to more fully understand the amount of force required to cause a neck injury in a neonate during a breech delivery.

Table 5.5 Summary of data from Duncan (1874)

Age (mo)	Gender	Mass (kg)	Strength (N) C-spine	Decapitation	Location
0	f	2.44	401	525	C5–C6
0	f	3.37	534	627	C6–C7
0	f	4.05	543	605	C4–C5
0	m	2.61	405	405	C4–C5
0.5	na	3.32	654	725	C3–C4

Duncan tested a total of five specimens, including four stillborn infants and one infant that died 2 weeks after birth. The cadavers were fully intact prior to testing with all soft tissue retained including the musculature and outer skin layers. The heads were fixed in a yoke and weights were added to the ankles to load the neck. The weights were added sequentially at 30 s intervals until decapitation occurred. The load at which the cervical spine failed was noted.

The tensile failure load of the four stillborn infants was measured as 471 ± 79 N (average ± standard deviation). The 2-week old infant had a tensile strength of 654 N. Failures in Duncan's sample were distributed throughout the lower cervical region from the C3–C4 joint to the C6–C7 joint. Duncan noted that the weights were applied to only one leg and that the leg was always stronger than the neck. However, there were multiple epiphyseal fractures in the loaded limb prior to "disseverment." A summary of all the data is presented in Table 5.5.

Duncan's work focused exclusively on neonates, which limits our ability to utilize the data to formulate scaling relationships across the spectrum of pediatric ages. Moreover, because the PMHS (postmortem human subjects) in Duncan's experiments were failed as whole specimens with all the skin and musculature included, direct comparisons with subsequent isolated cervical spine data are difficult. It is noteworthy that Duncan's tests produced primarily lower cervical spine injuries, which is in contrast with the epidemiologic observation of predominantly upper cervical spine injuries in the neonate.

Ouyang et al. (2005)

Ouyang et al. (2005) investigated the structural responses of whole pediatric PMHS cervical spines in sagittal plane bending and in tensile loading to failure. This study is important because it used methodologies previously published for whole adult spine testing, allowing some direct comparisons with adult data.

Ten cadaveric specimens of mixed gender ranging in age from 2 to 12 years (Table 5.6) were used. Testing fixtures were designed to support the head while applying both pure bending moments and tensile loads through the approximate head center of gravity. All testing, including failure tests, were conducted on whole cervical spines. Maximum moments applied during the bending protocol ranged from −2.4 Nm (extension) to 2.4 Nm (flexion). The bending protocol included three

Table 5.6 Anthropometry from Ouyang et al. (2005)

Subject	Age (years)	Gender	Mass (kg)	Stature (mm)	Head length (mm)	Head breadth (mm)	Neck circ. (mm)
1	2	f	13.0	970	160	135	220
2	2.5	m	10.5	875	165	140	220
3	3	m	10.0	910	155	145	210
4	3	f	10.5	850	160	145	205
5	4	m	14.0	1,090	160	145	240
6	5	m	13.0	1,010	165	140	230
7	6	m	16.5	1,080	170	145	245
8	6	m	20.0	1,090	175	150	230
9	7.5	f	17.0	1,170	175	135	230
10	12	f	20.0	1,400	170	135	255

discrete loading steps in both extension and flexion similar to methodology to previous studies (Camacho et al. 1997; Nightingale et al. 2002). The tension protocol included a viscoelastic battery followed by failure tests of the whole head-neck complex that was similar to the experiments by Van Ee et al. (2000) and Dibb et al. (2009). The specimens were failed with a free head end-condition, which permitted rotation and translation of the head in response to a tensile load. One specimen (number 6) was damaged during preconditioning and was not included in the results.

The bending stiffnesses of the pediatric specimens were compared in two age groups: 2–4 and 5–12 years old. The authors did not find a statistically significant difference in bending stiffness between the two groups; however, this may have been due to a lack of power. The mean bending stiffness of all the specimens was 0.041 N-m/degree. The average tensile failure load was 726 ± 171 N and the displacement at failure was reported to be 20.2 ± 3.2 mm (Table 5.7). The average tensile stiffness of the whole cervical spine specimens was 35 ± 6 N/mm. No statistically significant differences were observed between the age groups. The majority of the failures in the lower cervical spine were end-plate fractures occurring at the site of fixation.

A linear regression of the force at failure showed a statistically significant ($p = 0.024$) increase with age ($R^2 = 0.54$) (Fig. 5.4). The authors achieved better correlation with a log function ($R^2 = 0.85$) of the form:

$$F = a \ln(x+1) + b.$$

The displacement at failure did not show any clear trends (Fig. 5.5).

To date, Ouyang et al. (2005) have presented one of the more comprehensive studies on the biomechanics of the pediatric cervical spine. Ouyang provides data under tensile and bending loading for an age range between 2 and 12 years. Since the infant and teenage years were not available to them at the time, more data are needed for the development of scaling rules for the entire pediatric population. In addition, and similar to the work of Duncan, the majority of cervical injuries reported by Ouyang were in the lower cervical spine, which is not consistent with the epidemiology in which the majority of injuries in children 8 years and younger

Table 5.7 Failure testing results (Ouyang et al. 2005)

Subject	Age (years)	Age (months)	Force (N)	Disp. (mm)	Injury
1	2	24	494	14	C5–C6 level, fracture at superior end-plate of C6
2	2.5	30	531	19	C7–T1 level, fracture at inferior end-plate of C7
3	3	36	817	23	C5–C6 level, fracture at superior end-plate of C6
4	3	36	569	22	C6–C7 level, above the polymethylmethacrylate
5	4	48	634	17	C7–T1 level, fracture at superior end-plate of T1
7	6	72	912	24	C6–C7 level fracture at superior end-plate of C7
8	6	72	891	21	C4–C5 level, fracture at superior end-plate of C5
9	7.5	90	764	20	Fracture at T1–T2 level, above the polymethylmethacrylate
10	12	144	918	22	C7–T1 level, fracture at inferior end-plate of C7

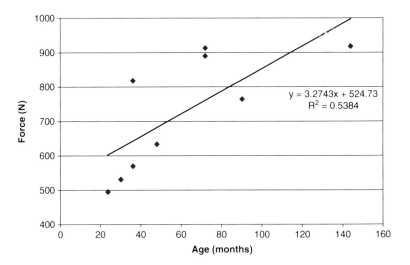

Fig. 5.4 Force at failure from Ouyang et al. (2005), showing a statistically significant increase with age

are in the upper cervical spine. Finally, Ouyang tested only whole cervical spines as opposed to sectioned motion segments. While the structural property data is invaluable, it is likely that the failures were influenced by the fixation at the T1 casting. Five of the nine specimens failed at the caudal fixation.

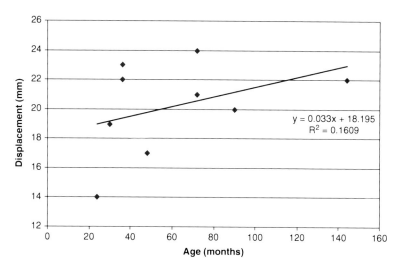

Fig. 5.5 Displacement at failure from Ouyang et al. (2005). The displacements at failure for the whole cervical spines were not statistically significant and were poorly correlated with age

Luck et al. (2008)

Luck et al. (2008) have conducted a comprehensive study on the tensile properties of the pediatric neck. The authors tested the strength and structural properties of both whole pediatric spines and pediatric spine motion segments, with the goal of providing quantitative values for model development, scaling, and injury assessment reference values.

Eighteen pediatric human cadaveric whole cervical spines ranging in age from 20 weeks gestation to 14 years old were tested in tension (Table 5.8). The mandible was disarticulated in order to facilitate good fixation in the testing frame (Luck et al. 2012). The testing included a full viscoelastic battery on the whole spines followed by dissection into the following three segments: O–C2 (including the skull to the second cervical vertebra), C4–C5, and C6–C7. The motion segments were subjected to the same battery of tests as the whole cervical spine and then they were failed with a fixed-fixed end condition. Cervical spine injuries were documented by dissection. For the nondestructive tests, linear stiffnesses were calculated based on the upper 50 % of the force–displacement curve, and the peak applied force was 10 % of the estimated strength of the specimen.

The data were grouped into two categories—neonatal (less than 1 month old) and specimens greater than 1 month old (Tables 5.9 through 5.12). The neonatal PMHS grouping ranged from 20 weeks gestation (prenatal) to 24 days old (postnatal). The neonatal group was the only population with enough samples for statistical analysis.

The tensile stiffness of the whole neonatal cervical spine was 6.8 ± 1.0 N/mm ($n = 6$). The neonatal tensile stiffness of the upper cervical spine, was 9.9 ± 2.1 N/mm ($n = 7$). The mean stiffness for the C4–C5 spinal segment was 47.5 ± 8.4 N/mm

Table 5.8 Anthropometry from Luck et al. (2008)

PMHS ID	Age	Sex	COD	Whole body Mass (kg)	Whole body Height (cm)	Head Mass (kg)	Head Breadth (cm)	Head Length (cm)	Spine Cervical Length (cm)	Spine T1 endplate Width (mm)	Spine T1 endplate Depth (mm)	Scale factor	Scaled loading rate (N/s)
Perinatal and neonatal PMHS													
02P	20 WKG [0 MN]	F	Heart failure	–	–	0.094	5.9	7.1	2.46	9.35	4.85	0.115	5.8
13P	29 WKG (5 WPB) [0 MN]	F	–	–	–	0.492	7.1	10.2	3.40	13.40	6.50	0.222	11.1
09P	33 WKG [0 MN]	M	Fetal demise	2.04	43.2	SC	9.1	–	3.61	11.05	7.00	0.197	9.9
07P	33 WKG [0 MN]	M	–	–	–	0.434	8.2	10.1	3.68	13.65	6.20	0.215	10.8
10P	35 WKG [0 MN]	M	Fetal demise	–	–	SC	7.9	10.0	4.12	15.60	8.15	0.323	16.2
08P	37.5 WKG [0 MN]	M	Pulmonary hypoplasia	–	50.8	0.780	9.4	11.4	4.21	15.00	8.00	0.305	15.3
05P	1 DY [0.03 MN]	F	Diaphragmatic hernia	2.75	–	0.665	9.1	11.7	4.14	13.00	6.00	0.198	9.9
03P	3 DY [0.1 MN]	M	Ischemic encephalopathy; cerebral infarction	–	–	0.492	8.5	10.3	3.61	12.25	6.75	0.210	10.5
06P	11 DY [0.37 MN]	F	Nonimmune hydrops fetalis and intercranial hemorrhage	2.02	44.5	0.702	10.4	11.2	3.83	13.40	6.60	0.225	11.3
11P	16 DY [0.53 MN]	F	Anencephaly	2.27	–	SC	6.3	6.1	3.66	15.35	6.55	0.256	12.8
04P	24 DY [0.8 MN]	F	Dandy-Walker syndrome	2.72	45.7	1.152	10.5	17.5	4.84	15.35	8.70	0.340	17.0
Pediatric PMHS > 1 month [older cohort]													
12P	5 MN	M	Respiratory failure	–	–	1.071	12.3	13.2	5.22	20.45	9.00	0.468	23.4
14P	9 MN	M	COPD	7.00	–	1.950	11.5	15.0	5.36	20.00	8.25	0.420	21.0
15P	11 MN	F	SIDS	8.16	71.1	1.570	11.9	14.8	6.12	21.70	10.85	0.599	29.9
16P	18 MN	M	Drowning	11.80	81.3	SC	13.5	15.2	7.17	21.90	12.20	0.680	34.0
17P	22 MN	F	Non-Hodgkins lymphoma	–	–	–	12.5	16.4	6.73	21.30	11.50	0.623	31.2
18P	9 YR [108 MN]	M	End-stage renal disease; hyperkalemia	–	–	2.440	13.1	16.3	8.66	23.60	15.10	0.906	45.3
01P	14 YR [168 MN]	F	Brain aneurysm	61.20	165.0	–	–	–	11.73	–	–	–	50

WKG weeks gestation, *WPB* weeks post birth, *DY* days, *MN* months, *YR* years, *SC* skull compromise. Equation for area of an ellipse $1/4\cdot(\pi\cdot W D)$ used to approximate area of T1 endplate for scaled loading rate

5 Experimental Injury Biomechanics of the Pediatric Neck

Table 5.9 Stiffness data for perinatal and neonatal specimens (Luck et al. 2008)

		Tensile stiffness and low-load displacement (LLD)							
		WCS		O–C2		C4–C5		C6–C7	
PMHS ID	Age (months)	LLD (mm)	Stiffness (N/mm)	LLD (mm)	Stiffness (N/mm)	LLD (mm)	Stiffness (N/mm)	LLD (mm)	Stiffness (N/mm)
02P	0	0.02[a]	5.4[a]	–	–	–	–	–	–
13P	0	1.28	7.7	0.29	12.2	0.06	61.4	0.10	39.4
09P	0	–	–	–	–	0.03	50.6	0.25	36.7
07P	0	0.69[b]	7.9[b]	0.61	11.9	0.12	46.1	0.08	44.4
10P	0	–	–	–	–	0.58	35.8	0.17	37.1
08P	0	1.98	6.0	0.23	10.5	–	–	–	–
05P	0.03	1.27	5.3	0.68	7.4	0.08	50.4	0.16	45.2
03P	0.1	0.49	7.7	0.72	11.2	0.01	54.2	0.00	42.2
06P	0.37	0.33	6.8	1.45	7.1	0.06	50.5	0.06	50.8
11P	0.53	–	–	–	–	0.01	35.5	0.02	61.5
04P	0.8	0.34	7.3	0.18	9.3	0.05	42.8	0.04	34.2
	Mean	0.95	6.8	0.59	9.9	0.11	47.5	0.10	43.5
	SD	0.67	1.0	0.44	2.1	0.18	8.4	0.08	8.5

– specimen not available, *grayed areas* PMHS used in two-way ANOVA
[a]Donor 02P tested from occiput to C6 (excluded from mean calculation)
[b]Donor 07P tested from occiput to C7 (excluded from mean calculation)

($n=9$) and for the C6–C7 spinal segment was 43.5 ± 8.5 N/mm ($n=9$). There was a statistically significant difference between spinal segments ($p<0.001$; $\alpha=0.05$). There were significant differences between the whole cervical spine and both lower cervical spine segments and between the upper cervical spine and both lower cervical spine segments ($\alpha=0.05$). A representative force–displacement response for the pediatric upper cervical spine (108 months) is provided in Fig. 5.6.

The neonatal tensile strength for the upper cervical spine was 230.9 ± 38.0 N ($n=7$). The mean strength for the C4–C5 spinal segment was 212.8 ± 60.9 N ($n=9$) and the strength of the C6–C7 spinal segment was 187.1 ± 39.4 N ($n=8$). Ultimate strength increased monotonically with age (Fig. 5.7, two-way ANOVA, $p<0.001$). No significant differences in the ultimate tensile strength by level were found within the neonatal cohort ($p=0.2334$, one-way ANOVA), whereas the upper cervical spine was significantly stronger than the lower cervical spine in the older cohort ($p<0.001$, two-way ANOVA).

The injuries produced by these tests were similar to those seen clinically. They included the following: physeal/endplate disruptions, cartilaginous synchondrotic disruptions, fractures at the neurocentral synchondroses and through the posterior synchondrosis. The upper cervical spines exhibited a variety of injuries including an occipitoatlantal dislocation, fracture/dislocations through the dentoneural/neurocentral synchondroses and dentocentral synchondrosis of the axis, and atlantoaxial dislocations.

Although the majority of the data is in the perinatal and neonatal range, enough specimens were available in the older ages to establish age dependent trends. In particular, the strength of the upper cervical spine increases with age very quickly,

Table 5.10 Stiffness data for pediatric specimens (Luck et al. 2008)

| PMHS ID | Age (months) | Tensile stiffness and low-load displacement (LLD) ||||||||||||
| | | WCS || O–C2 || C3–C4 || C4–C5 || C5–C6 || C6–C7 ||
		LLD (mm)	Stiffness (N/mm)	LLD (mm)	Stiffness (N/mm)	LLD (mm)	Stiffness (N/mm)	LLD (mm)	Stiffness (N/mm)	LLD (mm)	Stiffness (N/mm)	LLD (mm)	Stiffness (N/mm)
12P	5	0.48	9.7	0.73	14.5	–	–	0.02	58.2	–	–	0.02	43.5
14P	9	1.16	28.5	1.66	41.5	–	–	0.30	114.4	–	–	0.26	103.0
15P	11	1.74	31.3	1.15	54.4	–	–	0.16	141.8	–	–	0.17	107.6
16P	18	–	–	–	–	0.21	130.9	–	–	0.03	171.5	–	–
17P	22	2.14	36.5	1.05	64.0	–	–	0.09	153.1	–	–	0.07	93.1
18P	108	2.57	55.7	1.19	118.1	–	–	–	–	–	–	0.20	255.9
01P	168	0.82	70.1	0.22	199.0	–	–	–	–	–	–	–	–

– specimen not available, *grayed areas* PMHS used in two-way ANOVA

Table 5.11 Failure data for perinatal and neonatal specimens (Luck et al. 2008)

		Cervical segment initial failure and ultimate strength						
		O–C6	O–C2		C4–C5		C6–C7	
PMHS ID	Age (months)	Ultimate (N)	Initial (N)	Ultimate (N)	Initial (N)	Ultimate (N)	Initial (N)	Ultimate (N)
02P	0	57.2	–	–	–	–	–	–
13P	0	–	–	274.8	334.3	360.5	194.9	204.1
09P	0	–	–	–	178.2	180.4	124.8	154.1
07P	0	–	–	196.9	168.1	174.3	X	X
10P	0	–	–	–	188.4	207.5	177.8	209.7
08P	0	–	–	241.9	–	–	–	–
05P	0.03	–	188.3	208.9	–	225.5	172.6	181.3
03P	0.1	–	–	257.6	165.8	182.7	139.3	142.0
06P	0.37	–	–	173.6	173.8	176.0	140.2	151.9
11P	0.53	–	–	–	–	167.3	182.9	191.4
04P	0.8	–	260.0	262.3	237.2	240.8	250.4	262.2
	Mean		224.2	230.9	206.5	212.8	172.9	187.1
	SD		50.7	38.0	61.4	60.9	39.9	39.4

– specimen not available, *grayed areas* PMHS used in two-way ANOVA, X data loss

reaching values approximately 60 % of the adult strength by the age of two (Figs. 5.7 and 5.8). The stiffness data from the nondestructive tests provides low-load data while failure tests can provide high-load data. Another interesting finding was that the upper cervical spine of the infant was not significantly stronger than the lower cervical spine segments. This was not true of the older cohort, which showed the same trends in strength that have been documented in the adult (Dibb et al. 2009; Nightingale et al. 2007; Van Ee et al. 2000)

Arbogast et al. (2009)

Arbogast et al. (2009) investigated the kinematic responses of the head and spine of children and adults in low-speed noninjurious frontal sled tests. Twenty human volunteers ranging in age from 6 to 30 years were divided into four subcohorts: 6–8, 9–11, 12–14 years, and adults. They were subjected to acceleration pulses on the order of those experienced in amusement park bumper car-to-wall impacts. The kinematic response of the volunteers was quantified by tracking markers positioned at key landmarks on the head and spine. The excursion of these markers in the sagittal plane was normalized by subject-specific seated height in order to account for the effect of size in the response. Normalized forward excursion (x) was found to significantly decrease with age in all head markers (head vertex, opisthocranion, external auditory meatus, nasion) and in all spine markers (C4, T1, T4, T8). Similarly, normalized vertical (z) excursion for all head and spine markers were found to be greater in the younger subjects and at a minimum in the adults. For all ages, the head markers moved in a downward (z) direction (with the exception of the

Table 5.12 Failure data for pediatric specimens (Luck et al. 2008)

		Cervical segment initial failure and ultimate strength									
		O–C2		C3–C4		C4–C5		C5–C6		C6–C7	
PMHS ID	Age (months)	Initial (N)	Ultimate (N)	Initial (N)	Ultimate (N)	Initial (N)	Ultimate (N)	Initial (N)	Ultimate (N)	Initial (N)	Ultimate (N)
12P	5	382	462	–	–	–	330	–	–	242	293
14P	9	673	840	–	–	394	400	–	–	343	418
15P	11	153	1,019	–	–	406	631	–	–	489	492
16P	18	–	–	856	916	–	–	515	733	–	–
17P	22	1,097	1,231	–	–	467	845	–	–	646	807
18P	108	1,460	1,925	–	–	–	–	–	–	1,400	1,757
01P	168	495	2,960	–	–	–	–	–	–	–	–

– specimen not available, *grayed areas* PMHS used in two-way ANOVA

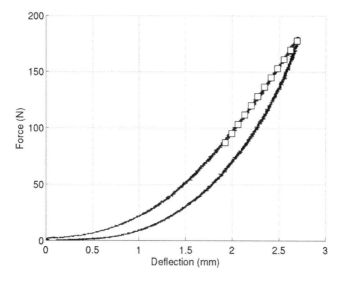

Fig. 5.6 A typical nonlinear force–displacement response for the upper cervical spine (O–C2) of a 108-month-old child. The tensile stiffness (N/mm) was determined by linear regression of the loading portion of the response from 50 to 100 % of the applied load (from Luck et al. 2008)

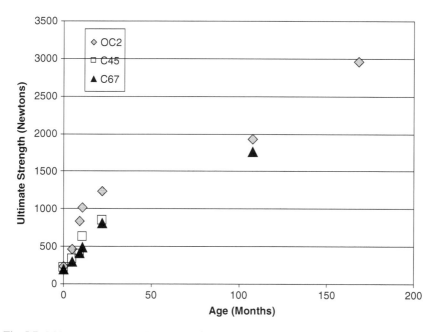

Fig. 5.7 Ultimate strength as a function of age and level (Luck et al. 2008)

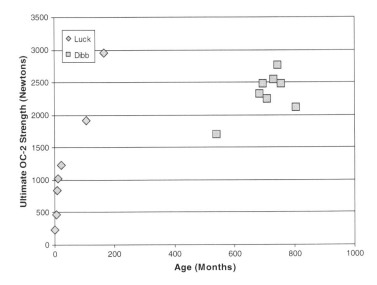

Fig. 5.8 Combined data on the strength of the upper cervical spine data from Luck et al. (2008) (pediatric), and from Dibb et al. (2009) (males with a mean age of 58 years). The data illustrate the rapid increase in upper cervical spine strength during early childhood and suggest that the upper cervical spine of the young adult may be considerably stronger than the older population

opisthocranion marker), while all of the spinal markers moved in a slightly upward (z) direction. Changes in the relative angle of the spinal segments, as defined by pairs of head and spinal markers (head and C4, C4–T1, T1–T4, and T4–T8), were reported to assess spinal mobility. The angle defined between the head markers and those of C4–T1 remained relatively constant for the adults and 12–14-year-old group, but exhibited slight neck extension for both the younger age groups at the point of maximum acceleration. The angle defined between C4–T1 and T1–T4 demonstrated the greatest change in magnitude, with the youngest ages experiencing the greatest change in angle.

These data provide an important component in furthering the understanding of the head and spine kinematics in low-severity frontal impact in both children and adults. The authors have also reported belt, seat pan, and footrest loads. These will be very useful in modeling the experiments both in FEA and with existing ATDs. Although the impacts are well below injury thresholds, they are likely to remain the best available data for the validation of scaling rules and for the development and validation of new computational models and ATDs.

Whole Pediatric PMHS Studies

A few studies have been conducted on whole pediatric post mortem subjects to simulate exposures of pediatric PMHS in child restraint systems (CRS). In addition to providing biomechanical data on the PMHS, the results were compared with

Table 5.13 A summary of pediatric PMHS sled tests (Dejeammes et al. 1984; Kallieris et al. 1976; Wismans et al. 1979)

Institution	Age, years	Length, cm	Weight, kg	Restraint system	Velocity, km/h	Decel., G	Head, G	HIC	Injury AIS
Heidelberg	2.5	97	16	Romer Vario	31	18	–	–	1
Heidelberg	6	125	27	Romer Vario	40	20	–	–	0
Heidelberg	6	124	30	Romer Vario	40	21	34	102	1
Heidelberg	11	139	31	Romer Vario	40	21	94	523	1
HSRI	6	109	17	5-Point CRS	48	20	41	347	2
APR1[a]	2	87	13	Integral 1	48	13	94	803	0
APR1[a]	2	87	13	Integral 1	50	13	53	207	0
APR1[a]	2	87	13	Integral 1	50	13	77	1,049	0
APR1[a]	2	87	13	Integral 1	50	13	109	559	0
APR2[a]	2	87	13	Tot guard	50	13	122	565	6
Heidelberg	13	162	39	3-Point Belt	49	15	102	800	0
Heidelberg	10	139	39	4-Point CRS	46	15	92	–	2
Heidelberg	12	144	52	3-Point Belt	–	–	–	–	1
Heidelberg	2.5	91	17	Romer Peggy	49	18	–	–	3

[a]All test were run using a single PMHS

identically restrained child dummies (Dejeammes et al. 1984; Kallieris et al. 1976; Wismans et al. 1979). This body of work is nicely summarized by Brun Cassan et al. (1993) and Klinich et al. (1996). The focus of these studies was on the kinematics of the head and torso, so there is only a limited amount of quantitative biomechanical data available for the cervical spine. However, some of the subjects were instrumented with head accelerometers, which may allow a rough estimate of neck loads.

The head acceleration data and the test conditions are all summarized in Table 5.13. Although the authors did not estimate neck loads in these studies, it is possible to do so using an estimate of 3.5 kg for the mass of the pediatric head (Irwin and Mertz 1997). This yields 1,167 N of neck tension (Heidelberg 6) on the low end for an injury of AIS 1. Since there was almost certainly head contact in the tests with the higher HIC values, such an estimate is not possible for the upper end of the spectrum. It may be possible to retrospectively examine the raw data to estimate peak neck loads prior to head contact; however, there is insufficient data in the referenced manuscripts to do so.

Other Work

Nuckley et al. (2005a; 2007) from the University of Washington have tested a number of human pediatric cervical spine specimens in a variety of loading modes. Their results have not been presented in a full length manuscript, so the results and the methodological details are sparse. In the 2005 abstract, the authors report results from compression tests to 75 % of body weight. Their data show an increasing trend in the C3–C5 compressive stiffness with age that is similar to trends seen in their prior

Table 5.14 Pediatric tolerance as a percent of adult (derived from Nuckley et al. 2007)

Age (years)	2	3	5	8	9	11	13	16	18	22	28
Tensile strength (C1–C2)	29	54	46	63	76	78	89	84	98	103	99
Compressive strength (C3–C5)	20	27	36	50	37	54	63	69	99	96	104
Extension strength (C6–C7)	27	26	31	37	24	69	46	79	–	92	109

baboon studies (Nuckley et al. 2002). The authors offer the following regression for predicting the compressive stiffness of the C3–C5 motion segment in N/mm:

$$\text{Stiffness}\left(\frac{N}{mm}\right) = -0.16(\text{years})^2 + 17(\text{years}) + 369.$$

Their 2007 abstract shows trends of increasing stiffness and decreasing range-of-motion with age for tension for C1–C2, and flexion and extension for C6–C7. Increasing trends in strength are also reported for tension, compression, and extension. Representative data for these modes of loading is presented; however, there is not enough information given in the abstract to allow the formulation of age scaling relationships for the mechanical responses. Motion segment strength by level and loading mode, as a percentage of adult, can be extracted from one of the figures (Table 5.14). These results should be used with caution, since it is not clear how the normative adult data was derived. In addition, the separation of C1 from the skull severely compromises the mechanical integrity of the upper cervical spine.

McGowan et al. (1993) conducted a quasi-static tensile test on a single fresh 8-h old PMHS. Specimen preparations included two cervical, one thoracic and one lumbar unit. Each unit consisted of three vertebrae and two intervertebral discs. The segments were tested in tension under displacement control at 1.25 mm/s. Apparently, fixation failures occurred in both the cervical and thoracic segments; therefore, neither the levels tested nor any biomechanical data were reported. For the lumbar spine, the specimen consisted of L3–L4–L5. Failure of the capsular ligaments of L3–L4 occurred at 260 N of tension and ultimate failure occurred at L4–L5 at approximately 265 N. Up until failure, the stiffness of the entire preparation was 94 N/mm. This stiffness was based on the series combination of the L3–L4 and L4–L5 segments.

Discussion

At this time, the biomechanics of the cervical spine in tension for perinatal and neonatal human can be considered well characterized. Although the number of samples is small relative to adults, it is a developmental age on which there is enough information to have confidence in our current estimates of stiffness and tolerance. This cannot be said for any other age. From the standpoint of ATD development, the important ages are 1, 3, 6, and 10 years. Although there is still almost no data for these specific ages, estimates can be made based on the trends observed in what is

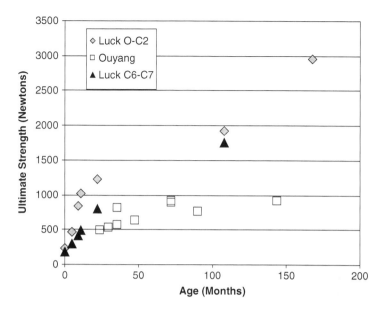

Fig. 5.9 Combined data from Luck et al. (2008) and Ouyang et al. (2005) (whole spine strength). The data are in reasonably good agreement on the strength of the lower cervical spine

currently available. However, as discussed below, any regression model needs to predict a rapid increase in strength during childhood and a slow decrease after skeletal maturity.

In upper cervical tension, sufficient data are available to illustrate these relationships (Fig. 5.8). It is noteworthy that the 9-year-old PMHS in the Luck study had 80 % of the adult strength (Dibb et al. 2009; Chancey et al. 2003), and that the 14-year-old PMHS was 20 % stronger than the mean adult (Fig. 5.8). The average age of the donors in the adult tests from Dibb was 58 years, which suggests that the older population may be significantly weaker than young adults in tension. The trends in strength with age appear to be different for the upper cervical spine than the lower. The Luck study suggests a rather dramatic increase in upper cervical spine strength during the first 3 years of life. The trend was much more gradual in the lower cervical motion segments. Both the Luck study and the Ouyang study show similar trends with age (Fig. 5.9). The difference in these trends might be explained by the complex developmental anatomy of the axis and atlas relative to the vertebrae of the lower cervical spine.

When the Duncan data are compared with the perinatal and neonatal data from Luck et al. (2008) (Fig. 5.10) it is clear that Duncan's specimens were stronger. This is likely due to the fact that Duncan used whole PMHS, while the Luck study tested only the osteoligamentous cervical spine. Unfortunately, the effects of post-mortem muscles on the structural properties of the neck are somewhat variable and unlike those of living muscle (Van Ee et al. 1998). However, it is apparent from Fig. 5.10 that the contribution of the passive soft tissues to ultimate strength must be considered.

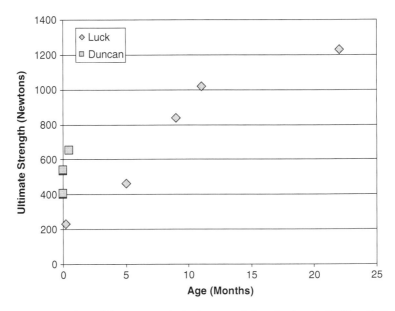

Fig. 5.10 A comparison of the upper cervical spine strength from Luck et al. (2008), with the data from Duncan (1874). The whole pediatric necks tested by Duncan appear to be stronger than the isolated osteoligamentous cervical spine segments

The data from Luck et al. (2008) and Ouyang et al. (2005) are in good agreement with regard to the strength of the lower cervical spine. Almost all of the failures of the whole spines in the Ouyang study occurred in the lower cervical spine, with almost all specimens failing at C4–C5 or lower. When the C6–C7 strength from Luck et al. (2008) are plotted with the data from Ouyang et al. (2005) (Fig. 5.9) there appears to be a continuous trend in the strength of the pediatric lower cervical spine, suggesting that these data sets can be combined.

Animal Surrogates

The difficulty of acquiring human pediatric tissue has fostered biomechanical research on juvenile animals as surrogates. The surrogates have included piglets (porcine), baboons (*Papio anubis*), and goats (caprine). Each was chosen based on developmental, anatomical, geometric, or weight bearing similitude between the juvenile animal and the human child.

Testing of porcine surrogates has formed the basis for the neck injury assessment reference values that are used in both the child and adult Hybrid III and CRABI dummies. Mertz et al. (1982) performed matched out-of-position airbag testing of piglets and an instrumented 3-year-old child dummy (Mertz et al. 1982; Mertz and Weber 1982). By comparing injuries in the animals with measurements from the dummy load cells, they estimated probabilities for significant neck injury as a function of the dummy neck load. An upper load limit of 1,350 N in the dummy was reported to correspond with a high probability of injury for the 3-year-old. Prasad and Daniel (1984) conducted similar piglet/dummy tests and reported injury reference values for the cervical spine that included the measured moment in addition to tension. Using a linear relationship between tension and moment, they proposed critical values for tension and bending of 2,000 N and 34 N-m respectively for the 3-year-old dummy. In order to estimate similar reference values for the other dummy sizes, scaling ratios were derived by Melvin based on geometric similitude and the strength and stiffness of human calcaneal tendon (Melvin 1995; Yamada 1970). Melvin estimated that the strengths of the 6- and 3-year-old are 41 and 32 % respectively compared to the 50th percentile male. Kleinberger et al. (1998) reanalyzed the Prasad and Daniel data using a multivariate logistic model and formulated new extension and tension values of 30 N-m and 2,500 N, respectively. Based on input from the American Automobile Manufacturers Association (AAMA), these values were further refined to 2,120 N for tension and 27 N-m for extension (Eppinger et al. 1999). Eppinger et al. (1999), used the methodology of Melvin, but with slightly different scale factors, to estimate injury assessment reference values (IARVs) for Federal Motor Vehicle Safety Standard 208 (FMVSS 208). Their analysis resulted in tension values of 45, 36, and 24 % for the 6-, 3-, and 1-year-old relative to the adult male. Values for the FMVSS 208 flexion criteria were based on previously reported ratios of 2.5 between flexion and extension. For compression, it was assumed that the critical values are the same as tension.

Despite their limited scope, the piglet studies of Mertz et al. (1982), and Prasad and Daniel (1984) are the foundation for the current Hybrid III 3-year-old ATD neck IARVs in all modes of loading, and the scaling of these values has formed the basis for neck injury criteria for the entire dummy family. While the IARVs discussed above provide some guidance for assessing injury risk, they are dummy based criteria and their relation to real human tolerance is still unknown. More recent baboon and caprine studies have attempted to validate these ratios in different animal models.

Several investigators have tested caprine cervical spines in both bending and tension. Scaling relationships were determined between juvenile and adult caprine specimens for tensile tolerance and stiffness (Pintar et al. 2000). This study recommended tolerances for the 12-, 6-, 3-, and 1-year-old of 78, 38, 20, and 12 % of the adult. Hilker et al. (2002) refined the tensile tolerance scaling ratios of Pintar et al. (2000) by examining the subset of the adult population that included a small adult, mid-sized adult and large adult. Comparisons were subsequently made between the mid-sized adult tolerance data and the younger subjects. The resulting ratios were slightly different than the initial estimates of Pintar et al. (2000). Hilker et al. (2002) reported scaling percentages of 62, 30, 16, and 10 % for the 12-, 6-, 3-, and 1-year-old, respectively, when compared to the mid-sized adult tensile tolerance data. Clarke et al. (2007) also examined cervical spine stiffness and age in caprine juveniles and adults; however, their study focused on the bending responses. They too found the immature motion segments to be significantly more compliant than the adult and suggested that these differences might be responsible for the different injury patterns in pediatric and adult spinal cord injury.

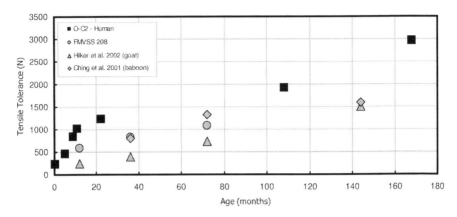

Fig. 5.11 Tensile tolerance of the pediatric upper cervical spine (O–C2) compared to scaled estimates from juvenile animal surrogate studies and from FMVSS 208 (Hilker et al. 2002; Ching et al. 2001; Eppinger et al. 1999). Tensile tolerances are higher than the predictions from both the animal studies and FMVSS 208. (From Luck et al. 2008)

Ching et al. (2001) performed tensile tests on juvenile baboons, and reported scaling ratios relative to the adult of 66, 55, and 33 % for the 12-, 6-, and 3-year-old, respectively. A follow-on study by Nuckley et al. (2005b) quantified the rate sensitivity of the tensile responses of the pediatric baboon specimens. A study by Nuckley and Ching (2006) using the juvenile baboon model examined the relative stiffnesses and strengths as a function of developmental age in human equivalent years and developed regression models for both. All the baboon models predicted increases in stiffness and strength with age; however, they should not be extrapolated beyond the age of skeletal maturity because they predict peak strengths in the eighth decade of life.

Neither the studies on juvenile animals nor the estimates in FMVSS 208 correctly predicted the tolerance of the human pediatric cervical spine in the Luck et al. (2008) study. Assuming an adult human male tolerance value of 2,417 N (Dibb et al. 2009), comparisons with the human pediatric data from Luck et al. (2008) can be made for the scale factors from FMVSS 208 (Eppinger et al. 1999; Melvin 1995), the caprine studies (Hilker et al. 2002), and the baboon studies (Ching et al. 2001). Both the animal models and the scale factors in FMVSS 208 underestimated the tolerance of the pediatric PMHS spine (Fig. 5.11). This was particularly true for ages less than 2 years. The human appears to strengthen dramatically in the first 2 years of life compared to the developmentally equivalent animal models. While the porcine-based FMVSS 208 scale factor predicts the tolerance of the human 1-year-old better than the caprine model, it still underestimates the human strength by almost 100 %.

In contrast to the tolerance results, the relationship between human pediatric stiffness and age was accurately described by the baboon and caprine studies (Fig. 5.12). Regression of the baboon and caprine data against the human data revealed correlation coefficients greater than 0.84. The mean absolute deviation between the animal models and the human pediatric data was 3 % for donors between the ages of 1–14 years. This suggests that the age-equivalent animal models can provide accurate scaling factors for stiffness at low loads.

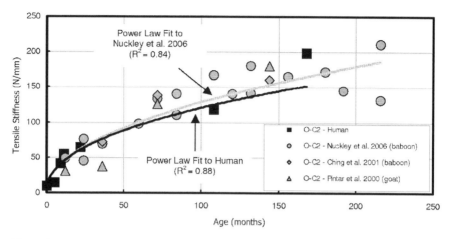

Fig. 5.12 Tensile stiffness (N/mm) of the human upper cervical spine (O–C2) compared to three previous juvenile animal surrogate studies (Pintar et al. 2000; Ching et al. 2001; Nuckley and Ching 2006). The stiffness trends and values for the age-equivalent juvenile animals compare well with the human. (From Luck et al. 2008)

Animal surrogate studies are, and have been, useful tools for understanding the biomechanics of the pediatric cervical spine. Still, there are important limitations to these surrogate studies. Experiments using juvenile animal models employ age estimation protocols to relate their respective models to a human equivalent age. These protocols rely on matching the developmental anatomy of the animal surrogates to known developmental anatomy of the maturing child. Based on the current data, it is unclear if the lack of predictive ability for tolerance is the result of differences in the structural properties or a result of the age-estimation techniques. The question of how to interpret tolerance and stiffness data for surrogates that have geometries different from the human is a confounding problem. Testing of additional pediatric PMHS is necessary not only to provide additional pediatric human data, but also to guide the interpretation of the many surrogate studies completed to date. If the biomechanical basis for the differences between surrogates and humans can be better understood, the animal studies could provide valuable complementary data to the scarce human data. These relationships will be of the utmost importance, as it is evident that surrogate studies will continue and the availability of human pediatric tissue will always be limited.

Material Properties of the Human Pediatric Neck

There are no data on the constitutive properties of pediatric cervical intervertebral discs, ligaments, or synchondroses. Investigators have used adult values for these tissues in finite element models and have scaled the constitutive properties (Kumaresan et al. 1997, 2000a, b; Yoganandan et al. 2001, 2002).

Yoganandan et al. (2001) offer a comprehensive review of all the materials and constitutive properties currently in use in FE models of the cervical spine. These are mostly adult values; however, they form the basis for the constitutive properties used in the models of Kumaresan et al. (1997; 2000a, b). One of the conclusions of the modeling effort by Kumaresan et al. (2000a, b) was that the developmental anatomic differences between the adult and child account for the majority of the difference in their structural responses and that it is imperative to include the growth regions and synchondroses for accurate prediction of pediatric responses.

Obviously there is a tremendous need for more data in this area in order to generate relationships between constitutive properties and age and to validate the existing scaling techniques. Ultimately many of our design specifications and performance criteria for the next generation of pediatric ATDs will come from finite-element and multibody dynamics modeling. While the geometries and inertial properties for these models are already well defined or within our grasp, many questions remain about the material properties: *if* they are different and *how* they may change as a function of age.

References

Adams VI (1992) Neck injuries: II. Atlantoaxial dislocation – a pathologic study of 14 traffic fatalities. J Forensic Sci 37:565–573
Angel CA, Ehlers RA (2001) Atloido-occipital dislocation in a small child after air-bag deployment. N Engl J Med 345:1256
Arbogast KB, Balasubramanian S, Seacrist T, Maltese MR, Garcia-Espana JF, Hopely T, Constans E, Lopez-Valdes FJ, Kent RW, Tanji H, Higuchi K (2009) Comparison of kinematic responses of the head and spine for children and adults in low-speed frontal sled tests. Stapp Car Crash J 53:329–372
Aufdermaur M (1974) Spinal injuries in juveniles. Necropsy findings in twelve cases. J Bone Joint Surg 56:513–519
Babcock JL (1975) Spinal injuries in children. Pediatr Clin N Am 22:487–500
Bailey DK (1952) The normal cervical spine in infants and children. Radiology 59:712–719
Bhattacharyya SK (1974) Fracture and displacement of the odontoid process in a child. J Bone Joint Surg 56:1071–1072
Blauth M, Schmidt U, Otte D et al (1996) Fractures of the odontoid process in small children: biomechanical analysis and report of three cases. Eur Spine J 5:63–70
Bodenham A, Swindells S, Newman RJ (1992) Permanent tetraplegia in an infant following improper use of a car seat restraint. Inj Br J Accid Surg 23:420–422
Bonadio WA (1993) Cervical spine trauma in children: part I. General concepts, normal anatomy, radiographic evaluation. Am J Emerg Med 11:158–165
Brown DK, Roe EJ, Henry TE (1995) A fatality associated with the deployment of an automobile airbag. J Trauma 39:1204–1206
Brown RL, Brunn MA, Garcia VF (2001) Cervical spine injuries in children: a review of 103 patients treated consecutively at a level 1 pediatric trauma center. J Pediatr Surg 36:1107–1114
Brun Cassan F, Page M, Pincemaille Y et al (1993) Comparative study of restrained child dummies and cadavers in experimental crashes. SAE Publication SP-986. Child Occupant Protection (SAE Technical Paper 933105). pp. 243–260. Warrendale, PA: Society of Automotive Engineers

Camacho DLA, Nightingale RW, Robinette JJ et al (1997) Experimental flexibility measurements for the development of a computational head-neck model validated for near-vertex head impact. In: Proceedings from the 41st Stapp car crash conference. Lake Buena Vista, FL. 41:473–486. Warrendale, PA: Society of Automotive Engineers, Inc

Cattell S, Filtzer DL (1965) Pseudosubluxation and other normal variations in the cervical spine in children. J Bone Joint Surg 47A(7):1295–1309

Chancey VC, Nightingale RW, Van Ee CA et al (2003) Improved estimation of human neck tensile tolerance: reducing the range of reported tolerance using anthropometrically correct muscles and optimized physiologic initial conditions. Stapp Car Crash J 47:135–153

Ching RP, Nuckley DJ, Hertsted SM, Eck MP, Mann FA, Sun EA (2001) Tensile mechanics of the developing cervical spine. Stapp Car Crash J 45:329–336

Clarke EC, Appleyard RC, Bilston LE (2007) Immature sheep spines are more flexible than mature spines: an in vitro biomechanical study. Spine 32:2970

Clasper JC, Pailthorpe CA (1995) Delayed diagnosis of an odontoid process fracture in an infant. Inj Intl J Care Inj 26:281–282

Dalmotas DJ, German A, Hendrick BE et al (1995) Airbag deployments: the Canadian experience. J Trauma 38:476–481

Dejeammes M, Tarriere C, Thomas C et al (1984) Exploration of biomechanical data towards a better evaluation of tolerance for children involved in automotive accidents. SAE Tech Paper #840530:427–441

Dibb AT, Nightingale RW, Luck JF, Chancey VC, Fronheiser LE, Myers BS (2009) Tension and combined tension-extension structural response and tolerance properties of the human male ligamentous cervical spine. J Biomech Eng 131(8):081008-1-081008-11

Duncan JM (1874) Laboratory note: on the tensile strength of the fresh adult fœtus. Br Med J 2:763

Eleraky MA, Theodore N, Adams M et al (2000) Pediatric cervical spine injuries: report of 102 cases and review of the literature. J Neurosurg 92:12–17

Eppinger R, Sun E, Bandak F et al (1999) Development of improved injury criteria for the assessment of advanced automotive restraint systems – II. Supplement to NHTSA Docket 9. Washington, D.C.: NHTSA

Ewald FC (1971) Fracture of the odontoid process in a seventeen-month-old infant treated with a halo. J Bone Joint Surg 53:1636–1640

Fesmire FM, Luten RC (1989) The pediatric cervical spine: developmental anatomy and clinical aspects. J Emerg Med 7:133–142

Fuchs S, Barthel MJ, Flannery AM, Christoffel KK (1989) Cervical spine fractures sustained by young children in forward-facing car seats. Pediatrics 84(2):348–354

Giguere JF, St-Vil D, Turmel A et al (1998) Airbags and children: a spectrum of c-spine injuries. J Pediatr Surg 33:811–816

Givens TG, Polley KA, Smith GF, Hardin WD Jr (1996) Pediatric cervical spine injury: a three-year experience. J Trauma 41(2):310–314

Graham JD, Goldie SJ, Segui-Gomez M et al (1998) Reducing risks to children in vehicles with passenger airbags. Pediatrics 102:1–7

Hadley MN, Zabramski JM, Browner CM, Rekate H, Sonntag VKH (1988) Pediatric spinal trauma: review of 122 cases of spinal cord and vertebral column injuries. J Neurosurg 68:18–24

Hamilton M, Myles S (1992a) Pediatric spinal injury: review of 174 hospital admissions. J Neurosurg 77:700–704

Hamilton MG, Myles ST (1992b) Pediatric spinal injury: review of 61 deaths. J Neurosurg 77:705–708

Hartemann F, Thomas C, Henry C et al (1977) Belted or not belted: the only difference between two matched samples of 200 car occupants. In: Proceedings from the 21st Stapp car crash conference, New Orleans, LA. 21:95–150. Warrendale, PA: Society of Automotive Engineers, Inc

Heilman CB, Riesenburger RI (2001) Simultaneous noncontiguous cervical spine injuries in a pediatric patient: case report. Neurosurgery 49:1017–1021

Henrys P, Lyne ED, Lifton C (1977) Clinical review of cervical spine injuries in children. Clin Orthop Relat Res 129:172–176

Hilker CE, Yoganandan N, Pintar FA et al (2002) Experimental determination of adult and pediatric neck scale factors. Stapp Car Crash J 46:323–351

Huelke DF, Mendelsohn RA, States JD (1978) Cervical fractures and fracture-dislocations sustained without head impact. J Trauma 18:533–538

Huelke DF, O'Day J, Mendelsohn RA (1981) Cervical injuries suffered in automobile crashes. J Neurosurg 54:316–322

Huelke DF, Mackay GM, Morris A (1992) Car crashes and non-head impact cervical spine injuries in infants and children. SAE Tech Paper 920562

Huelke DF, Mackay GM, Morris A et al (1993) A review of cervical fractures and fracture-dislocations without head impacts sustained by restrained occupants. Accid Anal Prev 25:731–743

Huelke DF, Mackay GM, Morris A (1995) Vertebral column injuries and lap–shoulder belts. J Trauma 8:547–556

Irwin AL, Mertz HJ (1997) Biomechanical bases for the CRABI and Hybrid III child dummies. SAE Tech Paper 106:3551–3562

Kallieris D, Barz J, Schmidt G et al (1976) Comparison between child cadavers and child dummy by using child restraint systems in simulated collisions. In: Proceedings from the 20th Stapp car crash conference. Dearborn, MI. 20:511–542. Warrendale, PA: Society of Automotive Engineers, Inc

Kasai T, Ikata T, Katoh S et al (1996) Growth of the cervical spine with special reference to its lordosis and mobility. Spine 21:2067

Keller J, Mosdal C (1990) Traumatic odontoid epiphysiolysis in an infant fixed in a child's car seat. Inj Br J Accid Surg 21:191–192

Kleinberger M, Sun E, Eppinger R et al (1998) Development of improved injury criteria for the assessment of advanced automotive restraint systems. Washington, D.C.: NHTSA

Klinich KD, Saul RA, Auguste G et al (1996) Techniques for developing child dummy protection reference values. NHTSA Event Report, Docket Submission, #74-14. Washington, D.C.: NHTSA

Kokoska ER, Keller MS, Rallo MC et al (2001) Characteristics of pediatric cervical spine injuries. J Pediatr Surg 36:100–105

Kuhns LR (1998) Imaging of spinal trauma in children: an atlas and text. BC Decker Inc, Hamilton

Kumaresan S, Yoganandan N, Pintar FA (1997) Age-specific pediatric cervical spine biomechanical responses: three-dimensional nonlinear finite element models. In: Proceedings from the 41st Stapp car crash conference. Lake Buena Vista, FL. 41:31–61. Warrendale, PA: Society of Automotive Engineers, Inc

Kumaresan S, Yoganandan N, Pintar F et al (1998) One, three and six year old pediatric cervical spine finite element models. In: Yoganandan N (ed) Frontiers in head and neck trauma. IOS Press, Amsterdam

Kumaresan S, Yoganandan N, Pintar FA et al (2000a) Biomechanical study of pediatric human cervical spine: a finite element approach. J Biomech Eng 122:60–71

Kumaresan S, Yoganandan N, Pintar FA et al (2000b) Biomechanics of pediatric cervical spine: compression, flexion and extension responses. Traffic Inj Prev 2:87–101

Luck JF, Nightingale RW, Loyd AM et al (2008) Tensile mechanical properties of the perinatal and pediatric PMHS osteoligamentous cervical spine. Stapp Car Crash J 52:107–134

Luck JF, Bass CRB, Owen SJ, Nightingale RW (2012) An apparatus for tensile and bending tests of perinatal, neonatal, pediatric and adult cadaver osteoligamentous cervical spines. J Biomech 45(2):386–389

Lui TN, Lee ST, Wong CW (1996) C1–C2 fracture–dislocations in children and adolescents. J Trauma 40:408–411

McGowan D, Voo L, Liu Y (1993) Distraction failure of the immature spine. Presented at ASME summer annual, Breckenridge, CO, pp 24–25

McGrory BJ, Klassen RA, Chao EYS et al (1993) Acute fractures and dislocations of the cervical spine in children and adolescents. J Bone Joint Surg 75:988–995

Melvin J (1995) Injury assessment reference values for the CRABI 6-month infant dummy in a rear-facing infant restraint with airbag deployment. SAE Tech Paper #950872:1–12

Mertz HJ, Weber DA (1982) Interpretations of the impact responses of a 3-year-old child dummy relative to child injury potential. In: Proceedings from the ninth international technical conference on experimental safety vehicles, Kyoto, Japan 9:368–376

Mertz HJ, Driscoll GD, Lenox JB et al (1982) Responses of animals exposed to deployment of various passenger inflatable restraint system concepts for a variety of collision severities and animal positions. In: Proceedings from the ninth international technical conference on experimental safety vehicles, Kyoto, Japan 9:352–367

Mikawa Y, Watanabe R, Yamano Y et al (1987) Fracture through a synchondrosis of the anterior arch of the atlas. J Bone Joint Surg 69:483

Myers BS, Winkelstein BA (1995) Epidemiology, classification, mechanism, and tolerance of human cervical spine injuries. Biomed Eng 23:307–410

NHTSA SCI (2009) Counts of frontal air bag related fatalities and seriously injured persons. DOT HS 811 104. Washington, D.C.: NHTSA. http://www-nrd.nhtsa.dot.gov/Pubs/811104.pdf. Last accessed 12 Jun 2012

Nightingale RW, Winkelstein BA, Knaub K et al (2002) Comparative bending strengths and structural properties of the upper and lower cervical spine. J Biomech 35:725–732

Nightingale RW, Chancey VC, Ottaviano D et al (2007) Flexion and extension structural properties and strengths for male cervical spine segments. J Biomech 40:535–542

Nitecki S, Moir CR (1994) Predictive factors of the outcome of traumatic cervical spine fracture in children. J Pediatr Surg 29:1409–1411

Nuckley DJ, Ching RP (2006) Developmental biomechanics of the cervical spine: tension and compression. J Biomech 39:3045–3054

Nuckley DJ, Hertsted SM, Ku GS, Eck MP, Ching RP (2002) Compressive tolerance of the maturing cervical spine. Stapp Car Crash J 46:431–440

Nuckley DJ, Yliniemi EM, Cohen AM, Harrington RM, Ching RP (2005a) Compressive mechanics of the maturing human spine. In: Proceedings from the ASB 29th annual meeting. Cleveland, OH

Nuckley DJ, Hertsted SM, Eck MP et al (2005b) Effect of displacement rate on the tensile mechanics of pediatric cervical functional spinal units. J Biomech 38:2266–2275

Nuckley DJ, Linders DR, Ching RP (2007) Human cervical spine mechanics across the maturation spectrum. In: Proceedings from the ASB 31st annual meeting. Palo Alto, CA

Odent T, Langlais J, Glorion C et al (1999) Fractures of the odontoid process: a report of 15 cases in children younger than 6 years. J Pediatr Orthop 19:51–54

Ogden JA (1984a) Radiology of postnatal skeletal development: XI the first cervical vertebra. Skeletal Radiol 12:12–20

Ogden JA (1984b) Radiology of postnatal skeletal development: XII the second cervical vertebra. Skeletal Radiol 12:169–177

Orenstein JB, Klein BL, Gotschall CS et al (1994) Age and outcome in pediatric cervical spine injury: 11-year experience. Pediatr Emerg Care 10:132–137

Ouyang J, Zhu Q, Zhao W et al (2005) Biomechanical assessment of the pediatric cervical spine under bending and tensile loading. Spine 30:716–723

Peacock A (1952) Observations on the postnatal structure of the intervertebral disc in man. J Anat 86:162

Pintar F, Mayer R, Yoganandan N et al (2000) Child neck strength characteristics using an animal model SAE Tech Paper #2000-01-SC06:44

Platzer P, Jaindl M, Thalhammer G, Dittrich S, Kutscha-Lissberg F, Vecsei V, Gaebler C (2007) Cervical spine injuries in pediatric patients. J Trauma 62(2):389–396

Prasad P, Daniel RP (1984) A biomechanical analysis of head, neck and torso injuries to child surrogates due to sudden torso acceleration. SAE Tech Paper 841656:25–40

Rachesky I, Boyce WT, Duncan B, Bjelland J, Sibley B (1987) Clinical prediction of cervical spine injuries in children. Am J Dis Child 141:199–201

Roche C, Carty H (2001) Spinal trauma in children. Pediatr Radiol 31:677–700

Sanderson SP, Houten JK (2002) Fracture through the C2 synchondrosis in a young child. Pediatr Neurosurg 36:277–278

Saveika JA, Thorogood C (2006) Airbag-mediated pediatric atlanto-occipital dislocation. Am J Phys Med Rehabil 85:1007–1010

Scheuer L, Black S (2004) The juvenile skeleton. Elsevier Academic Press, New York

Schippers N, Konings P, Hassler W et al (1996) Typical and atypical fractures of the odontoid process in young children: report of two cases and a review of the literature. Acta Neurochir 138:524–530

Seimon LP (1977) Fracture of the odontoid process in young children. J Bone Joint Surg 59:943–948

Shaw BA, Murphy KM (1999) Displaced odontoid fracture in a 9-month-old child. Am J Emerg Med 17:73–75

Smith JT, Skinner SR, Shonnard NH (1993) Persistant synchondrosis of the second cervical ertebra simulating a hangman's fracture in a child. J Bone Joint Surg 75:1228–1230

Swischuk LE (1977) Anterior displacement of C2 in children: physiologic or pathologic? A helpful differentiating line. Pediatr Radiol 122:759–763

Swischuk LE (1998) Five month old in a motor vehicle accident. Pediatr Emerg Care 14:299–301

Taylor JR (1975) Growth of human intervertebral discs and vertebral bodies. J Anat 120:49

Thakar C, Harish S, Saifuddin A et al (2005) Displaced fracture through the anterior atlantal synchondrosis. Skeletal Radiol 34:547–549

Van Ee CA, Chasse AL, Myers BS (1998) The effect of postmortem time and freezer storage on the mechanical properties of skeletal muscle. In: Proceedings from the 42nd Stapp car crash conference, Tempe, AZ. 42:169–178. Warrendale, PA: Society of Automotive Engineers, Inc

Van Ee CA, Nightingale RW, Camacho DL (2000) Tensile properties of the human muscular and ligamentous cervical spine. Stapp Car Crash J 44:85–102

Viano DC (1995) Restraint effectiveness, availability and use in fatal crashes: implications to injury control. J Trauma 38:538–546

Weiss MH, Kaufman B (1973) Hangman's fracture in an infant. Am J Dis Child 126:268–269

Wigren A, Amici F (1973) Traumatic atlanto-axial dislocation without neurological disorder. J Bone Joint Surg 55:642–644

Wismans J, Maltha J, Melvin JW (1979) Child restraint evaluation by experimental and mathematical simulation. In: Proceedings from the 23rd Stapp car crash conference, San Diego, CA. 23:383–415. Warrendale, PA: Society of Automotive Engineers, Inc

Yamada H (1970) Strength of biological materials. Williams and Wilkins Co, Baltimore, MD

Yoganandan N, Kumaresan S, Pintar FA (2001) Biomechanics of the cervical spine part 2. cervical spine soft tissue responses and biomechanical modeling. Clin Biomech 16:1–27

Yoganandan N, Kumaresan S, Pintar F et al (2002) Pediatric biomechanics. In: Naham A, Melvin JW (eds) Accidental injury: biomechanics and prevention. Springer-Verlag, New York

Chapter 6
Experimental Injury Biomechanics of the Pediatric Thorax and Abdomen

Richard Kent, Johan Ivarsson, and Matthew R. Maltese

Background and Introduction

Motor vehicle crashes are the leading cause of death and injury for children in the United States. Pediatric anthropomorphic test devices (ATD) and computational models are important tools for the evaluation and optimization of automotive restraint systems for child occupants. The thorax interacts with the restraints within the vehicle, and any thoracic model must mimic this interaction in a biofidelic manner to ensure that restraint designs protect humans as intended. To define thoracic biofidelity for adults, Kroell et al. (1974) conducted blunt impacts to the thoraces of adult postmortem human subjects (PMHS), which have formed the basis for biofidelity standards for modern adult ATD thoraces (Mertz et al. 1989). The paucity of pediatric PMHS for impact research led to the development of pediatric model biofidelity requirements through scaling. Geometric scale factors and elastic moduli of skull and long bone have been used to scale the adult thoracic biofidelity responses to the 3-, 6-, and 10-year-old child (Irwin and Mertz 1997; Mertz et al. 2001; van Ratingen et al. 1997). There is currently a need for data that apply to the child without scaling, both for validation of scaling methods used in the past and to confirm the validity of the specifications currently used to develop models of the child.

R. Kent, Ph.D. (✉)
Center for Applied Biomechanics, University of Virginia, 4040 Lewis & Clark Drive, Charlottesville, VA 22911, USA
e-mail: rwk3c@eservices.virginia.edu

J. Ivarsson, Ph.D.
Biomechanics Practice, Exponent, Inc., 23445 N 19th Avenue, Phoenix, AZ 85027, USA

M.R. Maltese, M.S.
Children's Hospital of Philadelphia, 34th and Civic Center Blvd, Suite 1150, Philadelphia, PA 19104, USA

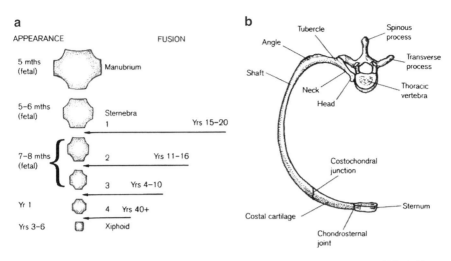

Fig. 6.1 Development of the sternum (a) and the typical thoracic rib (b) (Scheuer and Black 2000)

This chapter summarizes the available experimental data characterizing the injury and impact biomechanics of the pediatric torso. The torso includes the thorax and the abdomen. The thorax comprises the rib cage and the underlying soft tissue organs and is inferiorly bounded by the diaphragm. The rib cage consists of 12 pairs of ribs, 12 thoracic vertebrae, and the sternum. The underlying soft tissue organs are the lungs, heart, great vessels leaving and entering the heart, thymus gland, esophagus, lower portion of the trachea, thoracic duct, thoracic lymph nodes, and nerves passing into and through the thorax. The abdomen is bounded superiorly by the diaphragm and inferiorly by the pelvis and the muscles attached to it. There are two types of organs in the abdomen: the solid organs, which include the liver, spleen, kidneys, pancreas, adrenal glands, and ovaries and the hollow organs, which include the stomach, intestines, urinary bladder, and uterus. In general, the ribs are characteristic of other bones in that they are relatively soft and flexible in the child, becoming progressively more ossified and stiffer during development. The relative sizes of the thoracic and abdominal organs also change during development, with the heart and liver being relatively larger and the lungs being relatively smaller in children under age 3 than in the adult (Franklyn et al. 2007). Inspection of the torso maturation process suggests a complexity to scaling between adults and children due to the amount of time required for bones in the rib cage to appear and fuse. The sternum consists of six main bones—the manubrium superiorly, followed by sternebrae 1 through 4 and the xiphoid process (Fig. 6.1). The fourth sternebra appears at age 12 months, while the xiphoid process appears at 3–6 years. Fusing between sternebrae begins at age 4 years and continues through age 20 years. The sternum as a whole descends with respect to spine from birth up until age 2–3 years, causing the ribs to angle downward when viewed laterally, and the shaft of the rib to show signs of axial twist deformation (Scheuer and Black 2000). The costal cartilage also

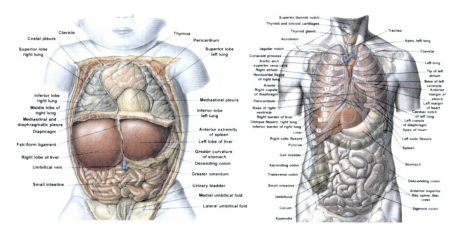

Fig. 6.2 Torso anatomy of a neonate (*left*) and an adult (*right*). Note the changes in relative organ size and apportionment of trunk volume between the thorax and the abdomen (Franklyn et al. 2007)

calcifies with age, likely influencing its flexibility. These human tissue changes during maturation likely influence the impact response of children in ways that are not considered in scaling.

The abdomen also makes up a larger proportion of the torso in young (neonatal) children (Fig. 6.2). This compresses the thorax, which is bounded by ribs that are oriented more horizontally (closer to the transverse plane) (Hammer and Eber 2005), and results in the ribs and sternum being positioned more superiorly than they are in the adult (Sinclair 1978). This anatomical state results in less bony protection around the abdomen, particularly the solid organs, of the child. The cross-sectional aspect ratio of the thorax and abdomen also change during pediatric development, becoming broader and flatter in the adult. This is the result of the pelvis becoming wider starting at about age 2, which allows the abdominal contents to descend.

The types of data contained in this chapter include subfailure, material-level (continuum) descriptions of tissue behavior, failure data at both tissue and structure levels, subinjury structural behaviors, and kinematic and injury descriptions of the torso in whole-body loading environments, including thoracic spine kinematics and their effect on whole-body behavior. Only data collected in a laboratory setting using a biological model of the human child are included in this summary. The chapter therefore includes only studies that employed human volunteers, pediatric human cadavers and tissue, or animal models or tissue intended to represent the human child. The many papers, reports, and standards that rely upon scaling of adult data, either human or animal, to represent the child have not been included, nor have studies that employ nonbiological models of the child, such as ATDs or computational models. Reconstruction of injurious events, such as car crashes or falls (e.g., Snyder 1963, 1969; Smith et al. 1975), has likewise been excluded. Finally, biomechanical studies that have limited relevance to an understanding of acute

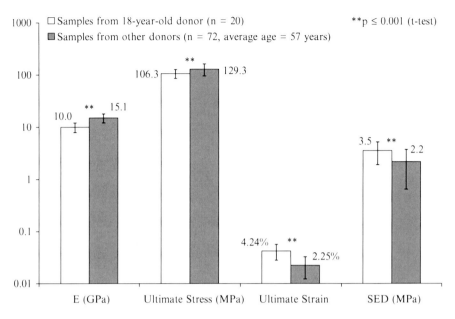

Fig. 6.3 Mean (±1 SD) dynamic tensile behavior of rib cortical bone samples from an 18-year-old donor compared to samples from older donors (data from Kemper et al. 2005) showing substantial reduction in ultimate strain and strain energy density (SED) with aging. Note logarithmic scale on ordinate

traumatic injury, such as biomechanical studies of gait or other daily activities (e.g., Stolze et al. 1997) or response to vibration (e.g., Giacomin 2005) have been excluded from this chapter. The definition of "pediatric" has intentionally been left ambiguous since, for example, the degree to which tissue from an 18-year-old differs from the tissue of an older population of donors (e.g., in Kemper et al. 2005) (Fig. 6.3) is valuable information given the general lack of pediatric data. It is also important to appreciate that the juvenile development of animals is different from humans. Developmental trends in juvenile animals have informed studies of human development and injury tolerance (e.g., this approach has been successfully applied to the cervical spine, as described in Chap. 5), but the interpretation of animal-based relationships between development and injury biomechanics can be difficult. Correlations have been made based on osseous development (e.g., Ching et al. 2001), size, anatomical arrangement, and other developmental comparisons (e.g., Kent et al. 2006; Franklyn et al. 2007), but the lack of a direct correlation between juvenile animal data and pediatric human data must be considered in any interpretation of experimental data using an animal model of the child.

The scope of the chapter does not cover all aspects of pediatric biomechanics. For example, the significance of the torso and thoracic spine in the biomechanics of locomotion and athletic performance are not included, nor are biomechanical aspects of tissue differentiation and remodeling. Rather, the focus is on injury tolerance and the subinjurious biomechanical behaviors of the human child during loading scenarios that occur in potentially injurious environments, especially automotive collisions. The reader will find that our knowledge of pediatric thoracic injury biomechanics is limited, but the amount and quality of work performed in recent years and in ongoing studies is encouraging. Novel approaches are being employed to address the significant experimental challenges hampering the study of pediatric injury, and the voids in knowledge are steadily being filled.

The chapter is organized by length scale from smallest to largest. The summary starts with tissue properties of various tissues of the torso, then proceeds to subfailure and failure properties of substructures such as whole organs and individual bones. The properties of the thorax and abdomen as composite structures are then presented followed finally by whole-body behavior and the kinematics of the thoracic spine.

Tissue Properties

A comprehensive review of the available literature on pediatric material properties, including those for the tissues of the thorax, has recently been published by Franklyn et al. (2007). Few studies have focused specifically on changes in thoracoabdominal tissue mechanical behavior during pediatric development. Much of the data available are isolated samples from larger studies that, often by happenstance, included a pediatric donor. In these cases, it is useful to compare that sample with the larger set of adult (often senescent) data, but care should be taken in the interpretation of these isolated data points since a single sample may not represent the pediatric population, particularly since pediatric tissue may often be harvested from cadaveric donors that do not represent the overall physical condition of living children. While this is obviously a limitation in any study of cadaveric tissue, this issue may be particularly important for the pediatric population since some pathologies leading to death may have a more pronounced effect on developing tissue than on the more stable tissues of the adult.

Cortical Bone

Changes in cortical bone during pediatric development have been described in a number of studies, most of which involve tissue samples from the extremities or from the skull. Those studies are summarized in Chap. 3 and were also discussed by Franklyn et al. (2007). Bending tests using pediatric rib samples have been published, as discussed later, but material-level tests of samples of cortical bone from the pediatric torso have been reported only by Kemper et al. (2005). That study did not focus specifically on pediatric properties, but the test population included an 18-year-old donor. Dynamic tensile tests at a nominal strain rate of 0.5/s were performed on cortical bone coupons harvested from anterior, lateral, and posterior locations on ribs 1 through 12 of this male cadaver. The entire test population included males and females up to age 67. Elastic modulus, E, yield stress and strain, ultimate stress and strain, and strain energy density (SED) up to failure were determined. The samples harvested from the 18-year-old donor had significantly ($p \leq 0.001$) lower elastic modulus and ultimate stress than the other (older) subjects, and significantly greater ultimate strain and strain energy density at failure. The mean elastic modulus was approximately 50 % greater in the pool of older samples (15.1 MPa vs. 10.0 MPa) and the ultimate stress was approximately 20 % greater (129.3 MPa vs. 106.3 MPa). The most important difference was in ultimate strain, with a mean value almost a factor of two greater for the younger samples (4.24 % vs. 2.25 %). The mean strain energy density at failure was approximately 60 % greater in the pool of 18-year-old samples (3.5 MPa vs. 2.2 MPa). As illustrated in Fig. 6.4, even a middle-aged donor exhibited remarkably less strain at failure than the 18-year-old.

Trabecular Bone

Weaver and Chalmers (1966) reported compression tests of trabecular bone samples harvested from the third lumbar vertebrae of human cadavers, six of which were under the age of 20 years. The compressive strength of those pediatric samples was found to be similar to the group of middle-aged cadavers (around age 40), but lower than the group of samples of age 20–30. The compressive strength of the pediatric samples ranged from 2.4 to 3.7 MPa (see Franklyn et al. 2007). Mosekilde et al. (1985, 1987) and Mosekilde and Mosekilde (1986) reported compression tests on trabecular bone samples from several thoracic and lumbar vertebrae of human cadavers, including two of age 15 and 19 years. Maximum compressive stress, energy absorption, and other properties were measured. The maximum compressive stress ranged from 4.2 to 4.7 MPa for trabecular bone cored samples and was 5.85 MPa for a whole vertebra tested from the 15-year-old. Energy absorption ranged from 0.64 to 0.85 mJ/mm^2. Kleinberger et al. (1998) summarized data originally collected by McElhaney et al. (1970) and concluded that the ultimate compressive strength of human lumbar vertebral cancellous bone increased throughout pediatric development and decreased with age above approximately 20 years (Fig. 6.5).

Costal Cartilage

Yamada (1970) reported the results of tensile tests of samples of costal cartilages from 28 subjects grouped into age categories. The ultimate tensile strength was reported to be 0.46 MPa and invariant with age up to age 19, while the ultimate percent elongation was reported to decrease slightly during pediatric development, from 31.2 % in the samples of age 0–9 years to 28.2 % in the samples of age

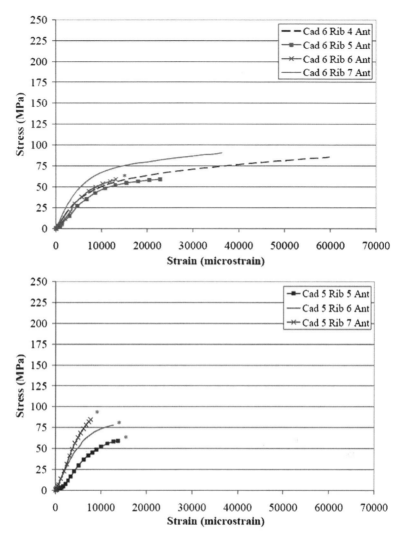

Fig. 6.4 Dynamic tensile behavior of cortical rib bone from an 18-year-old donor (*top*) compared to a 46-year-old (*bottom*) illustrating significant reduction in failure strain and SED, but less pronounced reduction in elastic modulus or yield stress (Kemper et al. 2005)

10–19 years. Oyen et al. (2006) presented the results of spherical indentation experiments on juvenile porcine costal cartilage samples and developed a linear viscoelastic model of the tissue using a correspondence technique developed by Mattice et al. (2006). That study found that the cartilage midsubstance exhibited a time-zero elastic modulus, E_0, of 5.34 ± 1.59 MPa, a long-time modulus, E_0, of 0.73 ± 0.27 MPa and that the assumption of linear viscoelasticity was reasonable over the strain magnitudes considered. The authors also concluded that those modulus values were similar to those seen for adult humans.

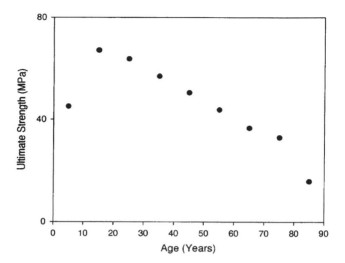

Fig. 6.5 Ultimate compressive strength of human lumbar vertebral trabecular bone as a function of age showing increase during development followed by reduction through senscense (from McElhaney et al. 1970 as reported by Kleinberger et al. 1998)

Lung

The lungs are not solid organs but instead consist mainly of tiny air-containing sacs called alveoli, which number approximately 150 millions per lung. The mechanical behavior of lung tissue is complex, exhibiting pronounced effects from nonlinearity, rate sensitivity, surface tension, transpulmonary pressure (P_{tp}), and heterogeneity, and the degree to which it has been characterized even for adults is limited (e.g., Vawter et al. 1978; Vawter 1980). For the child, the state of knowledge is even more limited, with linear mechanical properties applicable for studying respiration being the only information currently available (Lai-Fook et al. 1976; Hajji et al. 1979; Stamenovic and Yager 1988; Mansell et al. 1989; Tepper et al. 1999; Okazawa et al. 1999; Lai-Fook and Hyatt 2000). The nonlinear and rate-sensitive behaviors likely to dominate the behavior during potentially injurious loading remain unknown for the child.

Mansell et al. (1989) used a punch indentation technique to investigate the mechanical properties of lungs removed from piglets aged 6–12 h ($N=6$), 3–5 days ($N=6$), 25–30 days ($N=5$), and 80–85 days ($N=5$). For the three older groups of animals, the bulk modulus increased approximately linearly with increasing P_{tp}. In the 3- to 5-day-old group, the slope of this linear relationship was 6.8, which was significantly greater than the corresponding slopes in the 25- to 30-day-old group (slope 4.9, $p<0.001$) and in the 80- to 85-day-old group (slope 5.1, $p<0.005$). The bulk modulus in the three groups varied from approximately 10 cmH$_2$O at a P_{tp} of 2 cmH$_2$O to 100–145 cmH$_2$O at a P_{tp} of 20 cmH$_2$O. The bulk modulus of the lungs from the 6- to 12-h-old piglets was lower (approximately 8 cmH$_2$O at a P_{tp} of 2–80 cmH$_2$O at a P_{tp} of 20 cmH$_2$O) and did not increase linearly with increasing P_{tp}.

Values of the shear modulus were similar for the three older groups at most transpulmonary pressures between 5 and 20 cmH$_2$O and increased approximately linearly with increasing P_{tp} with a slope of 0.6. However, the shear modulus at $P_{tp} = 5$ cmH$_2$O increased progressively with postnatal age. The measured shear modulus at transpulmonary pressures between 5 and 20 cmH$_2$O in the three older groups were in the range of 4.5–12.5 cmH$_2$O. Shear modulus in the newborn group varied more than in the older lungs and did not show a linear relationship with P_{tp}. Values were higher at $P_{tp} = 5$ cmH$_2$O ($p < 0.05$) and at $P_{tp} = 15$ cmH$_2$O ($p < 0.05$) in the newborn lungs compared with the 3- to 5-day-old lungs.

Except at $P_{tp} = 5$ cmH$_2$O, Poisson's ratio for the three older groups were virtually identical and increased slightly with P_{tp} (from approximately 0.32 at $P_{tp} = 5$ cmH$_2$O to 0.4 at $P_{tp} = 20$ cmH$_2$O). However, the Poisson's ratio at $P_{tp} = 5$ cmH$_2$O was greatly reduced to nearly zero for the oldest lungs. In the newborn lungs, Poisson's ratio was reduced at all P_{tp} levels and was close to zero at $P_{tp} = 5$ cmH$_2$O.

Lai-Fook and Hyatt (2000) assumed that the lungs could be approximated as perfectly elastic homogeneous continua and investigated the effects of age on bulk modulus, shear modulus and Poisson's ratio of human lungs at transpulmonary pressures of 4, 8, 12, and 16 cmH$_2$O. The 20 isolated lungs tested were obtained from human cadavers aged 10–78 years at the time of death. The bulk modulus was measured from incremental changes in P_{tp} and lung volume, whereas the shear modulus and Poisson's ratio were determined from indentation testing and the measured values of the bulk modulus. Linear regression demonstrated that the bulk modulus increased significantly ($p < 0.05$) with age at each of the four P_{tp} values. Corresponding analyses demonstrated that the shear modulus increased significantly with age at all tested P_{tp} values except at 4 cmH$_2$O, whereas the Poisson's ratio increased significantly with age at all tested P_{tp} values except at 16 cmH$_2$O. Plots of the linear functions are shown in Fig. 6.6.

Tepper et al. (1999) used techniques very similar to Lai-Fook and Hyatt (2000) to estimate the bulk modulus and the shear modulus of immature (3–4 weeks) and mature (6 months) rabbit lungs. For both the mature and immature lung specimens, the bulk modulus increased significantly with increasing P_{tp} (from approximately 10 cmH$_2$O at a P_{tp} of 2 cmH$_2$O to approximately 70 cmH$_2$O at a P_{tp} of 10 cmH$_2$O). However, there was no significant difference in bulk modulus between the immature and mature rabbit lungs. For both the immature and mature rabbit lungs, the shear modulus increased significantly with increasing P_{tp}. In contrast to the bulk modulus, the shear modulus demonstrated age dependence being significantly lower for the immature than for the mature specimens at transpulmonary pressures of 6 and 10 cmH$_2$O. The measured average shear modulus for the immature and mature lungs at transpulmonary pressures of 2, 4, 6, 8, and 10 cmH$_2$O are shown in Table 6.1 below.

During a normal respiration cycle, the transpulmonary pressure increases from approximately 5.4 to 9.5 cmH$_2$O during inspiration and then drops back to 5.4 cmH$_2$O during expiration. If it is assumed that the average transpulmonary pressure throughout the respiration cycle can be calculated as the mean of the lowest (5.4 cmH$_2$O) and highest pressure (9.5 cmH$_2$O), then the material properties at pressure values of 7.5–8 cmH$_2$O should be used in a model that does not incorporate pressure-varying properties for the lung.

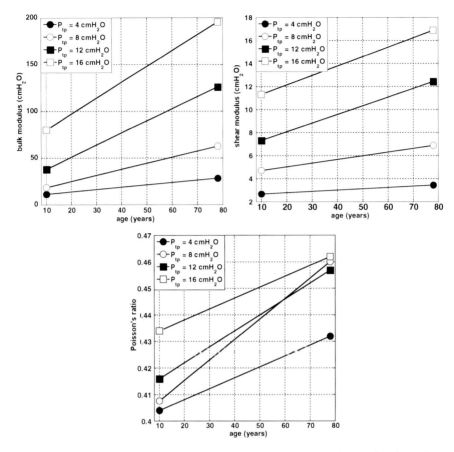

Fig. 6.6 Bulk modulus, shear modulus and Poisson's ratio of the human lung as functions of age at transpulmonary pressures of 4, 8, 12, and 16 cmH$_2$O as reported by Lai-Fook and Hyatt (2000)

Table 6.1 Shear modulus versus transpulmonary pressure for mature and immature rabbit lung as measured by Tepper et al. (1999)

P_{tp} (cmH$_2$O)	Shear modulus (cmH$_2$O)	
	Mature	Immature
2	3.3	3.0
4	4.0	3.3
6	4.5	3.8
8	5.3	4.7
10	6.2	5.0

Among the studies reviewed above, the one by Lai-Fook and Hyatt (2000) is the only one that used human specimens and those results therefore appear as the most relevant ones to use as a basis for estimating the mechanical properties of the lungs in children. However, as Lai-Fook and Hyatt emphasize, all but one subject from which the lung specimens were removed were older than 17 years, so the estimated age-dependent functions for the moduli and Poisson's ratio were dominated by the effects of aging through adulthood rather than with changes due to pediatric lung development. Consequently, using the linear age-dependent expressions derived by Lai-Fook and Hyatt for determining the moduli and Poisson's ratio of the lungs in the pediatric population might lead to incorrect estimates.

Considering instead the study on piglet lungs by Mansell et al. (1989), the lung bulk moduli at a transpulmonary pressure of 8 cmH_2O for the 3–5, 25–30, and 80–85 days old piglets from the graph in Fig. 6.2a in the original paper are 40.0, 36.7, and 27.1 cmH_2O, respectively. The shear modulus at the same transpulmonary pressure is, according to Fig. 6.3a in the original paper, 7.3 cmH_2O in the 3- to 5-day-old age group and 8.1 cmH_2O in the 25–30 and 80–85-day-old age groups, whereas the corresponding values of the Poisson's ratio are, according to Fig. 6.4a in the original paper, 0.39, 0.37, and 0.32. These ranges of bulk modulus (27.1–40.0 cmH_2O) and shear modulus (7.3–8.1 cmH_2O) are within the ranges of the bulk and shear moduli predicted by Lai-Fook and Hyatt's linear age-dependent equations valid for a transpulmonary pressure of 8 cmH_2O suggesting that human and pig lungs have similar mechanical properties. Consequently, it seems reasonable to assume that the mechanical properties of the human lung at a transpulmonary pressure of 8 cmH_2O can be approximated by the corresponding properties of the pig lung at the same transpulmonary pressure. Since the equivalent human lung ages of 3–5, 30–35, and 80–85-day-old piglets are not known, it is assumed that the bulk and shear moduli of the lung tissue in children of various ages can be approximated by the averages of the corresponding values from the three age groups of pigs.

In summary, then, it appears that there is reasonable evidence to suggest that lung changes substantially in the period shortly after birth, and that these changes are primarily related to the bulk or volumetric properties of the tissue rather than the deviatoric properties. Whether pediatric tissue well into childhood is substantially different than adult lung tissue is, however, not clear. Furthermore, it is important to realize the potential limitations in a linear elastic description of lung tissue. The behavior of lung tissue at injurious levels of strain during high-rate loading may not be adequately described with two material constants, and the mechanical parameters (shear modulus, bulk modulus, and Poisson's ratio) presented in these studies focused on respiratory behavior may be of limited utility for modeling the complexity of lung tissue, especially in an injurious loading regime.

The failure characteristics of pediatric lung are likewise poorly understood. Stitzel et al. (2005) published a correlation between first principal strain and area of contusion following a direct impact to exposed lung in an in vivo rat model. This study was not focused on pediatric lung, but in the absence of any data indicating a difference in the failure tolerance of pediatric and adult lung tissue, their value of 3.5 % strain corresponding to the area of contusion measured at 24 h post impact is a reasonable threshold, though subsequent work by the same group (Gayzik et al. 2007) indicates that this threshold may depend upon the strain rate. This is an important area for further study since lung injuries are the dominant thoracic injury in children.

Liver, Spleen, Kidney, and Hollow Organs

The literature on the material properties of the liver, spleen, and kidney of children is very sparse. Some failure properties of the pediatric solid organs have been summarized by Franklyn et al. (2007). It also appears that in some cases the material properties of pediatric and adult organ tissues (e.g., Mattice 2006 for the kidney, Stingl et al. 2002 for the liver and spleen) may not be substantially different.

Citing Ohara (1953), Yamada (1970) reported on the tensile properties of the liver parenchyma in rabbits. The ultimate tensile strength and elongation were said to be 23.5 ± 4.9 kPa and 46 ± 2.3 %, respectively.

Fazekas et al. (1971) investigated the pressure stability of $N=11$ intact human livers removed from cadavers aged 10–86 years (52.7 ± 24.2) at the time of death. The necessary force and pressure (mean±SD) for laceration (capsule laceration + slight parenchymal lesion) were 1.65 ± 0.63 kN and 154 ± 71.6 kPa, respectively. Multiple ruptures of the livers occurred at 3.14 ± 0.89 kN and 153 ± 53.0 kPa. The highest laceration force was recorded for the liver from the 10-year-old subject (the only immature specimen) that lacerated at a force 70 % higher than the average laceration force for the ten adult specimens. Fazekas et al.'s results were briefly summarized in English (Fazekas et al.'s article is in German) by Schmidt (1979).

Melvin et al. (1973) conducted 17 constant velocity compression tests on the livers from three Rhesus monkeys in vivo. The Rhesus subject was anesthetized whereupon the liver was surgically mobilized and laid onto a load cell while still being perfused by the living animal. The specimens were impacted by a flat plate with a cross-sectional area of 11.6 cm^2. Impact velocity and depth of penetration were varied in order to yield various types of injuries and severity levels. Tests were performed at constant speeds of 5, 250, and 500 cm/s. The maximum compressive strain level was varied from 40 to 75 % to produce injuries of various severities. The load deflection data from the tests were converted to stress–strain curves (stress was calculated as if deformation took place in uniaxial compression, i.e., by dividing the reaction force by the cross-sectional area of the impact plate). The results of the tests are shown in Table 6.2.

Based on the results of Melvin et al. (1973), Miller (2000a, b) derived a constitutive expression for the in vivo deformation of the liver under compression. The constitutive model was based on a hyperviscoelastic strain energy function W with time dependent constants, written in the form of a convolution integral. By adopting a number of assumptions and simplifications, Miller derived an expression for the only nonzero Lagrange stress and identified the material constants (see original article for details). A comparison between the predicted and experimentally obtained stress–strain curves results demonstrated that the theoretical model was very good in predicting the constitutive behavior of the liver for compression levels reaching 35 % and for loading velocities varying over two orders of magnitude (approximately 0.225–22.5 1/s). As an additional test of the model, Miller used the model to predict the values of strain energy density and compared with the corresponding values obtained by Melvin et al. (1973) for maximum stress. The predicted strain

Table 6.2 Results from Melvin et al.'s (1973) impact tests on rhesus monkey livers in vivo

Test velocity (cm/s)	Max stress (kPa)	Strain at max stress (%)	Max strain (%)	Strain energy density at max stress (J/cm^3)	Max strain energy density (J/cm^3)
5	145	58.5	58.5	0.026	0.026
	262	41.5	41.5	0.024	0.024
	228	55.0	55.0	0.042	0.042
	293	59.0	59.0	0.046	0.046
250	234	44.0	44.0	0.025	0.025
	297	43.0	50.0	0.042	0.054
	318	57.0	57.0	0.055	0.055
	310	53.0	60.0	0.060	0.079
	310	42.0	62.1	0.037	0.095
	328	47.0	59.0	0.055	0.089
	374	56.0	75.0	0.077	0.134
	407	50.0	66.0	0.055	0.109
500	265	41.6	48.6	0.037	0.053
	234	36.0	49.0	0.041	0.069
	524	48.8	56.0	0.075	0.106
	455	47.0	58.0	0.061	0.100
	662	51.0	57.0	0.111	0.148

energy densities for the low, medium, and high speed tests were 66.1, 36.1, and 2.1 % higher, respectively, than the corresponding experimentally obtained values. Despite the difference between the theoretical and experimental strain energy densities in case of the low and medium speed tests, the agreement has to be considered as acceptable since maximum stress occurred at average strain levels of 48.9–53.5 %. Miller concluded by stating that before the finite element simulation of liver deformation in vivo can be carried out, further research is needed to determine the way this organ is attached to the body such that proper boundary conditions can be formulated mathematically.

Wang et al. (1992) subjected blocks of hepatic tissue removed from pigs to indentation testing and obtained force relaxation curves at various indentation depths. By use of classical elastic theory and nonlinear regression, Wang and coworkers estimated the elastic modulus and Poisson's ratio of the pig hepatic tissue. The measured elastic moduli (mean±SD) of specimens 1, 2, and 3 were 2.55±0.03, 2.38±0.04, and 2.18±0.04 kPa, respectively, while the Poisson's ratios measured for the same three specimens were 0.399±0.008, 0.402±0.011, and 0.403±0.014.

Uehara (1995) (as cited in Tamura et al. 2002) subjected porcine liver to tensile testing and found that the ultimate tensile strength increased with increasing loading rate while the maximum extension ratio decreased with increasing loading rate. The ultimate tensile strength at a loading rate of 8.33 mm/s was reported to be 205±28 kPa.

Seki and Iwamoto (1998) investigated the mechanical strength of liver samples removed from five 6-month-old swine. The organ specimen to be tested was positioned on a table that moved upwards towards the flat end of a circular detection bar of diameter 5 mm that contacted the organ. The table moved with a constant velocity of 20 mm/min. The force loaded on the organ from the bar was recorded and the breaking force was defined as the first peak or notch on the load–displacement curve. The breaking stress was calculated as the breaking force divided by the cross-sectional area of the detection bar (19.6 mm^2). The breaking stress was determined in two groups of liver specimens; specimens with intact serous membranes and specimens with the serous membranes removed. Five specimens with and five specimens without serous membranes were obtained and tested from each liver. The breaking stresses (mean ± SEM) of the specimens with and without serous membrane were 451 ± 29 and 353 ± 29 kPa, respectively.

Liu and Bilston (2000) investigated the rheological properties of bovine liver tissue in the linear viscoelastic domain. Liver samples in the form of disks of 30 mm diameter and thicknesses of 2 mm were cut from the surface near the edge of the liver lobule and tested in shear strain sweep oscillation, shear stress–relaxation and shear oscillation. A shear strain sweep oscillation test conducted at the constant frequency of 1 Hz gave initial values of the storage shear modulus and loss shear modulus of approximately 2 and 0.8 kPa, respectively. However, the sampled dynamic properties started to change significantly at strain amplitude of approximately 0.2 %, which consequently corresponds, to the linear strain limit of the tissue (larger shear strains result in nonlinear tissue behavior). Data from two relaxation tests for which the average strain amplitude was 0.13 % demonstrated that the relaxation modulus had not reached a constant plateau value after ~250 s of relaxation, which was said to indicate that liver tissue is a fluid-like or very soft viscoelastic material. Data from two oscillatory shear tests in which the frequency was varied from 0.01 to 20 Hz and the average strain was 0.11 % demonstrated that that the storage shear modulus increased with frequency from approximately 1 kPa at 0.01 Hz to 6 kPa at 11 Hz and then decreased slightly with frequency up to 20 Hz, whereas the loss shear modulus increased with frequency from approximately 0.7 kPa at 0.01 Hz to 1 kPa at 1 Hz and then decreased with frequency.

Ishihara et al. (2000) (as cited in Tamura et al. 2002) subjected fresh porcine liver with and without the visceral and diaphragmatic covering to tensional testing and reported that these membranous tissues affect the tensile properties of the liver. The ultimate compressive stress of the liver with capsule at a loading rate of 10 mm/min was reported to be 123 kPa.

Yeh et al. (2002) measured the elastic modulus of fresh human liver samples from 19 patients (mean age 55 years, 18 adults and 1 child aged 10 months) with various liver diseases. Cubic liver specimens of approximate size 1 cm^3 were subjected to cyclic compression-relaxation. The elastic modulus for each specimen was computed from the slope of the stress–strain curve, where the strain was defined as the ratio of total deformation to the initial height. The samples used in the tests were graded according to their degree of fibrosis. Liver fibrosis was scored from 0 to 4 according to the following definition: 0 = no fibrosis, 1 = portal fibrosis without septa, 2 = septa fibrosis, 3 = numerous septa without cirrhosis, and 4 = cirrhosis. In cirrhotic livers, the widths of fibrotic bands varied and the score was further divided so that a score of 4 represented thin fibrotic bands and a score of 5 was for thick fibrotic bands. The measured elastic modulus was presented as functions of fibrosis score and preload strain and is shown in Table 6.3.

Table 6.3 Results from Yeh et al.'s (2002) tests with human liver samples

Fibrosis score	N	Modulus at 5 % preload strain (mean ± SD) (Pa)	Modulus at 10 % preload strain (mean ± SD) (Pa)	Modulus at 15 % preload strain (mean ± SD) (Pa)
0	3	640 ± 80	1,080 ± 160	2,000 ± 630
1	1	1,260	2,770	6,800
2	2	1,050 ± 240	1,790 ± 670	5,400 ± 3,470
3	3	870 ± 170	1,820 ± 290	5,390 ± 2,780
4	7	1,110 ± 170	2,370 ± 360	6,270 ± 1,020
5	3	1,650 ± 110	4,930 ± 930	19,980 ± 6,950

Table 6.4 Average (mean ± SD) ultimate stress, ultimate strain, and SED (strain energy density) to failure as obtained by Tamura et al. (2002) in their compressive tests on samples of porcine liver

	0.005/s	0.05/s	0.5/s
Ultimate stress (kPa)	123.4 ± 31.4	135.2 ± 17.0	162.5 ± 27.5
Ultimate strain	0.432 ± 0.026	0.420 ± 0.038	0.438 ± 0.040
SED to failure (kJ/m^3)	18.19 ± 5.11	19.73 ± 4.58	23.83 ± 5.83

Tamura et al. (2002) conducted compressive relaxation and failure tests on samples of porcine liver (20 mm × 20 mm × 10 mm). The relaxation tests were conducted with the specimens placed on an aluminum plate in an environmental chamber of constant temperature (36 °C) filled with physiological saline solution. A total of ten liver samples were compressed a distance of 4 mm at a constant rate of 60 mm/s and held in the deformed state for 120 s. The failure tests were conducted at three different compressive loading rates, 0.05, 0.5, and 5.0 mm/s, which approximately corresponded to strain rates of 0.005, 0.05 and 0.5/s. Prior to compression to failure, all samples were preconditioned for five cycles. The rate and amplitude of the preloading cycles were 0.5 mm/s and 20 % sample strain, respectively. The specimens were assumed to be incompressible which allowed calculation of instantaneous cross-sectional area of the samples by dividing the constant specimen volume by the instantaneous specimen height. The instantaneous cross-sectional area was then used to determine true specimen stress. Compressive failure was defined as the point on the force–displacement curve at which the force dropped abruptly. From the results of the compressive relaxation and failure tests, Tamura and coworkers determined both the failure properties as well as a QLV model. The ultimate stress, ultimate strain, and SED (strain energy density) to failure obtained at the different loading rates are shown in Table 6.4.

Liu and Bilston (2002) and Nasseri et al. (2002) measured the viscoelastic properties of bovine liver and porcine kidney in shear. While neither of these studies described the tissue as juvenile, the ages of the animals are preadult (Franklyn et al. 2007). Both of these studies identified 0.2 % strain as the linear viscoelastic limit. The kidney model was developed within this linear regime using oscillation and stress–relaxation experiments. A multimode upper convected Maxwell model with variable viscosities and time constants, with a Mooney hyperelastic function, both multiplied by a nonlinear damping function was found to model the experimental data well with a single set of parameters. The liver model was a nonlinear viscoelastic differential model previously developed for brain tissue (see Bilston et al. 2001) and fit to experimental data up to 5.5 % Cauchy strain.

Mattice (2006) performed spherical indentation stress–relaxation experiments on kidney parenchyma from 13 female swine aged from 28 to 162 days. He found that any trends with age in either linear or quasilinear viscoelastic models of the tissue were small compared to the inherent scatter in the data. The mean values of the short-time and long-time elastic moduli were 17.66 and 4.21 kPa in low-strain tests, but pronounced nonlinearity in the elastic response was found as the strain increased.

Fazekas et al. (1972) investigated the pressure stability of $N=9$ intact human spleens removed from cadavers aged 10–83 years (46.1±21.9) at the time of death. The necessary force and pressure (mean±SD) for laceration (capsule laceration + slight parenchymal lesion) were 432 ± 212 N and 82.4 ± 42.2 kPa, respectively. Multiple ruptures of the spleens occurred at 859 ± 235 N and 152 ± 88.3 kPa. The highest laceration force was recorded for the spleen from the 10-year-old subject (the only immature specimen), which lacerated at a force 82 % higher than the average laceration force for the eight adult specimens.

Uehara (1995) (as cited in Tamura et al. 2002) reported tensile tests on porcine spleen and found that the ultimate tensile strength increased with increasing loading rate, while the maximum extension ratio decreased with increasing loading rate. The ultimate tensile strength at a loading rate of 8.33 mm/s was reported to be 101 ± 21 kPa.

Seki and Iwamoto (1998) investigated the mechanical strength of spleen samples removed from five 6-month-old swine. The organ specimen to be tested was positioned on a table that moved upwards towards the flat end of a circular detection bar of diameter 5 mm that contacted the organ. The table moved with a constant velocity of 20 mm/min. The force loaded on the organ from the bar was recorded and the breaking force was defined as the first peak or notch on the load–displacement curve. The breaking stress was calculated as the breaking force divided by the cross-sectional area of the detection bar (19.6 mm^2). The breaking stress was determined in two groups of spleen specimens; specimens with intact serous membranes and specimens with the serous membranes removed. Five specimens with and five specimens without serous membranes were obtained and tested from each spleen. The breaking stresses (mean±SEM) of the specimens with and without serous membrane were 882 ± 88 and 176 ± 10 kPa, respectively.

Ishihara et al. (2000) (as cited in Tamura et al. 2002) subjected fresh porcine spleen with and without the visceral and diaphragmatic covering to tensional testing and reported that these membranous tissues affect the tensile properties of the spleen. The ultimate compressive stress of the spleen with capsule at a loading rate of 10 mm/min was reported to be 100 kPa.

Tamura et al. (2002) conducted compressive relaxation and failure tests on samples of porcine spleen (20 mm×20 mm×10 mm). The relaxation tests were conducted with the specimens placed on an aluminum plate in an environmental chamber of constant temperature (36 °C) filled with physiological saline solution. A total of nine spleen samples were compressed a distance of 7 mm at a constant rate of 90 mm/s and held in the deformed state for 120 s. The failure tests were conducted at three different compressive loading rates, 0.05, 0.5, and 5.0 mm/s, which approximately

Table 6.5 Average (mean ± SD) ultimate stress, ultimate strain, and SED (strain energy density) to failure as obtained by Tamura et al. (2002) in their compressive tests on samples of porcine spleen

	0.005/s	0.05/s	0.5/s
Ultimate stress (kPa)	107.5 ± 12.2	114.6 ± 12.5	146.3 ± 14.9
Ultimate strain	0.825 ± 0.041	0.809 ± 0.040	0.834 ± 0.012
SED to failure (kJ/m^3)	22.51 ± 3.55	24.46 ± 3.17	32.55 ± 3.44

Table 6.6 Stomach failure data adapted from Yamada (1970) as summarized by Franklyn et al. (2007)

	Lunar month	Age (years)	
Property	6th–10th	0–9	10–19
Longitudinal tensile breaking load per unit width (g/mm)	42	132	172
Transverse tensile breaking load per unit width (g/mm)	34.2	113	142
Longitudinal ultimate tensile strength (g/mm^2)	29.2	47	#
Transverse ultimate tensile strength (g/mm^2)	23	40	#
Longitudinal ultimate percentage elongation	65.8	106	
Transverse ultimate percentage elongation	90.8	124	

The 10–19 years data was amlgamated with the 10–39 years data (so the age group was 10–39 years) for this parameter, so not included.

corresponded to strain rates of 0.005, 0.05, and 0.5/s. Prior to compression to failure, all samples were preconditioned for five cycles. The rate and amplitude of the preloading cycles were 0.5 mm/s and 40 % sample strain, respectively. The specimens were assumed to be incompressible which allowed calculation of instantaneous cross-sectional area of the samples by dividing the constant specimen volume by the instantaneous specimen height. The instantaneous cross-sectional area was then used to determine true specimen stress. Compressive failure was defined as the point on the force–displacement curve at which the force dropped abruptly. From the results of the compressive relaxation and failure tests, Tamura and coworkers determined both the failure properties as well as a QLV model (see original article for details on this model). The ultimate stress, ultimate strain, and SED (strain energy density) to failure obtained at the different loading rates are shown in Table 6.5.

Yamada's (1970) studies of stomach wall tissue included some neonatal and pediatric tissue. Longitudinal and transverse failure characteristics were summarized throughout development by Franklyn et al. (2007), who averaged data from the greater curvature, the lesser curvature, and the pyloric region (Table 6.6). The data indicated an increase in the strength of the stomach wall during development, with substantial increases occurring between neonatal and older pediatric samples.

Heijnsdijk et al. (2003) included some pediatric swine (body mass 19–45 kg) in their study of bowel tolerance to pinching from a laparoscopic bowel grasper. They found that the tolerance of the pediatric swine bowel was not different than the adult human bowel, and that the large bowel had a higher force tolerance (13.5 ± 3.7 N) than the small bowel (11.0 ± 2.5 N). The pinch interface was two 1.5-mm diameter hemispheres, but the experimental protocols and data reporting are insufficient to estimate a reliable stress from the measured force values.

Vasculature and Cardiac Tissue

The failure properties of the thoracoabdominal vasculature have been relatively well studied. Yamada (1970) included the tensile breaking load (TBL), ultimate percent elongation, ultimate tensile stress from samples of thoracic and abdominal aorta and coronary artery. These data were summarized by Franklyn et al. (2007) in a table that is reproduced here (Table 6.7). The tensile breaking load was found to increase with age, but this appears to be primarily a structural phenomenon related to the increase in size with age. Khamin (1975) reported significant decreases with aging (after age 11 years) in the ultimate tensile stress of the aortic wall (Fig. 6.7). Subfailure behavior of pediatric vasculature has not been well quantified. Iannuzzi et al. (2004) estimated the elastic modulus of abdominal aorta in normal children to be 0.24 ± 0.10 Pa for very small deformations and Roach and Burton (1959) found the elastic modulus of the internal iliac artery to be relatively lower in samples from donors younger than age 20 compared to those from older donors, and to increase with increasing stretch. As with most tissues, however, the severe limitations imposed by the assumptions related to linearity and elasticity of these tissues make it difficult to draw firm conclusions about the true relationships between development and mechanical behavior of the vasculature.

Substructure Properties

Ribs

In his review on biomechanical data of children, Stürtz (1980) cites one publication of interest for the tolerance of the child thorax to blunt impact—the dissertation of Theis (1975). Theis (as cited in Stürtz 1980) reported on results from quasistatic 3-point bending tests on ribs 6 and 7 from children under the age of 14 years ($n=3$) and from dynamic bending tests on ribs 6 and 7 from children under the age of 14 years ($n=17$, bending rate=2.9 m/s), persons 15–64 years of age ($n=65$, bending rate=2.0 m/s), and persons older than 65 years ($n=31$, bending rate=1.7 m/s). The average breaking load for the three pediatric ribs that were loaded quasistatically was 240 N, whereas the average dynamic breaking loads for the ribs from the children under the age of 14 years, persons 15–64 years of age, and persons older than 65 years were 234, 372, and 173 N, respectively. To give the reader some perspective on the dynamic rates used by Theis (1975), Kent et al. (2004) found that the peak sternal displacement rate experienced by restrained adult cadavers in 48-km/h frontal sled tests was approximately 1.0–1.5 m/s.

An unpublished study by Pfefferle et al. (2007) reported 3-point bending tests of 48 rib specimens from 13 pediatric human subjects (age 1 day to 6 years). Force and applied displacement were measured. The ratio of force to displacement during a linear regime of the loading curve was defined as stiffness and beam theory was used

Table 6.7 Mechanical properties of the vasculature (adapted from Franklyn et al. 2007)

Artery	Region/direction	Property	Age in years				Author
			Fetal[a]	Infant/child	Adolescence	Young adult	n	
Aorta	Thoracic	TBL longitudinal (g/mm)	35–80	116.5±14	137±7	146.5±7		Yamada
Aorta	Thoracic	TBL transverse (g/mm)	58–105	157±19	204±15	213±15		Yamada
Aorta	Thoracic	UPE longitudinal (%)	109	67.5±4	98±5	90±8		Yamada
Aorta	Thoracic	UPE transverse (%)	99	69.5±5	104.5±8	101±8		Yamada
Aorta	Thoracic	UTS longitudinal (MPa)	0.78–1.14	1.02±0.08	0.92±0.06	0.87±0.06		Yamada
Aorta	Thoracic	UTS transverse (MPa)	1.27–1.46	1.40±0.10	1.31±0.10	1.22±0.05		Yamada
Aorta	Thoracic	UTS longitudinal (MPa)			2.44±0.14	1.79±0.04	109	Khamin
Aorta	Thoracic	UTS transverse (MPa)			4.40±0.12	3.57±0.06	107	Khamin
Aorta	Abdominal	TBL longitudinal (g/mm)	NA	163	163	132		Yamada
Aorta	Abdominal	TBL transverse (g/mm)	NA	154	154	154		Yamada
Aorta	Abdominal	UPE longitudinal (%)	NA	102	102	94		Yamada
Aorta	Abdominal	UPE transverse (%)	NA	103	103	114		Yamada
Aorta	Abdominal	UTS longitudinal (MPa)	NA	1.59	1.59	1.10		Yamada
Aorta	Abdominal	UTS transverse (MPa)	NA	1.49	1.49	1.19		Yamada
Aorta	Abdominal	UTS longitudinal (MPa)			2.03±0.14	1.79±0.11	62	Khamin
Aorta	Abdominal	UTS transverse (MPa)			3.54±0.24	2.86±0.16	67	Khamin
Internal iliac	C	E at zero wall stretch		0.7×10^3 Pa[a]	1.0×10^3 Pa[a]		8	Roach and Burton
Internal iliac	C	E at 100 mm wall stretch[b]		1×10^4 Pa[a]	2.5×10^4 Pa[a]		8	Roach and Burton
Internal iliac	C	E at high wall stretch		4×10^4 Pa[a]	4.5×10^4 Pa[a]		8	Roach and Burton
Internal iliac	NA	Collapse pressure		35/10/45 mmHg	20 mmHg[a]		4	Roach and Burton
Coronary	L	TBL per unit width			85±3.1 g/mm			Yamada
Coronary	L	UTS			140±3.0 g/mm^2			Yamada
Coronary	L	UPE			99±2.4 %			Yamada

Note: Yamada, Khamin, and Roach and Burton used slightly different age ranges: Yamada used 0–9, 10–19, and so forth, while Khamin and Roach and Burton used 0–10, 11–20, and so forth. For Yamada's data, the thoracic aorta is the average of the ascending and descending parts
TBL tensile breaking load, UPE ultimate percent elongation, UTS ultimate tensile strength, E Young's modulus
[a]Graphs only were presented in the paper, so data points were measured from the graphs
[b]Physiological pressure

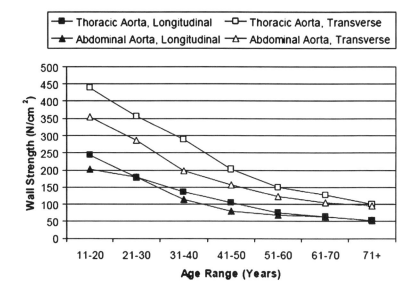

Fig. 6.7 Aorta wall strength decreasing with age (from Khamin 1975 as summarized in Franklyn et al. 2007)

to estimate the elastic modulus and yield stress, and these characteristics of the ribs were reported as a function of biological age and a "developmental age" based on typical growth charts. Rotation of the curved rib segment on the test fixture during loading was noted, however, so there is uncertainty in the interpretation of this study's results. The "developmental age" was found to correlate quite well with increasing rib stiffness, presumably since this is a structural measurement and developmentally older children would be expected to have larger ribs, but this conclusion is confounded by the authors' apparent use of variable span distances. The average stiffness in the neonatal samples was approximately 7 N/mm and an exponential regression to the entire dataset indicated a stiffness of approximately 52 N/mm at a developmental age of 6 years. The yield force was also highly correlated with developmental age, ranging from approximately 10 N at age 1 day to 90 N at age 6 years. The material property estimates from these experiments are not considered reliable due to the limitations in the beam theory approximation, but were estimated to be below approximately 5 GPa for all ages, which is consistent with the rib cortical bone samples from the 18-year-old subject tested by Kemper et al. (2005).

The available literature is consistent in the finding that the bending strength of pediatric ribs is lower than the corresponding strength of ribs from persons aged 15–64 years, and thus, the pediatric rib cage should provide a less stiff shell for protecting the underlying soft tissue organs than does the adult rib cage. The increased failure strain of younger ribs, however, means that underlying organ injury from the fractured surfaces of ribs should be less common in children. The net effect of these competing aspects of thoracic injury risk in children relative to adults is not known.

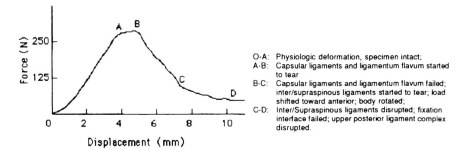

Fig. 6.8 Force–displacement curve for an infant lumbar spine segment (L3-4-5) tested in distraction (tension) to failure (modified from McGowan et al. 1993)

Spinal Components

Motion segments from the lumbar spines of 42 human cadavers (age 18–74 years) were tested in combined compression and bending by Adams and Dolan (1991) and by Adams et al. (1994). Each motion segment contained two adjacent vertebrae with the associated disk and intervertebral ligaments. These studies were focused on the identification of injurious conditions for manual lifting tasks, and as such are limited in their utility to predict acute injury in an impact event, but they are useful in that the peak bending moment did not exhibit a clear trend with age. Individual specimen variability dominated any age effect. The peak bending moment was in the range of 60–70 Nm for the two subjects under age 20.

McGowan et al. (1993) reported distraction data for a single lumbar spine segment (L3-4-5) harvested from an 8-h-old infant human cadaver. The third and fifth lumbar vertebrae were potted and a tensile force was applied to failure at a displacement rate of 1.25 mm/s. Initial failure was noted at the capsular ligaments and the ligamentum flavum at L3-4, followed by tearing of the intrasupraspinous ligaments at L4-5, after which a pronounced drop in force was measured with continued displacement. The functional elements started to fail at a tensile force of approximately 260 N, and the stiffness during the linear portion of the initial loading curve was approximately 94 N/mm (Fig. 6.8).

Thorax as a Structure

Very few published studies have addressed the issues of the strength and blunt impact tolerance of the intact pediatric thorax. The only results from dynamic tests on intact child cadaver thoraces are those from Kent et al. (2009) and from Ouyang et al. (2006) (Fig. 6.9). The Ouyang series involved nine cadavers of children aged 2–12 years that were subjected to frontal thoracic impact by means of an impactor weighing 2.5 kg (cadavers of 2–3 year-old children) or 3.5 kg (cadavers of 5–12 year-old children)

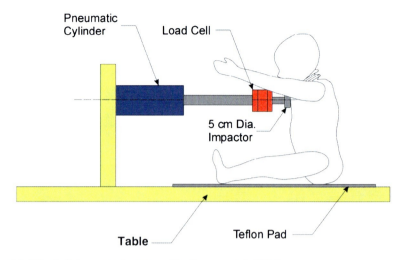

Fig. 6.9 Blunt hub impact test setup used by Ouyang et al. (2006)

Table 6.8 Test matrix from Ouyang et al. (2006) blunt hub thoracic impact tests

Subject	Impactor mass (kg)	Impactor diam. (cm)	Impact speed (m/s)	Age/gender	Wt. (kg)	Ht. (cm)	Cause of death
1	2.5	5.0	5.9	2/F	13.0	97.0	Fluorocetamide poisoning
2	2.5	5.0	5.9	2.5/M	10.5	87.5	Cerebral edema
3	2.5	5.0	6.0	3/M	13.5	93.0	Brain tumor
4	2.5	5.0	6.0	3/F	10.5	85.0	Heart disease
5	3.5	7.5	6.4	5/M	13.0	101.0	Cerebritis
6	3.5	7.5	5.9	6/M	16.5	108.0	Leukemia
7	3.5	7.5	6.5	6/M	20.0	109.0	Mediterranean anemia
8	3.5	7.5	6.0	7.5/F	17.0	117.0	Acute urinemia
9	3.5	7.5	6.2	12/F	29.0	142.5	Leukemia

at impact speeds ranging from 5.9 to 6.5 m/s (Table 6.8). Following impact, six of the impacted subjects had pneumothorax, one had a bleeding thymus gland and two had no signs of thoracic injury (Table 6.9). None of the subjects sustained a rib fracture. Time histories of the viscous criterion, acceleration, force and deformation were recorded in all the tests. The subjects were divided into two age cohorts and force–deformation cross-plots were generated and reduced to impact response corridors (Fig. 6.10). The authors reported good qualitative agreement between the thoracic force and deflection data of the older pediatric PMHS cohort and a scaled biofidelity corridor developed from adult PMHS impacts, but quantitatively the pediatric PMHS cohort exhibited a lower force–deflection onset rate, higher maximum forces, and lower maximum deflection than the scaled corridor.

Parent (2008) questioned the accuracy of the Ouyang chest deformation measurements and reported corrected force–deflection response corridors. For six of the

Table 6.9 Test results from Ouyang et al. (2006) blunt hub thoracic impact tests

Subject	Max. ext. chest def. (mm, % of undeformed)[a]	Max. viscous criterion (m/s)	Max. accel. at T4 (g)	Max. applied force (N)	Observed trauma
1	57.7, 46.1 %	1.9	73.2	740	Rt. pneumothorax
2	45.0, 40.9 %	2.7	63.9	790	Rt. pneumothorax
3	44.8, 40.7 %	2.1	63.9	825	None
4	44.9, 35.9 %	1.5	124.8	750	None
5	55.7, 46.4 %	2.1	62.4	900	Pneumothorax
6	44.5, 37.1 %	0.7	71.2	1,200	Pneumothorax
7	31.5, 24.2 %	1.0	91.7	1,560	Rt. pneumothorax
8	48.7, 40.6 %	1.6	64.0	900	Bleeding thymus gland
9	72.3, 48.2 %	4.5	35.6	1,130	Lt. pneumothorax

[a]Posterior displacement of the anterior chest relative to the posterior as measured with a chestband wrapped around the external thorax at the level of the fourth thoracic vertebra. Maximum deformation and maximum viscous criterion were very near the mid-sternum in all tests

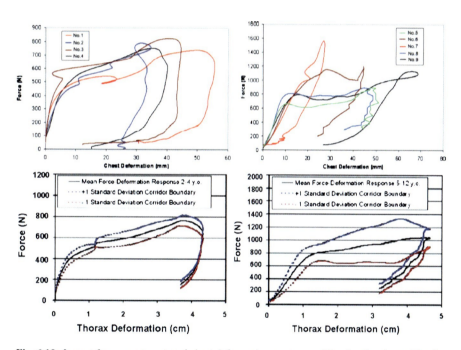

Fig. 6.10 Impact force versus external chest deformation as measured by chestbands or video for the young cadaver cohort (age 2–4 years, *upper left*) and the older cohort (age 5–12 years, *upper right*) of Ouyang et al. (2006), and corridors developed by those authors to describe the behavior (*lower plots*). (Note that the plots do not have the same range on either axis)

nine subjects tested by Ouyang, chestbands with between 18 and 26 active gauges at one inch resolution were used to measure external chest deflection. Such low-resolution chestbands may underestimate the true deflection (Hagedorn and Pritz 1991; Shaw et al. 1999). Parent (2008) confirmed the apparent under-estimation by Ouyang by assessing the chestband measurements in identical blunt hub impact tests using the Hybrid III 3-year-old child dummy. The chest deflection measured with chestbands in those dummy tests was less than that measured using more reliable methods in similar tests of lesser severity reported by Saul et al. (1998). Parent furthermore notes that the corridors presented by Ouyang (and reproduced in Fig. 6.10) were developed from unscaled data and were defined by the average force plus or minus one standard deviation for each point of average deflection. Such a corridor development technique can result in regions of unreasonably small corridor width (cf. Lessley et al. 2004), such as in the young cohort corridor at about 12 mm and in the unloading portions of both corridors.

As a result of these experimental and analytical limitations in the Ouyang study, Parent (2008) reinterpreted the pediatric PMHS blunt thoracic impact response data and corridor development. The external deflection for each subject was determined by measuring the distance between the impactor and the spine of the subject using high-speed video captured during the event. The deflection was then synchronized with the mass-compensated force by minimizing the error between the impactor motion measured from the high-speed video and the impactor motion calculated by double integration of the impactor accelerometer. This process was intended to correct for the potential error of misjudging impact time if using high-speed video alone to determine external deflection and thereby to improve synchronization of the deflection and force time-histories. For each of the two impactor masses, a corridor was developed by first calculating corridors for the force and deflection individually. This was done by taking the average of the individual responses and offsetting the result by ±1 SD. The standard deviation was determined by fitting a smoothing spline to the standard deviation at each point in time to remove local discontinuities. The union of the four possible combinations of standard deviation-offset force and deflection time-histories defined the force–deflection response corridors (Fig. 6.11). Parent also excluded the data from one 5-year-old cadaver in his determination of the older-cohort corridor, as its mass placed it below even the average 3-year-old.

Kent et al. (2009) reported a series of diagonal-belt and distributed loading tabletop experiments on a 7-year-old female cadaver (Fig. 6.12). Repeated, subinjurious loading was performed at three frequencies (0.5, 1, 4 Hz) and with a high-rate (~2 m/s peak) "ramp-hold" waveform. These experiments used the same loading environment used earlier (Kent et al. 2004) to characterize the thoracic response of 15 adult cadavers and thus are useful as an assessment of methods for scaling adult thoracic response data for application to the child. Scaling of the adult responses using any of four published techniques did not successfully predict the pediatric behavior. All of the scaling techniques intrinsically reduce the stiffness of the adult response, when in reality the pediatric subject was as stiff as, or slightly stiffer than, published adult corridors (Fig. 6.13).

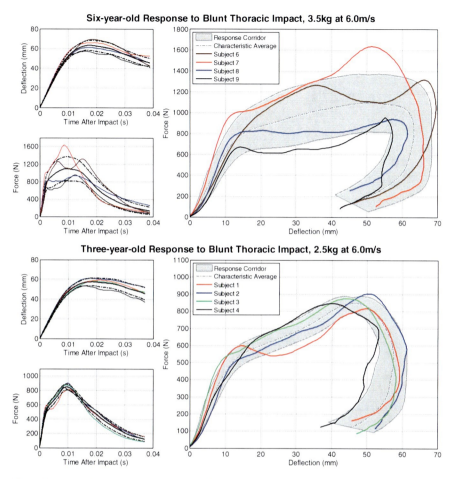

Fig. 6.11 Applied force, external sternal deflection, and response corridors developed by Parent (2008) after reanalyzing data from the tests of Ouyang (2008, personal communication)

The pronounced differences in thoracic injury mechanics between children and adults are reflected in the results of Ouyang et al. (2006), who reported that six of the nine subjects received pneumothoraces in the absence of rib fracture. Clinically, children frequently receive pulmonary injury in the absence of rib fracture (which is rare in adults, Holmes et al. 2002), and when rib fractures are present in children they are associated with severe trauma (Garcia et al. 1990). This has implications for development of thoracic injury criteria for children. Criteria and thresholds developed from adult PMHS impact data will predict best the injuries present in those experiments—rib fractures with the occasional soft tissue injury (Eppinger et al. 1999; Kallieris et al. 1981; Kroell et al. 1974), but thoracic injury criteria for children must do the opposite—predict primarily soft tissue injury, with the occasional rib fracture. It is not clear that the current methods of developing thoracic

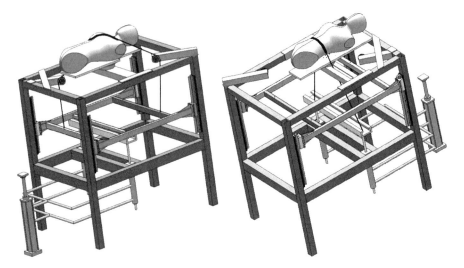

Fig. 6.12 Table-top loading configuration used by Kent et al. (2009) to characterize the thorax (*left*) and abdomen (*right*) of a 6-year-old cadaver

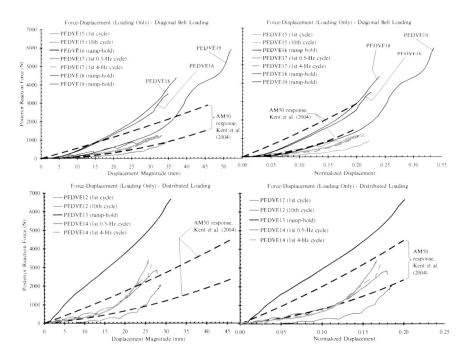

Fig. 6.13 Response of a 6-year-old cadaver to diagonal belt (*top*) and distributed (*bottom*) loading compared to adult corridors. Note that the pediatric subject is at least as stiff as the adult corridors. (Figure modified from Kent et al. 2009)

injury criteria for children, which involve scaling adult-based criteria that are often based on rib fracture severity in adult cadaver tests, are a valid method for arriving at the appropriate injury criteria or thresholds for children.

Given the small number of pediatric PMHS subjects available for impact testing, alternative methods for obtaining pediatric thoracic force and deflection data are necessary to further quantify any potential differences between actual pediatric thoracic response and the scaled pediatric thoracic biofidelity corridors. In particular, cardiopulmonary resuscitation (CPR) involves the deflection of the sternum toward the spine in a manner not dissimilar to the PMHS experiments previously described. Clinical resuscitation guidelines prescribe targets for CPR chest compressions: 38–51 mm of chest compression for the adult, and one-third to one-half the anterior–posterior (AP) chest depth for the child (American Heart Association 2006). In terms of absolute deflection, the CPR compression target for the 6 year old child is 47–72 mm, assuming an AP chest depth of 143 mm (Irwin and Mertz 1997). For comparison, chest deflections in hub impact testing with PMHS range from approximately 50 to 70 mm in adults (Kroell et al. 1974) and from 31.5 to 73 mm in children (Ouyang et al. 2006).

Various electromechanical devices have been developed over the past three decades to improve the quality of CPR and to study the effect of the mechanics of thoracic compression on clinical outcomes (Gruben et al. 1990; Tsitlik et al. 1983; Vallis et al. 1979). Recently, a load cell and accelerometer force–deflection sensor (FDS) (Philips Healthcare, Andover, MA) has been integrated into a patient monitor–defibrillator to provide visual and audio feedback on the quality of CPR chest compressions (Maltese et al. 2008). The FDS is interposed between the hand(s) of the person administering CPR and the sternum of the patient, and it records force and acceleration data during each compression cycle. The FDS has become the standard of care for patients aged 8 years or older in the Emergency Department and Pediatric Intensive Care Unit at the Children's Hospital of Philadelphia, providing the opportunity to measure the force–deflection properties of children undergoing CPR. Maltese et al. (2008) reported on 18 subjects (11 females) aged 8–22 years who received CPR chest compressions. The relationship between applied force and sternal displacement for a typical CPR cycle to a child is shown in Fig. 6.14.

These data become most useful when compared with other similar studies of adults in CPR. In particular, Tsitlik et al. (1983) instrumented a mechanical chest compressor device used in clinical chest compressions on ten patients aged 17–72 receiving CPR chest compressions. In this study, only the maximum force and maximum deflection during a compression cycle were measured. However, with each compression cycle, the compressor device was designed to compress the chest at sequentially increasing deflections up to maximum limit, followed by sequentially decreasing deflections down to a minimum deflection, and then repeat the process. Since force and deflection data are recorded at the time of maximum compression when the chest velocity is zero, the viscous forces are minimal and thus the elastic force is directly measured. Thus, Tsitlik's measurement techniques are particularly advantageous to compare to the Maltese et al. 2008 study. Of note, subjects of similar age in the two studies have comparable elastic forces at 15 %

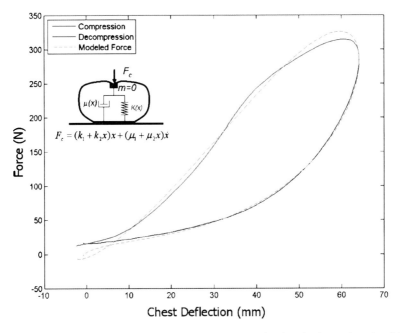

Fig. 6.14 Applied force versus sternal displacement (chest deflection) in the tenth cycle of CPR to a human child, and the model developed by Maltese et al. (2008) to describe this behavior

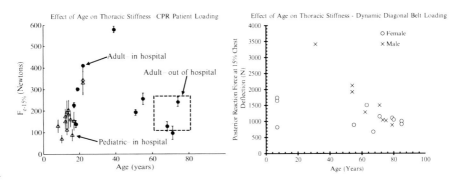

Fig. 6.15 Structural stiffness of the chest as a function of age in CPR patients (*left*) and in cadavers loaded dynamically by a diagonal belt (*right*). (Figure modified from Kent et al. 2009)

compression (Fig. 6.15). In addition, Arbogast et al. (2006a) analyzed force–deflection data from 91 adult subjects who received out-of-hospital CPR chest compressions using force-accelerometer technology similar to that which was used by Maltese et al. 2008. Individual subject age was not available in this study, though the interquartile range (IQR) and median age are provided. Elastic force at 15 % compression for this study is 188 N (SD 80) for subjects of median age 70 (IQR 61–81).

When all three CPR datasets are taken together, it can be seen that the elastic stiffness of chest climbs from age 8 to age 40, but falls after age 40 to force levels comparable to pediatric patients. Interestingly, the dynamic diagonal belt loading experiments of Kent et al. (2004, 2009) showed the same trend (Fig. 6.15). Aging bone shows a decrease in elastic modulus beyond adult middle age, and ribs alone in bending demonstrate decreased breaking strength with increased age. The CPR and belt data, although sparse (note that the single data point around age 30 in both studies is the primary basis for the trend), suggest a reduction in thoracic stiffness beyond middle age that supports the material and rib strength findings. Indeed, it is likely that the rise in stiffness to age 40 occurs in the absence of rib fractures, and reduction in stiffness after age 40 is related to the increased incidence of rib fracture. For example, in the young population of CPR patients, three of the patients reported herein received autopsies, and no rib or sternal fractures were found, whereas Tsitlik et al (1983) found rib fractures in the 55- and 68-year-old subjects in Fig. 6.15. In general, rib fractures are rare in children secondary to CPR (0–2 %), but are more common in adult (13–97 %) (Hoke and Chamberlain 2004). This pattern extends to the impact environment—no rib fractures were found in a recent series of blunt impacts into the thoraces of nine PMHS ages 2–12 years (Ouyang et al. 2006), yet the same type of test performed on adults produced rib fractures in 18 of 22 subjects (Kroell et al. 1974). This pattern is likely due to several aspects of pediatric development in the rib cage, including the large sections of cartilaginous tissue in the costal and intrasternebrae space, which ossify with maturity (Scheuer and Black 2000) and thus provide greater flexibility of the pediatric chest compared to that of the adult.

Abdomen as a Structure

The subjects reported by Ouyang et al. (2006) (see above) were also subjected to frontal, mid-sagittal abdominal impacts using the same pneumatic system described in their study of thoracic response [Ouyang J, 2008, personal communication]. The cadavers' heads were supported in an upright position using a cervical collar and tape. The impact location was selected as the position one-third of the distance from the umbilicus to the bottom of the sternum. However, if the impact location resulted in impactor contact closer than 1 cm from the bottom of the rib cage, the impact location was moved downward until there was a 1 cm distance from the bottom of the rib cage and the impactor.

The cadaver was positioned to limit the free stroke of the piston to 65 % of the measured abdominal depth at the umbilicus. This step was taken to limit the potential for spinal injuries from high velocity contact with the impactor. As with the thoracic impacts, a 2.5 kg, 5.0-cm-diameter impactor or a 3.5 kg, 7.5-cm-diameter impactor was used, depending on the age cohort to which the subject belonged. The impactor speed at impact was 6.3 ± 0.3 m/s, regardless of cohort. Abdominal testing on each test subject occurred before thoracic tests to minimize the possibility that structural instabilities caused by thoracic injuries would compromise the abdominal tests.

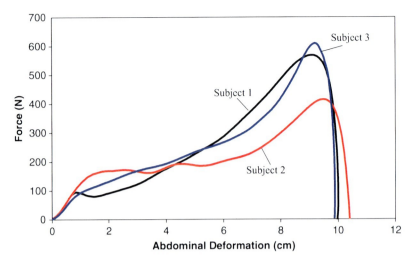

Fig. 6.16 Inertially compensated impactor force versus abdominal deformation in a series of nominally 6 m/s impact tests to a cohort of young (age 2–4 years) pediatric human cadavers (Ouyang J, 2008, personal communication)

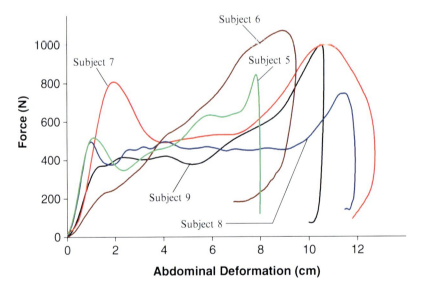

Fig. 6.17 Inertially compensated impactor force versus abdominal deformation in a series of nominally 6 m/s impact tests to a cohort of older (age 5–12 years) pediatric human cadavers (Ouyang J, 2008, personal communication)

Inertially compensated impactor force is cross-plotted against abdominal deformation for the younger cadaveric cohort (ages 2–4 years) in Fig. 6.16, and for the older cadaveric cohort (ages 5–12 years) in Fig. 6.17. The profiles for the young cohort show qualitatively and quantitatively similar behavior among the tests.

Table 6.10 Injury outcomes and maximum values measured in abdominal impact tests of Ouyang (2008, personal communication)

Subject	Abdominal injury	Peak abdominal compression (%)	Peak $V \times C$ (m/s)	Peak force (N)
4	Colon perforation, kidney hemorrhage	N/A	N/A	N/A
1	Stomach perforation	59	2.1	568.3
2	Small intestine perforation	83	3.2	414.1
3	Small intestine perforation, bile found outside gall bladder (no gross rupture found)	76	2.3	608.5
9	Transverse colon hemorrhage	57	1.7	993.9
7	None	67	2.6	997.0
8	Colon perforation, kidney hemorrhage	75	2.2	743.1
6	Kidney hemorrhage, liver hemorrhage	53	1.7	1,070.3
5	Retroperitoneal hematoma, transverse colon perforation	61	1.7	841.5

For the older cohort, there is evidence of possible contact between the rib cage of subject 7 and the impactor.

The abdominal injuries sustained by the pediatric cadavers subjected to abdominal impacts by Ouyang (2008, personal communication) ranged from none to perforations of the hollow organs and hemorrhaging of the solid organs (Table 6.10).

In 2006, Kent et al. reported the development of a porcine (sus scrofa domestica) (Fig. 6.18) model of the 6-year-old human's abdomen and defined the biomechanical response of this abdominal model. First, a detailed abdominal necropsy study was undertaken, which involved collecting a series of anthropometric measurements and organ masses on 25 swine, ranging in age from 14 to 429 days (4–101 kg mass). These were then compared to the corresponding human quantities to identify the best porcine representation of a 6-year-old human's abdomen. This was determined to be a pig of age 77 days, and whole-body mass of 21.4 kg. The subinjury, quasistatic response to belt loading of this porcine model compared well with pediatric human volunteer tests performed with a lap belt on the lower abdomen by Chamouard et al. (1996) (Fig. 6.19). A test fixture was designed to produce transverse, dynamic belt loading on the porcine abdomen (Fig. 6.20). A detailed review of field cases by Arbogast et al. (2006b) defined the specifications for the testing, and the following test variables were studied: loading location (upper/lower), penetration magnitude (23–68 % of initial abdominal depth), muscle tensing (yes/no), and belt penetration rate (quasistatic, dynamic 2.9–7.8 m/s). Dynamic tests were performed on 47 postmortem subjects. Belt tension and dorsal reaction force were cross-plotted with abdominal penetration to generate structural response corridors (Figs. 6.21 and 6.22). Subcutaneous stimulation of the anterior abdominal muscle wall stiffened the quasistatic response significantly, but was of negligible importance in the dynamic tests. The upper abdomen exhibited stiffer response quasistatically, and also was more sensitive to penetration rate, with stiffness increasing significantly over the range of dynamic rates tested here. In contrast, the lower abdomen was relatively rate insensitive.

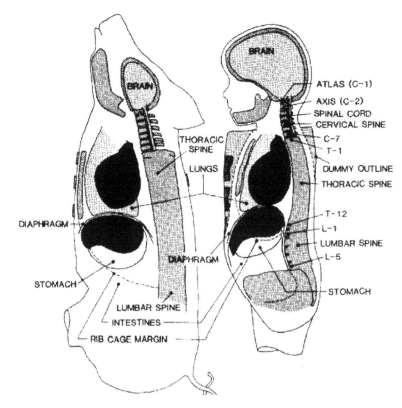

Fig. 6.18 Schematic depiction of the size and location of the major organs of the juvenile swine and the pediatric human (adapted from Prasad and Daniel 1984)

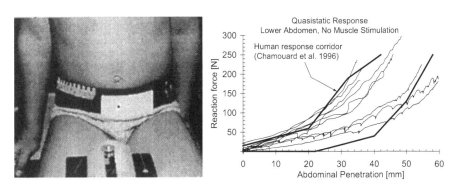

Fig. 6.19 Chamouard et al. (1996) pediatric volunteer abdominal belt loading method (*left*) and quasistatic response of pig model developed by Kent et al. (2006) compared to corridors based on human child volunteer tests (*right*)

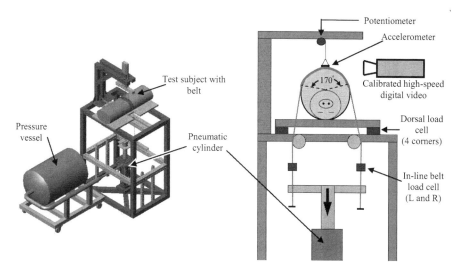

Fig. 6.20 Schematic drawings of the test fixture of Kent et al. (2006) showing major components and instrumentation

Kent et al. (2008) followed up their earlier study with a paper describing the injuries sustained by the tested swine and assessing various mechanical parameters as injury predictors. They found that the maximum measured values of penetration depth, $(d_n)_{max}$, and belt tension, $(F_b)_{max}$, were reasonable discriminators of injurious loading conditions and published probability curves relating the risk of moderate (MAIS 2+) and serious (MAIS 3+) injury to these and other parameters (Fig. 6.23). This study also found that the rate of penetration did not modulate the penetration tolerance over the range of rates from 2.9 to 7.3 m/s.

The 2009 Kent et al. study discussed above (Fig. 6.12) included a series of abdominal loading experiments on the 6-year-old cadaver using the same protocol used in the swine series rate bin 1 (lowest rate dynamic tests). That study also presented an analytical model to estimate the human's response in the two higher-rate bins. The pediatric cadaver exhibited abdominal response similar to the swine, including the degree of rate sensitivity. The upper abdomen of the PMHS was slightly stiffer than the porcine behavior, while the lower abdomen of the PMHS fit within the porcine corridor (Fig. 6.24).

In his review on biomechanical data of children, Stürtz (1980) considers two publications of interest for the biomechanical load limits of the overall abdomen in children: the studies by Kallieris et al. (1976) and Gögler et al. (1977). Kallieris et al. (1976) ran four frontal sled tests with four cadavers of children aged 2.5, 6, 6 and 11 years. The cadavers were restrained by means of a lap-belt (18 % elongation at 9.8 kN) put around a semi-cylindrically shaped safety table that supported the abdominal region during impact. Thus, the neck–head region and the upper part of the body remained free. The preimpact sled velocity was 40 km/h in the three tests with the older cadavers and 30 km/h in the test with the cadaver of the 2.5-year-old. In all four tests, the sled was brought to rest during approximately 70 ms of

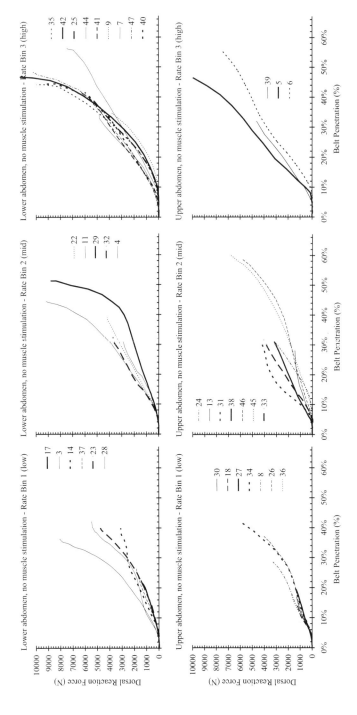

Fig. 6.21 Dorsal reaction force versus normalized belt penetration at three rates (*left* to *right*) and for the lower and upper abdomen (*top* and *bottom*) by Kent et al. (2006). No muscle stimulation. The *numbers* in the legends refer to the test numbers

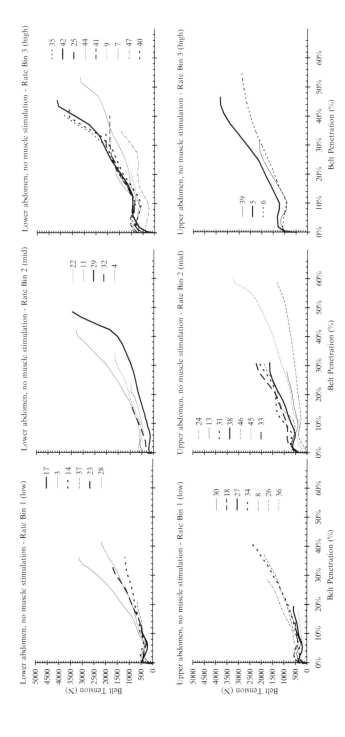

Fig. 6.22 Belt tension versus normalized belt penetration at three rates (*left* to *right*) and for the lower and upper abdomen (*top* and *bottom*) by Kent et al. (2006). No muscle stimulation. The *numbers* in the legends refer to the test numbers

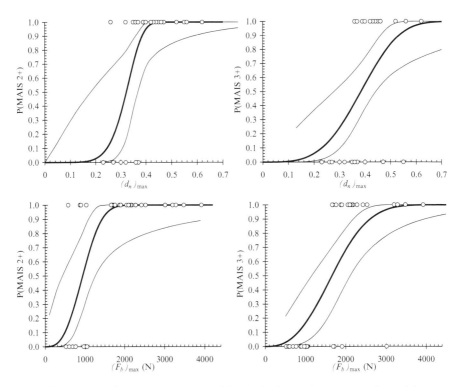

Fig. 6.23 Injury risk functions and 95 % confident limits for maximum penetration and force as determined by Kent et al. (2008). Experimental data points shown as *circles*

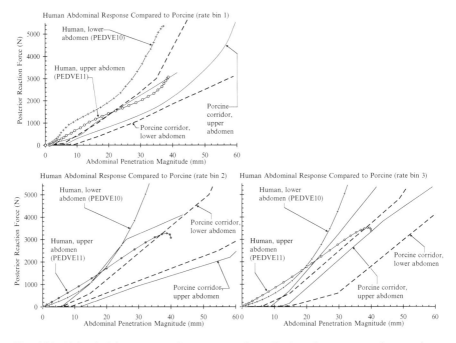

Fig. 6.24 Abdominal force–penetration response of a pediatric cadaver compared to envelopes bracketing the porcine responses reported by Kent et al. (2006) for three rate bins

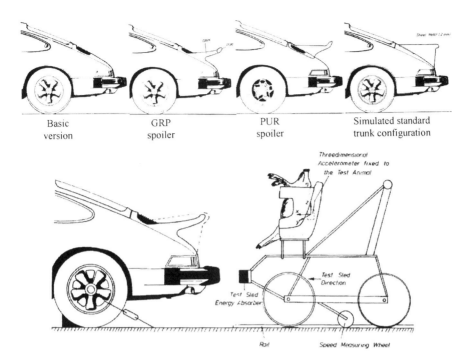

Fig. 6.25 Abdominal impact tests performed by Gögler et al. (1977) involving juvenile swine propelled into four rear-end variants of a vehicle. Rear-end variants shown on *top*, test method *below*. Selena, this is a composite of several screen captures from their paper

deceleration with corresponding stopping distances of 21 cm for the 30 km/h test and 29–32 cm for the 40 km/h tests. The peak resultant belt forces recorded in the three 40 km/h tests were 5.7–6.6 kN. Several muscular hemorrhages as well as hemorrhages of disks and ligaments were noticed but none of the tested cadavers showed injuries to the internal abdominal organs. Gögler et al. (1977) subjected 12 Göttingen mini pigs to abdominal impact by means of projecting them into various rear end configurations of a Porsche 911 (Fig. 6.25). The mini pigs used were said to correspond to an 8–12 year old child as far as weight, height, and age were concerned. Applied force, resultant spinal acceleration, and injury outcomes were reported. Thoracoabdominal injury severity ranged from MAIS 2 to 4, but the peak trunk acceleration was relatively low, ranging from 10.7 to 39.0 g (one subject experienced 54.5 g, but that may be artifactually related to a fall after the test). The results from the tests led Gögler et al. to suggest a force threshold of approximately 981 N for AIS 3 abdominal injury in 8- to 12-year-old children. The thresholds were found to be highly sensitive to the shape of the impacted surface and comparison with the results of Kent et al. (2008) suggests that nonimpact abdominal loading (such as that from a belt) may have a much lower risk of injury at a given force level than impact abdominal loading, such as that generated by Gögler et al. The probability of MAIS 3+ injury in the belt loading study using a 6-year-old model was less than 10 % at 1,000 N.

Thoracoabdominal Response in Whole-Body Loading Environments

Child-size adult cadavers, pediatric cadavers, and animal models of the human child have been tested in a variety of whole-body loading environments, including airbag out-of-position (OOP) studies and full-scale sled impact tests with different restraint configurations (Patrick and Nyquist 1972; Aldman et al. 1974; Backaitis et al. 1975; Schreck and Patrick 1975; Kallieris et al. 1976; Gögler et al. 1977; Wismans et al. 1979; Mertz et al. 1982; Robbins et al. 1983; Dejeammes et al. 1984; Prasad and Daniel 1984; Brun-Cassan et al. 1993; Lopez-Valdes et al. 2009a, b). A small minority of these (18 tests) have involved pediatric human cadavers or adult cadavers the size of a child, with various animal models making up the bulk of the data. Animals used to represent the human child in whole-body impact environments include juvenile swine, baboons, and chimpanzees. These tests have represented children from age 2 to 13 years with subjects ranging from 10 to 52 kg whole-body mass. Sled tests have been run at impact speeds ranging from 24 to 57 km/h. Both lateral and frontal impacts have been performed. Injury outcomes ranged from none (AIS 0) to fatal (AIS 6). Some representation of torso acceleration was measured in many of the experiments. The reported values ranged from 11 to 312 g. Pressure transducers were also placed into the vascular system near the heart in some subjects, and the measured pressure peaks ranged from 5.3 to 171.7 kPa. Head trajectories and acceleration were documented in a limited number of studies. Head excursion ranged from 35.3 to 90 cm, and head acceleration from 33.5 to 798 g. HIC values ranged from 42 to 22,800. Unfortunately, the technology, methods, and inconsistency of protocols inherent to the era and type of this testing make it difficult to interpret findings across studies and render much of the data unreliable. Injury coding is also inconsistent, making any combined analysis of this dataset difficult. Subject to the limitations imposed by these factors, some general observations can be made. First, if the thoracoabdominal injury severity data and peak torso acceleration are combined from the studies where both are documented, the correlation between the two is found to be very weak ($r^2=0.13$) (Fig. 6.26). In fact, all subjects that sustained no injury (AIS=0) experienced acceleration above 52 g and one subject tolerated 184 g with only a moderate (AIS 2) injury. This data, from test 7 in Prasad and Daniel (1984) involved an airbag deployment into a pig. The animal sustained subendocardial left ventricular internal hemorrhage that was of moderate severity and extent, as well as a transient, nonspecific change in cardiac rhythm. In contrast, the subject with the lowest acceleration (11 g) sustained an AIS 3/4 injury. This subject (number 10 in Gögler et al. 1977) sustained an abdominal blow from the rear end of a Porche 911 without a spoiler at a closing speed of 14.4 km/h. The total force applied to this subject was relatively low (450 N), but the degree of trauma was extensive: severe acute shock, incomplete rupture of the aortic wall, and a kidney contusion. Though neither abdominal penetration nor chest deflection was measured in any of the available tests, this biomechanical pattern is consistent with

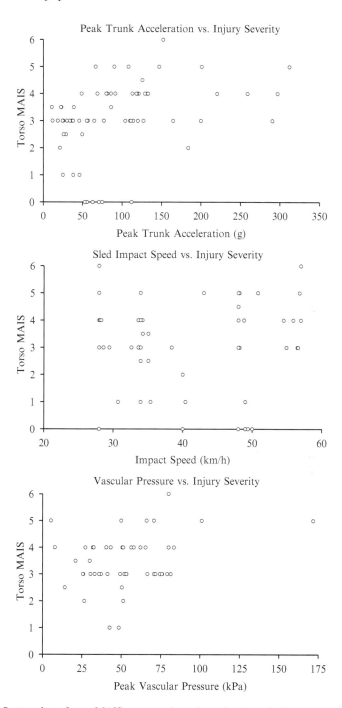

Fig. 6.26 Scatter plots of torso MAIS versus peak trunk acceleration, sled impact speed, and peak vascular pressure from 12 different studies involving torso loading via a range of mechanisms

a deformation-related injury. These results reinforce the limitations of acceleration or force as a thoracic or abdominal injury criterion since its correlation with injury is highly sensitive to changes in the loading environment, and acceleration is rarely the mechanism of injuries observed after an impact. The data do, however, show that all subjects with acceleration above 200 g sustained at least an AIS 3 thoracoabdominal injury. There are seven subjects with acceleration levels of that magnitude (PIG18 and BAB7 from Mertz et al. 1982; A-878, A-879, 76T004, and 76T005 from Robbins et al. 1983; and 5 from Prasad and Daniel 1984). Some of the injuries to those subjects appear to have an inertial mechanism (e.g., the lung "tore loose" from the chest wall in 76T005 and there was "severe hemorrhage" of the omentum and portal vein in BAB7), but some of the subjects do not appear to have any injuries caused by inertial loading. Interestingly, there were no aortic injuries in these seven subjects, which finding supports recent studies questioning the role of whole-body acceleration as a mechanism of traumatic aortic rupture in the field (e.g., Melvin et al. 1998; Hardy et al. 2006; Forman et al. 2008; Lee and Kent 2007).

Impact speed and peak vascular pressure also exhibit weak correlation with the severity of injury outcome, and all three parameters (impact speed, torso acceleration, vascular pressure) have very little ability to discriminate among test conditions that generated AIS 3+ thorcoabdominal injuries and those which did not (Kruskal's gamma = 0.31 for speed, 0.28 for acceleration, and 0.32 for pressure). These findings reflect the degree to which restraint or loading condition and the mechanical behavior of the impact surface influence injury risk, and probably also the large degree to which the experimental methods used in these studies differed.

Cadaver Sled Tests

A review of the literature identified 15 full-scale sled tests performed with 11 pediatric human cadavers (Kallieris et al. 1976, 1978; Wismans et al. 1979; Dejeammes et al. 1984; Brun-Cassan et al. 1993; Mattern et al. 2002) and three full-scale sled tests performed with an adult cadaver having the size of a child (Lopez-Valdes et al. 2009a, b). The ages of the pediatric subjects range from 2.5 to 13 years, the test speeds from 31 to 50 km/h, and the restraint conditions include shields with a lap belt, child seats, forward-facing child seats with 4- and 5-point harnesses, and 3-point belt systems. The child-size adult was tested with a force-limiting, pretensioned belt and with a standard belt, in a booster seat in all cases. Synthesis of the data to form a complete picture of the findings is challenging since incomplete and inconsistently documented results are reported in multiple publications, with different identification systems used by the different authors and repeated trials on one subject. An accurate and concise summary of the tests, including cross-references to numbering systems used in different studies, is shown in Table 6.11. Additional information and results are summarized in Figs. 6.27 through 6.30.

Belt tension has been reported for some of these sled tests, and the reported values range from 1.6 to 6.8 kN. The reported peak accelerations of the trunk range

6 Experimental Injury Biomechanics of the Pediatric Thorax and Abdomen

Table 6.11 Summary of pediatric cadaver sled tests in the literature

Test[a]	Speed (km/h)	Avg. sled decel. (g)	Restraint	Age (years)	Gender	Mass (kg)	Stature (cm)	Ref.[b]	Injury summary	Notes
HD 36-75	31	18	Lap belt with shield	2.5	M	16	97	1	AIS 1 hemorrhages in spine	Originally reported in ref. 1
HD 38-75	40	20	Lap belt with shield	6	F	27	125	1	None	Originally reported in ref. 1
HD 39-75	40	21	Lap belt with shield	6	M	30	124	1	AIS 1 hemorrhages in spine	Originally reported in ref. 1
HD 41-75	40	21	Lap belt with shield	11	M	31	139	1	AIS 2 hemorrhages in spine	Originally reported in ref. 1
HD 5	46	15	4-pt harness	10	M	39	139–150	3	AIS 2 spine injuries	#77 in ref. 3. Stature is reported inconsistently between ref. 3, 4, and Mattern et al. (2002). Heidelberg test number is 77/01
APR 1	48	13	Shield (Int. 2)	2	F	13	87	3	None	This subject was tested five times (see also APR 2). These four tests are #'s 78, 79, 80, 82 in ref. 3
	50	13							None	
	50	13							None	
	50	13							None	
HSRI	48	20	5-pt harness child restraint	6	M	17	109	2	Shoulder abras. Liver contusion of unknown severity	Originally reported in ref. 2
HD 8	49	15	6-kN load-limited 3-pt belt	13	M	39	162	3	None	#71 in ref. 3. Ref. 4 cites ref. 1 for methods
HD 9	49	25	3-pt belt	12	F	52	144	3	AIS 1 in thoracic spine	#73 in ref. 3. Ref. 4 cites ref. 1 for methods
APR 2	50	13	Shield (tot guard)	2	F	13	87	3	C1 fx w/AO disloc	Same subject as APR 1 tests. #84 in ref. 3

(continued)

Table 6.11 (continued)

Test[a]	Speed (km/h)	Avg. sled decel. (g)	Restraint	Age (years)	Gender	Mass (kg)	Stature (cm)	Ref.[b]	Injury summary	Notes
HD 89-12	49	18	Shield with shell (Romer Peggy)	2.5	F	17	91	4	AIS 3 hemorrhages in spine, laceration of ligamentum flavum, dens fx	Ref. 4 cites ref. 1 for methods
76/41	50	20	3-pt belt (17 % elongation)	12	M	41	147	1	AIS 1 skin abrasions over L shoulder and both hips	Limited results published by Kallieris et al. (1978), Sherwood et al. (2003), and Ash et al. (2009). Detail in Mattern et al. (2002)

[a]Numbering used in Brun-Cassan et al. (1993) except test 76/41, which was not discussed in that study. Different authors used different numbering schemes (see notes)

[b]1: Kallieris et al. 1976, 1978; 2: Wismans et al. 1979; 3: Dejeammes and Quincy 1974; Dejeammes et al. 1984; 4: Brun-Cassan et al. (1993). The reference given is the first reference in which the test was reported, but unique results from some of these tests are reported in more than one of these six references and others

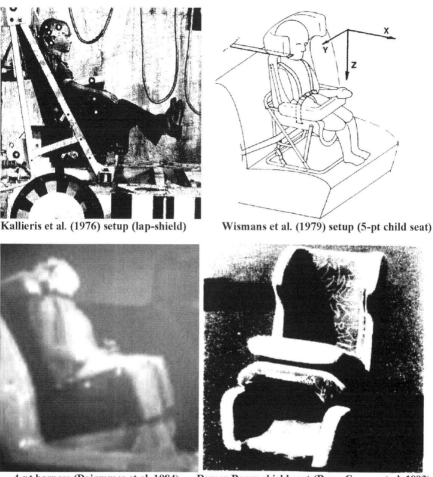

Fig. 6.27 Restraint conditions used in some of the tests listed in Table 6.11

from 40 to 112 g, though differences in instrumentation mounting, data processing, and other factors make direct comparisons and interpretation difficult. The head trajectory that results from the restraint-to-chest interaction is quite variable, with the lowest measured value of 37 cm occurring in the test of a 4-year-old in a 5-point child seat at 50 km/h (Wismans et al. 1979) and the highest value of 90 cm being measured in two of the tests (38/75 and 41/75) by Kallieris et al. involving 6- and 11-year-old subjects restrained by a lap belt with a shield in a 40 km/h impact. Head trajectory plots developed by Brun-Cassan et al. (1993) for a forward-facing child restraint are presented in Fig. 6.28. Similar head trajectory plots from Kallieris et al. (1976) are excluded from this review, since the lap-shield restraint condition is no longer in use. Interested readers are referred to the original paper.

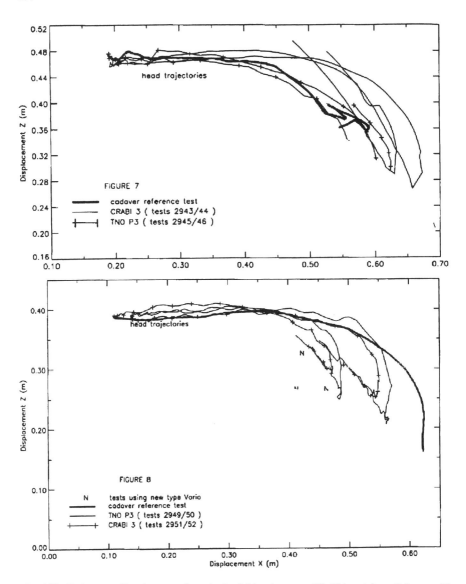

Fig. 6.28 Trajectory of head center of gravity in child cadaver test 89–12 involving a 2.5-year-old in a forward facing child seat at 49.4 km/h impact speed (*upper plot*) and test 36–75 involving a 2.5-year-old in a shield booster at 30.7 km/h (*lower plot*). Cadaver trajectories are the *thick lines*. Modified from Brun-Cassan et al. (1993)

The sled system used in the Heidelberg tests (Kallieris et al. 1976, 1978; Dejeammes et al. 1984; Brun-Cassan et al. 1993; Mattern et al. 2002; Sherwood et al. 2002, 2003) is shown in Fig. 6.29 along with some key dimensions and the seat and anchor locations used in some of the tests. Selected results from test 76/41, which involved a 3-point belt system and a 12-year-old subject, are presented in Fig. 6.30.

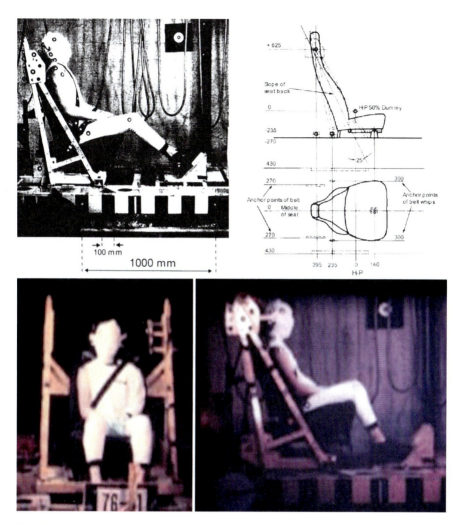

Fig. 6.29 Sled system used in the pediatric cadaver tests performed at the University of Heidelberg (*upper left*) and the seat and belt anchor locations used in at least tests 76/41 and 77/01 (*upper right*). Also shown is initial position of test subject in test 76/41 (*lower left* and *right*) (adapted from Mattern et al. 2002)

Interestingly, no serious thoracoabdominal trauma has been reported in any of these pediatric cadaver sled tests. In fact, 8 of the 15 tests resulted in no trauma documented in the torso. The torso injuries observed in these subjects included AIS 1 and 2 hemorrhages in the thoracic spine, primarily involving the musculature and ligaments. A shoulder abrasion, a liver abrasion of unknown severity on the anterior aspect of the right lobe, and "increased bilateral mobility" of the rib cage was reported in the test of Wismans et al. (1979), but no rib fractures were found. These findings are consistent with the low risk of thoracoabdominal injury experienced by

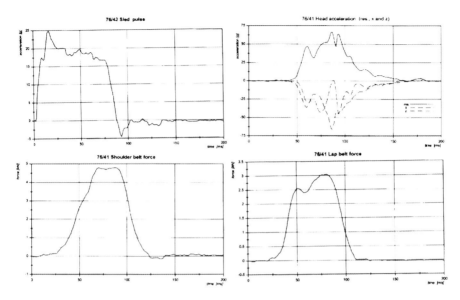

Fig. 6.30 Available data from Heidelberg test 76/41, which involved a 12-year-old cadaver restrained by a 3-point seatbelt at 50 km/h that has been analyzed by Kallieris et al. (1978) and Sherwood et al. (2002, 2003) (adapted from Mattern et al. 2002)

children in the field and support the assertion that the primary motivations for creating biofidelic physical and computational models of the pediatric chest are (1) to provide a realistic restraint interaction so that the spinal kinematics and head trajectory represent an actual child and (2) to provide the appropriate boundary condition at the base of the neck so that the neck loads measured in the model accurately reflect those that would be present in an actual child.

Sherwood et al. (2002, 2003) reanalyzed the 3-point belted, 12-year-old cadaver (test 76/41 in Table 6.11) specifically with respect to the thoracic spine kinematics and their influence on neck injury risk. These authors identified significant differences in the spinal kinematics of the cadaver compared to a Hybrid III 6-year-old dummy in nominally matched test conditions (Fig. 6.31). Specifically, the thoracic spine of the cadaver underwent significant forward flexion during the impact and had significant kyphotic curvature at maximum head excursion. In contrast, the dummy's thoracic spine, which is virtually rigid, exhibited no curvature. The resulting difference in the boundary condition at T1 generated loads in the dummy's neck that were considered by the authors to be unrealistically high (218 Nm in the lower neck) and to overstate neck injury risk in that impact environment (Sherwood et al. 2002, 2003).

Ash et al. (2009) quantified the differences in thoracic spinal kinematics first observed qualitatively by Sherwood. The horizontal displacement of the cadaver's T1 was found to be substantially greater (16 cm at $t=0$) than that of the dummy (3 cm at $t=0$) (Figs. 6.32 and 6.33), which the authors concluded to result from the pronounced bending in the thoracic region of the cadaver's spine.

In addition to the pediatric cadaver sled tests described above, Kallieris et al. 1982 presented a series of cadaver sled tests that included an 18-year-old (run H7917, 77 kg,

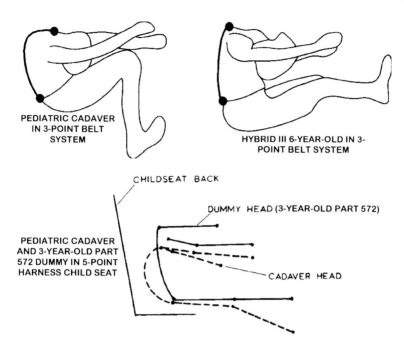

Fig. 6.31 Schematic depiction of thoracic and cervical spine kinematics in nominally matched tests of a 3-point belted Hybrid III 6-year-old (*left*) and a pediatric cadaver (HD 9) as described by Sherwood et al. (2002, 2003) (*top*, see Fig. 6.29) and a similar phenomenon documented by Wismans et al. (1979) with 5-point harness child seat (*bottom*, see Fig. 6.27). The distributed curvature of the human thoracic spine and the resulting boundary condition at T1 is hypothesized to reduce neck loads substantially compared to what is measured in the dummy

176 cm) and a 19-year-old (H8008, 66 kg, 180 cm, Fig. 6.34), both male. These tests were performed at 48 km/h and the subjects were restrained with a 3-point seatbelt. The kinematics of these two subjects and their interaction with the restraint system were not clearly different from the older subjects included in this study, with the 3-ms clip acceleration at T1 between 30.4 and 44.6 g and at T12 between 38.6 and 41.2 g. Maximum head, chest, and pelvis displacement of the 18-year-old were 63, 37, and 42 cm (data not reported for the 19-year-old). Peak lap belt tension was 4.6 kN for the 18-year-old and 6.0 kN for the 19-year-old. The distribution of trauma to these younger subjects was, however, different than older cadavers tested in the same conditions. Subject H7917 had a skin abrasion on the left shoulder, hemorrhages bilaterally in the subscapularis muscles, a fracture of the sternum between the fourth and fifth ribs, and spinal trauma including hemorrhages in the cervical, thoracic, and lumbar spine, laceration of the ligamentum flavum around C6/7 and subluxation at that level. This subject sustained no rib fractures. The 19-year-old also had skin abrasions in the regions of belt loading, in addition to hemorrhages in the cervical and thoracic spine and two rib fractures (left ribs 6 and 7 in the costal cartilage). There is a clear relationship between age and the number of rib fractures sustained by subjects in that study, with the two oldest subjects (both age 51) sustaining 12 and 14 rib fractures despite being subjected to nominally the same loading environment.

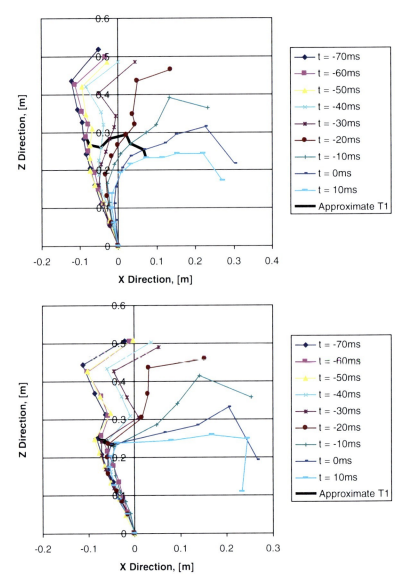

Fig. 6.32 Spinal contours with approximate location of the first thoracic vertebra (T1) of a pediatric PMHS (76/41) (*top*) and the Hybrid III 6-year-old (*bottom*) in a 50 km/h frontal impact with a 3-point belt restraint, as determined by Ash et al. (2009)

Lopez-Valdes et al. (2009a, b) presented the response of a child-size adult cadaver in a series of frontal impact sled tests. Specific focus was on the whole-body kinematics and resulting head trajectories under two different restraint conditions (booster seat and standard belt, booster seat and force-limiting pretensioning belt) in a rear seat environment. Two speeds were considered (nom. 29 and 48 km/h).

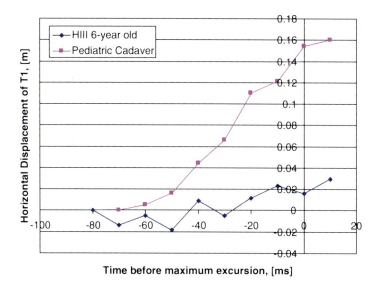

Fig. 6.33 Comparison of horizontal displacement of T1 as determined by Ash et al. (2009)

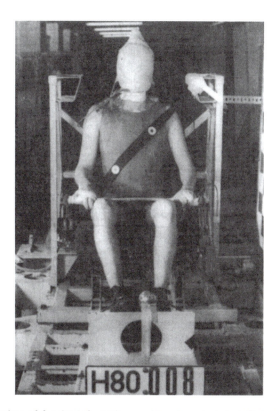

Fig. 6.34 Front view of 3-point-belted 19-year-old cadaver tested at 48 km/h as reported by Kallieris et al. (1982)

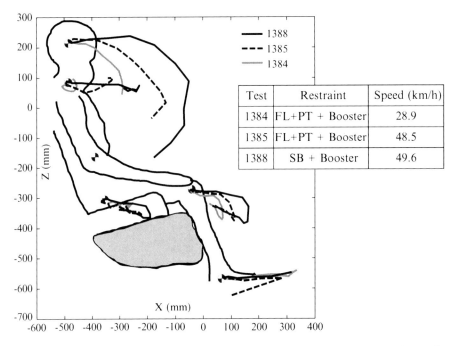

Fig. 6.35 Trajectories of the head, shoulder, h-point, knee, and foot relative to the vehicle buck in sled tests reported by Lopez-Valdes et al. (2009a, b). (Table inside figure)

At 48 km/h, the pretensioning, force-limiting seatbelt reduced the forward excursion of both the head (353 mm vs. 424 mm) and the h-point (120 mm vs. 152 mm) compared to the standard system (Fig. 6.35). Maximum torso pitch was similar for both seatbelts (Fig. 6.36).

In a follow-up to their cadaver test paper, Lopez-Valdes et al. (2009b) presented a comparison of the Hybrid III 6-year-old dummy and the child-size adult PMHS in matched impact conditions. Data from the cadaver were scaled using the characteristic length of a 50th percentile 6-year-old using the erect sitting height. The dummy predicted the peak values of the scaled displacements of the cadaver, but important differences were found in the torso angle and the seatbelt forces. Poor biofidelity in spinal flexion was identified as a major limitation of the dummy.

Animal Models in Sled Tests

Numerous studies have reported results from sled tests using a variety of animal models to study restraint interaction, thoracoabdominal impact response, and injury. Perhaps the largest of these studies was reported by Robbins et al. (1983). Twenty-three thoracic impact tests were performed at the University of Michigan using juvenile male baboons. These were sled tests with frontal impacts involving a lap

6 Experimental Injury Biomechanics of the Pediatric Thorax and Abdomen 271

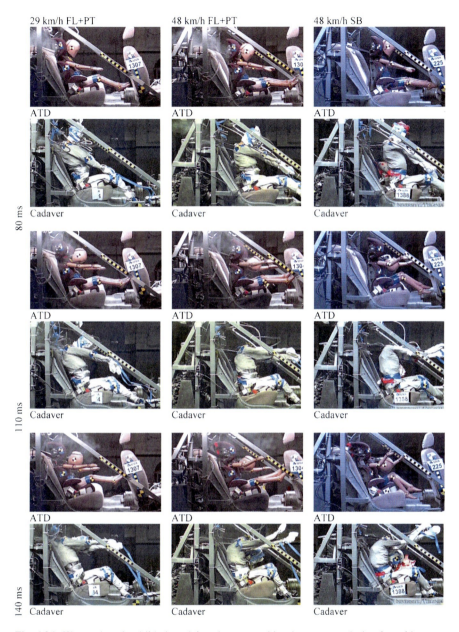

Fig. 6.36 Kinematics of a child-size adult cadaver seated in a booster seat during frontal impacts at 29 and 48 km/h with a force-limiting, pretensioned seatbelt and a standard seatbelt compared to the Hybrid III 6-year-old dummy (from Lopez-Valdes et al. 2009a, b).

belt, an energy-absorbing column, and various interaction surfaces (e.g., pads, airbags) at the contact point between the animal and the end of the column (Fig. 6.37). In some tests, a scaled steering wheel was used. Frontal impact sled speed ranged from 33.9 to 48.8 km/h. Two lateral impacts were also reported, with impact speeds

Fig. 6.37 Frontal and lateral impact sled test setup in juvenile baboon tests reported by Robbins et al. (1983). Unpadded column with scaled steering wheel shown in *upper left*, padded column in *upper right*, and lateral impact due to sliding into rigid wall shown on *bottom*

of 29.4 and 43.0 km/h. The test subjects' body weight ranged from 10.0 to 17.8 kg and seated height ranged from 58.4 to 68.6 cm. According to the age trends published in Kent et al. (2006), both of these values represent children between approximately age 2 and 6 years. Thoracoabdominal injury outcomes ranged from AIS 1 ($n=3$, slight intestinal bleeding, insignificant lung hemorrhages, minor lung bruises) in lower-speed tests with foam pads or an airbag on the column to AIS 5 ($n=4$, pericardial hematoma and liver lacerations, extensive hemorrhage of pulmonary artery, significant lung hemorrhages with vascular involvement and cardiac lacerations, extensive pancreatic and pericardial hemorrhaging). One of the AIS 5-injured subjects experienced a 43 km/h lateral impact into a rigid wall that generated lateral torso acceleration of 312 g and vascular pressure of 171.7 kPa. Two of the three AIS 5-injured subjects in frontal impacts experienced test speeds above 43 km/h, one without any padding on the column.

Backaitis et al. (1975) reported 13 sled tests of juvenile baboons in three different child seats mounted to a standardized automobile seat (the "GM Love seat", the "Peterson seat", and the "Mopar seat"). The GM Love seat, which is also called the

6 Experimental Injury Biomechanics of the Pediatric Thorax and Abdomen   273

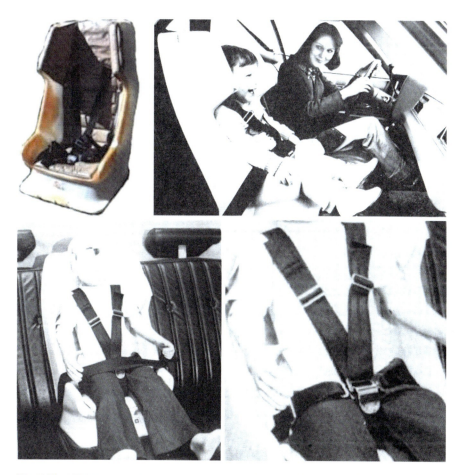

Fig. 6.38 "GM Love Seat" harness system (*top*) and arranged as used by Schreck and Patrick (1975) in tests with juvenile chimpanzees (*bottom*)

"Graco-Century" love seat, was discontinued before 1990. It is a 5-point harness child seat, as shown in Fig. 6.38 and has been tested extensively in dummy-based studies (see Weber and Melvin 1982). Each subject was tested two or three times. Subjects faced either laterally or forward and tests were performed at speeds ranging from 24 to 48 km/h with peak sled deceleration from 21 to 30 g. The animals were intended to represent human age 3, with a seated height of approximately 56 cm and a body weight of approximately 11 kg. A single larger subject was also tested (seated height 62 cm, body mass 14 kg). Tests were not intended to produce injury but rather to quantify kinematics and restraint interaction for assessing dummies, so no injuries were documented in the study. An attempt was made to measure trunk acceleration in this study using an accelerometer strapped to the chest, but the measurement was considered unreliable due to motion of the instrumentation during impact. Peak head excursion ranged from 35.3 to 60.7 cm and was found to depend on the impact speed and particular child seat used. Peak excursion was also correlated with HIC ($r=0.654$, $p=0.021$).

Schreck and Patrick (1975) reported seven sled tests involving two sedated juvenile chimpanzees in the GM forward-facing 5-point harness "Love Seat" child restraint system at impact speeds increasing from 32.2 to 48.3 km/h. One subject was 3.5 years old and the other was 5 years old. Body mass was 15.3 and 18.0 kg, and seated height was 51 and 59 cm, corresponding to human age of less than 4 years. The GM harness system, which is attached to the vehicle via the lap belt passing over the seat, is shown in Fig. 6.38. The kinematics of the subject in a 46.1 km/h, 22 g-peak deceleration test are shown in Fig. 6.39.

There was no evidence of bone fracture or soft tissue injury to the chimpanzee subjects following the test series, though serum levels of various enzymes and X-ray were used rather than autopsy to identify injury. Both animals tolerated the tests well, recovered, and were donated to the Detroit Zoological Society at the conclusion of the study.

Out-of-Position and Dynamic Airbag Tests

Patrick and Nyquist (1972) reported 22 static airbag deployments into four anesthetized juvenile baboons (average body mass of 14.4 kg is equivalent to approximately age 3 in the human) in five distinct positions (viz. normally seated, crouching on seat, standing, kneeling on floor, sitting on floor). Two different airbag characteristics and two instrument panel types were used, but very little information is given about the airbag system and other test parameters. Furthermore, no instrumentation was attached to the test subjects. The study is notable for the relatively few injuries generated, though little information is given about injury diagnostic techniques used or their sensitivity.

Aldman et al. (1974) deployed airbags either statically or during a frontal sled impact into 14 swine intended to represent children aged 3–6 with a mean reported body mass of 14.5 kg. Specific information about the deployment characteristics of the airbags is not reported, but the paper states that the airbags had been shown to be appropriate for protecting mid-size male adults in 48 km/h impacts. The dynamic tests involved a sled deceleration pulse with 12 g peak and 28 km/h change in velocity. The fully deployed airbags had a volume of 190 L. Three different inflation pulses having different peak pressures, onset times, and waveform shapes were considered (see Fig. 6.7 in original paper). The swine were placed 10–15 cm from the center of the airbag outlet area and most were suspended with their right sides toward the deploying bag (Fig. 6.40), though two subjects were tested with their ventral aspect toward the bag. The apex of the heart was at the level of the horizontal center line of the airbag outlet. Four subjects were tested at 48 km/h in a rear-facing child seat with no airbag deployment as a baseline for comparing injury outcome. The airbag interaction was found to be quite chaotic, with subjects

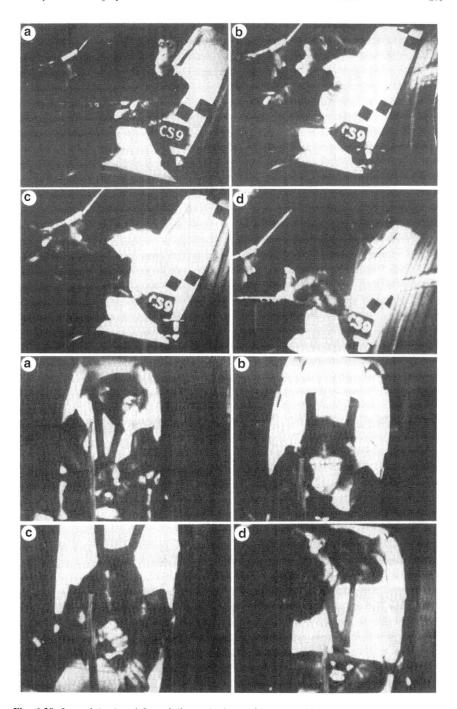

Fig. 6.39 Lateral (*top*) and frontal (*bottom*) views of occupant kinematics in a 46.1 km/h test showing pretest position (**a**), forward excursion (**b**), maximum forward head excursion (**c**), and maximum rebound (**d**) (Schreck and Patrick 1975)

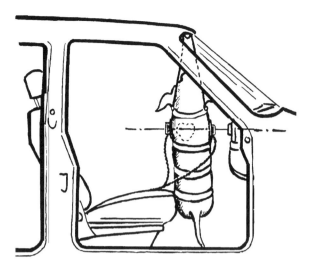

Fig. 6.40 Experimental setup used by Aldman et al. (1974) in airbag deployment tests using juvenile swine

displacing laterally or vertically during loading from the deploying bag. This motion was considered by authors to be related to injury outcome (see detailed discussion in original paper). Injury outcome ranged from AIS 0 ($n=7$, including all subjects tested in the rear-facing child seat) to AIS 6 ($n=8$, primarily contusions and lacerations to the heart, lung, and liver). Unfortunately, no instrumentation was mounted to the test subjects.

Mertz et al. (1982) subjected 43 juvenile pigs and three juvenile baboons to deploying airbags in simulated frontal impacts. Three sled deceleration pulses and seven animal positions were tested (two of the positions, 6 and 7, were preliminary and not fully documented), as well as ten different types of inflators (different gas flow characteristics) in combination with eight types of airbags, four folding patterns, and three types of covers. Head acceleration, spine and sternum acceleration, and blood pressure near the heart were documented along with head, neck, and thoracic trauma. The trauma descriptor used in this study was a "threat-to-life" (TL) value, which appears to be similar to AIS. The stated target human age was 3 years. The swine were on average 10 weeks old, the baboons were 4.5 years old, and both had body mass of approximately 15 kg. Sled impact speed ranged from 28.1 to 56.8 km/h, peak trunk acceleration from 23 to 297 g, and peak vascular pressure from 14.7 to 83.3 kPa. Thoracoabdominal trauma TL values ranged from 0 to 6 ($n=1$, significant lung and heart trauma, disrupted heart function, extensive rib fractures, liver trauma, and diaphragm injury, animal died 24 min post impact of blood loss associated with severely lacerated liver). The apparent use of so many independent variables (airbag characteristics, speeds, etc.) and the lack of reported information regarding these variables make interpretation of the study findings difficult. Figure 6.41 depicts the positions of the animal relative to the deploying airbag and Table 6.12 summarizes key outcomes by position. Position 1 was tested at a higher impact speed on average than position 2, but generated lower thoracic acceleration, vascular pressure, and thoracoabdominal injury severity. Position 3 was tested at a similar average impact

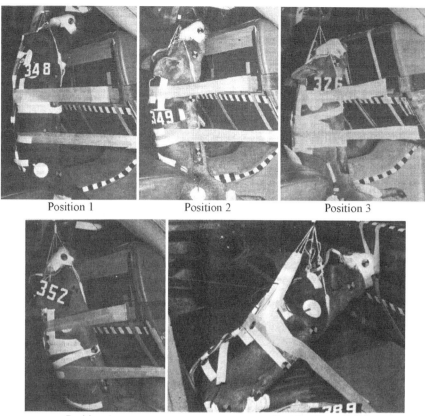

		ANIMAL POSITION	ANIMAL TO INSTRUMENT PANEL REFERENCE		SELECTION RATIONALE
POS. NO.	FIG. NO.	DESCRIPTION	REFERENCE POINTS	DISTANCE (mm)	
1	1	SPINE VERTICAL, HEAD ABOVE I.P. BROW, XIPHOID 50 MM ABOVE CENTERLINE OF INFLATOR.	XIPHOID TO INFLATOR CENTERLINE	325 REAR 50 ABOVE	• EXPOSE ABDOMEN TO DEPLOYING CUSHION/COVER. • EXPOSE HEAD AND NECK TO UPWARD DEPLOYING L-CUSHION DESIGN.
2	2	SPINE VERTICAL, SNOUT ON I.P. BROW, XIPHOID AT SAME LEVEL OF INFLATOR CENTERLINE.	XIPHOID TO INFLATOR CENTERLINE	300 REAR 0 ABOVE	• EXPOSE THORAX TO DEPLOYING CUSHION/COVER.
3	3	SPINE VERTICAL, SNOUT AT I.P. BROW, PRESTERNAL PROTUBERANCE AT LEVEL OF TOP OF COVER.	PRESTERNAL PROTUBERANCE TO INFLATOR CENTERLINE	290 REAR 115 ABOVE	• EXPOSE HEAD/NECK TO DEPLOYING CUSHION/COVER.
4	4	SIMILAR SNOUT & PRESTERNAL PROTUBRANCE LOCATIONS AS POSITION 3, LOWER TORSO ROTATED FORWARD PARALLEL TO COVER.	PRESTERNAL PROTUBERANCE TO INFLATOR CENTERLINE	240 REAR 115 ABOVE	• EXPOSE HEAD/NECK AND TORSO TO DEPLOYING CUSHION/COVER.
5	5	SPINE 30° TO HORIZONTAL, SNOUT MIDWAY BETWEEN TOP OF COVER AND I.P. BROW.	TIP OF SNOUT AND I.P. PANEL	AGAINST I.P. PANEL	• EXPOSE HEAD/NECK TO UPWARD DEPLOYING L-CUSHION DESIGN.

Fig. 6.41 Animal positions tested by Mertz et al. (1982) and the selection rationale

Table 6.12 Summary of Mertz et al. (1982) dynamic OOP airbag tests with juvenile swine. Mean values reported by animal position

Body position	n	Impact speed (km/h)	Thoracoabdominal injury severity (TL)	Peak trunk acceleration (g)	Peak blood pressure near heart (kPa)	Peak head acceleration (g)	HIC
1	10	45.7	2.6	80.3	42.9	282.6	5,863.7
2	15	37.4	2.8	109.6	45.4	163.2	1,821.5
3	14	37.0	1.9	59.3	39.7	221.0	3,496.6
4	4	29.5	3.3	151.5	64.5	253.0	3,416.7
5	2	28.6	2.5	26.5	34.7	171.5	1,832.5

speed as position 2, yet resulted in substantially less thoracoabdominal trauma, trunk acceleration, and vascular pressure. Position 4, which was similar to position 3 except the lower torso was rotated ventrally to be closer to the instrument panel, generated the most thoracoabdominal trauma and the greatest trunk acceleration and vascular pressure despite being tested at a much lower impact speed than positions 1, 2, and 3. Position 5 generated thoracoabdominal trauma similar to that generated by position 1 despite being performed at a lower impact speed and resulting in substantially lower trunk acceleration and vascular pressure.

Prasad and Daniel (1984) reported 15 dynamic airbag tests with 10–14 weeks old swine. Instrumentation included accelerometers on the snout, sternum, ventral abdominal wall, and thoracic spine, as well as a pressure measurement in the vasculature near the heart. The test protocol is described as similar to that used by Mertz et al. (1982), but details of the individual test conditions for each subject are not reported. Detailed injury information is provided and the traumatic outcomes are correlated with various mechanical measurements throughout the body. Only the thoracoabdominal findings are reviewed here. Torso injury outcomes ranged from MAIS 2 to MAIS 5, but no rib fractures were found. Transient heart arrhythmia and hemorrhages of the pulmonary parenchyma were the most serious thoracic injuries observed, while the abdominal injuries ranged from none to a critical (AIS 5) rupture of the gall bladder. Tears between the lobes of the liver and of the falciform ligament were observed, as were hepatic contusions. Maximum torso acceleration, as measured at T6, ranged from 38 to 113 g with durations of 6–18 ms. This study found no apparent relationship between the torso 3-ms clip acceleration and thoracic trauma. High sternal acceleration levels seemed to correlate with heat arrhythmia, with transient arrhythmias being observed at maximum sternal acceleration levels of approximately 1,200 g (Fig. 6.42), though the authors caution that the arrhythmia may also be related to the "unnatural" (presumably upright) posture of the swine when tested. The authors also observed a weak correlation between peak aortic blood pressure and the severity of abdominal injury, with all AIS 3+ injuries occurring at pressures above 53 kPa. It is worth noting, however, that two tests that generated no abdominal trauma had aortic pressures above this level. The magnitude of aortic blood pressure was found to correlate with the peak displacement rate of the sternum.

Conclusions and Recommendations

A review of the literature on pediatric thoracoabdominal injury biomechanics reveals a large amount of disparate data that have been collected using a range of biological models, including human cadavers and several animal species. From these data, some aspects of pediatric torso injury biomechanics can be stated with some certainty. For example, the cortical bone of the pediatric ribs has a substantially greater failure strain than that of adult ribs. Unfortunately, as with many analyses of published literature, the inconsistency in measurement methods, data

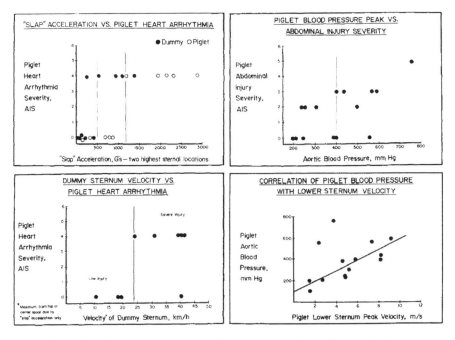

Fig. 6.42 Relationships between sternal acceleration and velocity and heart arrhythmia (*upper and lower left*), aortic blood pressure and abdominal injury severity (*upper right*), and aortic blood pressure and estimated sternal displacement rate (*lower right*) in dynamic airbag tests of juvenile swine (Prasad and Daniel 1984)

processing, and, particularly, documentation make firm conclusions difficult. Despite several decades of research, for example, it is still not clear whether pediatric and adult organs have substantially different constitutive behavior, or whether the failure tolerance of some pediatric tissues (particularly soft tissues) differs substantially from that of adults. Further, the anatomical morphology of the child as it relates to injury tolerance of the torso is virtually unexplored. The historical data compiled by Yamada (1970) form much of the foundation of our knowledge on changes in failure tolerance during pediatric development, but most of those findings are insufficiently documented and have not been reproduced. Thus, they require independent confirmation using contemporary experimental methods and documentation standards before they can be considered reliable.

From the standpoint of automotive safety system design, future work in this area should occur in both the short term and the long term. In the short term, an improved understanding of the spinal behavior of the human child in various impact environments (frontal and side impacts, seatbelted vs. harness restrained, etc.) is imperative. The head trajectories and neck loads measured by current crash test dummies are unreliable since their spinal biomechanics are not human-like. Future development of pediatric models, both physical and computational, should include biofidelic compliance in the thoracic spine, especially during bending. The head is the most frequently and most seriously injured body region in children, so torso research should focus, at least in the near term, on how the biomechanics of a restraint's interaction with the torso creates load paths on the spine that dictate the path of the head and the risk and severity of a head contact. Longer-term areas for focus include injury tolerance of the torso per se, in particular with regard to the types of soft-tissue injuries sustained by children. For example, injury criteria based on rib fractures have some justification for use in studies of adult crash protection, especially for the elderly, but the adaptation of these criteria to children is inadequate, since the mechanisms of injury in the child's thorax may be entirely different.

References

Adams MA, Dolan P (1991) A technique for quantifying the bending moment acting on the lumbar spine in vivo. J Biomech 24:117–126

Adams MA, McNally DS, Chinn H (1994) Posture and the compressive strength of the lumbar spine. Clin Biomech 9:5–14

Aldman B, Andersson A, Saxmark O (1974) Possible effects of airbag inflation on a standing child. In: Proceedings from IRCOBI, Zurich, pp 194–215

American Heart Association (2006) 2005 American Heart Association guidelines for cardiopulmonary resuscitation and emergency cardiovascular care. Circulation 112:Suppl

Arbogast K, Kent R, Ghati Y et al (2006a) Mechanisms of abdominal organ injury in seat belt restrained children. J Trauma 62:1473–1480

Arbogast KB, Maltese MR, Nadkarni VM et al (2006b) Anterior-posterior thoracic force–deflection characteristics measured during cardiopulmonary resuscitation: comparison to post-mortem human subject data. Stapp Car Crash J 50:131–145

Ash J, Sherwood S, Abdelilah Y et al (2009) Comparison of anthropomorphic test dummies with a pediatric cadaver restrained by a three-point belt in frontal sled tests. In: Proceedings from 21st enhanced safety of vehicles (ESV) conference, Stuttgart, Germany, 2009

Backaitis S, Medlin J, Radovich V et al (1975) Performance evaluation of child dummies and baboons in child restraint systems in a systemized crash environment. SAE Tech Paper #751153

Bilston L, Liu L, Phan-Thien N (2001) Large strain behaviour of brain tissue in shear: some experimental data and differential constitutive model. Biorheology 38:335–345

Brun-Cassan F, Page M, Pincemaille Y et al (1993) Comparative study of restrained child dummies and cadavers in experimental crashes. SAE Tech Paper #933105

Chamouard F, Tarriere C, Baudrit P (1996) Protection of children on board vehicles influence of pelvis design and thigh and abdomen stiffness on the submarining risk for dummies installed on a booster. In: Proceedings from 15th conference on the enhanced safety of vehicles (ESV), Paper 96-S7-O-03

Ching R, Nuckley D, Hertsted S et al (2001) Tensile mechanics of the developing cervical spine. Stapp Car Crash J 45:329–336

Dejeammes M, Quincy R (1974) Recherche sur la conception de systemes de retention d'enfants a l'aide d'un modele animal. In: Proceedings from IRCOBI, Lyon, France, pp 260–277

Dejeammes M, Tarriere C, Thomas C et al (1984) Exploration of biomechanical data towards a better evaluation of tolerance for children involved in automotive accidents. SAE Tech Paper #840530

Eppinger R, Sun E, Bandak F et al (1999) Development of improved injury criteria for the assessment of advanced automotive restraint systems. NHTSA USDOT Docket # NHTSA-1999-6407-0005

Fazekas IG, Kosa F, Jobba G et al (1971) Die druckfestigkeit der menschlichen Leber mit besonderer hinsicht auf die verkehrsunfälle. Z Rechtsmed 68:207–224

Fazekas IG, Kosa F, Jobba G et al (1972) Beiträge zur druckfestigkeit der menschlichen milz bei stumpfen krafteinwirkungen. Arch Kriminol 149:158–174

Forman J, Stacey S, Evans J et al (2008) Posterior acceleration as a mechanism of blunt traumatic injury of the aorta. J Biomech 41:1359–1364

Franklyn M, Peiris S, Huber C et al (2007) Pediatric material properties: a review of human child and animal surrogates. Crit Rev Biomed Eng 35:197–342

Garcia VF, Gotschall CS, Eichelberger MR et al (1990) Rib fractures in children: a marker of severe trauma. J Trauma Inj Infect Crit Care 30:695–700

Gayzik F, Hoth J, Daly M, Meredith J, Stitzel J (2007) A finite element-based injury metric for pulmonary contusion: investigation of candidate metrics through correlation with computed tomography. Stapp Car Crash J 51:189–209

Giacomin J (2005) Absorbed power of small children. Clin Biomech 20:372–380

Gögler E, Best A, Braess HH et al (1977) Biomechanical experiments with animals on abdominal tolerance levels. In: Proceedings from 21st Stapp car crash conference, pp 713–751

Gruben KG, Romlein J, Halperin HR et al (1990) System for mechanical measurements during cardiopulmonary resuscitation in humans. Biomed Eng 37:204–210

Hagedorn AV, Pritz HB (1991) Evaluation of chest deflection measurement band. NHTSA USDOT Technical Report, Washington, DC

Hajji MA, Wilson TA, Lai-Fook SJ (1979) Improved measurements of shear modulus and pleural membrane tension of the lung. J Appl Physiol 47:175–181

Hammer J, Eber E (2005) The pecularities of infant respiratory physiology. Prog Respir Res Basel Karger 33:2–7

Hardy WN, Shah CS, Kopacz JM et al (2006) Study of potential mechanisms of traumatic rupture of the aorta using in situ experiments. Stapp Car Crash J 50:247–266

Heijnsdijk EAM, van der Voort M, de Visser H, Dankelman J et al (2003) Interand intraindividual variabilities of perforation forces of human and pig bowel tissue. Surg Endosc 17:1293–1296

Hoke RS, Chamberlain D (2004) Skeletal chest injuries secondary to cardiopulmonary resuscitation. Resuscitation 63:327–338

Holmes JF, Sokolove PE, Brant WE (2002) A clinical decision rule for identifying children with thoracic injuries after blunt torso trauma. Ann Emerg Med 39:492–499

Iannuzzi A, Rosaria Licenziati M, Acampora C et al (2004) Preclinical changes in the mechanical properties of abdominal aorta in obese children. Metabolism 53:1243–1246

Irwin A, Mertz HJ (1997) Biomechanical basis for the CRABI and Hybrid III child dummies. In: Proceedings from 41st Stapp car crash conference, pp 261–272

Ishihara T, Nakahira Y, Furukawa K (2000) Measurement of the mechanical properties of the pig liver and spleen. In: Proceedings from Japan society of mechanical engineers (JSME) conference, 2:209–210

Kallieris D, Barz J, Schmidt G et al (1976) Comparison between child cadavers and child dummy by using child restraint systems in simulated collisions. SAE Tech Paper #760815:513-542

Kallieris D, Schmidt G, Barz J et al (1978) Response and vulnerability of the human body at different impact velocities in simulated three-point belted cadaver tests. In: Proceedings from IRCOBI, pp 105–209

Kallieris D, Mattern R, Schmidt G et al (1981) Quantification of side impact responses and injuries. In: Proceedings from 25th Stapp car crash conference, pp 329–368

Kallieris D, Mellander H, Schmidt G et al (1982) Comparison between frontal impact tests with cadavers and dummies in simulated true car restrained environment. In: Proceedings from 26th Stapp car crash conference, pp 353–367

Kemper A, McNally C, Kennedy E et al (2005) Material properties of human rib cortical bone from dynamic tension coupon testing. Stapp Car Crash J 49:199–230

Kent R, Lessley D, Sherwood C (2004) Thoracic response to dynamic, non-impact loading from a hub, distributed belt, diagonal belt, and double diagonal belts. Stapp Car Crash J 48:495–519

Kent R, Stacey S, Kindig M et al (2006) Biomechanical response of the pediatric abdomen, part 1: development of an experimental model and quantification of structural response to dynamic belt loading. Stapp Car Crash J 50:1–26

Kent R, Stacey S, Kindig M et al (2008) Biomechanical response of the pediatric abdomen, part 2: injuries and their correlation with mechanical parameters. Stapp Car Crash J 52:135–166

Kent R, Salzar R, Kerrigan J, Parent D, Lessley D, Sochor M, Luck J, Loyd A, Song Y, Nightingale R, Bass C, Maltese M (2009) Pediatric thoracoabdominal biomechanics. Stapp Car Crash J 53:373–402

Khamin N (1975) Strength properties of the human aorta and their variation with age. Mech Compos Mater 13:100–104

Kleinberger M, Yoganandan N, Kumaresan S (1998) Biomechanical considerations for child occupant protection. Ann Proc Assoc Adv Auto Med 42:115–136

Kroell C, Schneider D, Nahum A (1974) Impact tolerance and response of the human thorax II. SAE Tech Paper #741181:201–282

Lai-Fook S, Hyatt R (2000) Effects of age on elastic moduli of human lungs. J Appl Physiol 89: 163–168

Lai-Fook SJ, Wilson TA, Hyatt R et al (1976) Elastic constants of inflated lobes of dog lungs. J Appl Physiol 40:508–513

Lee SH, Kent RW (2007) Blood flow and fluid–structure interactions in the human aorta during traumatic rupture conditions. Stapp Car Crash J 51:211–233

Lessley D, Crandall J, Shaw C (2004) A normalization technique for developing corridors from individual subject responses. SAE Tech Paper #004-01-0288

Liu Z, Bilston L (2000) On the viscoelastic character of liver tissue: experiments and modelling of the linear behaviour. Biorheology 37:191–201

Liu Z, Bilston LE (2002) Large deformation shear properties of liver tissue. Biorheology 39:735–742

Lopez-Valdes F, Forman J, Ash J et al (2009) The frontal-impact response of a booster-seated child-size PMHS. In: Proceedings from IRCOBI (in press)

Lopez-Valdes F, Forman J, Kent R et al (2009) A comparison between a child-size PMHS and the Hybrid III 6 YO in a sled frontal impact. In: Annual Proceedings/Association for the Advanced Automotive Medicine (in press)

Maltese M, Castner T, Niles D et al (2008) Methods for determining pediatric force–deflection characteristics from cardiopulmonary resuscitation. Stapp Car Crash J 52:83–105

Mansell A, Moalli R, Calist C et al (1989) Elastic moduli of lungs during postnatal development in the piglet. J Appl Physiol 67:1422–1427

Mattern R, Kallieris D, Riedl H (2002) Reanalysis of two child PMHS-tests. Final report. Univer Heidelberg, Heidelberg, Germany

Mattice J (2006) Age-dependent changes in the viscoelastic response of the porcine kidney parenchyma using spherical indentation and finite element analysis. Master of Science Thesis. University of Virginia, Virginia

Mattice J, Lau A, Oyen M et al (2006) Spherical indentation load-relaxation of soft biological tissues. J Mater Res 21:2003–2010

McElhaney J, Alem N, Roberts V (1970) A porous block model for cancellous bones. ASME Paper # 70-WA/BHF-2

McGowan D, Voo L, Liu Y (1993) Distraction failure of the immature spine. Proc ASME Bioeng Conf 24:24–25

Melvin JW, Stalnaker RL, Roberts VL et al (1973) Impact injury mechanisms in abdominal organs. In: Proceedings from 17th Stapp car crash conference, pp 115–126

Melvin JW, Baron KJ, Little WC et al (1998) Biomechanical analysis of indy race car crashes. Stapp Car Crash J 42:247–266

Mertz HJ, Driscoll G, Lenox J (1982) Responses of animals exposed to deployment of various inflatable restraint system concepts for a variety of collision severities and animal positions. In: Proceedings from 9th ESV, Kyoto, Japan, pp 352–368

Mertz HJ, Irwin A, Melvin JW et al (1989) Size, weight and biomechanical impact response requirements for adult size small female and large male dummies. SAE Auto Front Impact SP-782:133–144

Mertz HJ, Jarrett K, Moss S et al (2001) The Hybrid III 10-year-old dummy. Stapp Car Crash J 45:316–326

Miller K (2000a) Constitutive modelling of abdominal organs. J Biomech 33:367–373
Miller K (2000b) Biomechanics of soft tissues. Med Sci Monit 6:158–167
Mosekilde L, Mosekilde L (1986) Normal vertebral body size and compressive strength: relations to age and to vertebral and iliac trabecular bone compressive strength. Bone 7:207–212
Mosekilde L, Viidik A, Mosekilde L (1985) Correlation between the compressive strength of iliac and vertebral trabecular bone in normal individuals. Bone 6:291–295
Mosekilde L, Mosekilde L, Danielsen CC (1987) Biomechanical competence of vertebral trabecular bone in relation to ash density and age in normal individuals. Bone 8:79–85
Nasseri S, Bilston LE, Phan-Thien N (2002) Viscoelastic properties of pig kidney in shear, experimental results and modelling. Rheol Acta 41:180–192
Ohara T (1953) On the comparison of strengths of the various organ-tissues. J Kyoto Prefect Univ Med 53:577–597
Okazawa M, D'Yachkova Y, Pare PD (1999) Mechanical properties of lung parenchyma during bronchoconstriction. J Appl Physiol 86:496–502
Ouyang J, Zhao W, Xu Y et al (2006) Thoracic impact testing of pediatric cadaveric subjects. J Trauma 61:1492–1500
Oyen M, Lau A, Kindig M et al (2006) Mechanical properties of structural tissues of the pediatric thorax. J Biomech 39:156
Parent DP (2008) Scaling and optimization of thoracic impact response in pediatric subjects. Master's Thesis. Mechanical and Aerospace Engineering, University of Virginia, Virginia
Patrick L, Nyquist G (1972) Airbag effects on the out-of-position child. SAE Tech Paper #720442
Pfefferle K, Litsky A, Donnelly B et al (2007) Biomechanical properties of the excised pediatric human rib. In: Proceedings from 35th international workshop on injury biomechanical research, NHTSA, Washington, DC
Prasad P, Daniel R (1984) A biomechanical analysis of head, neck, and torso injuries to child surrogates due to sudden torso acceleration. SAE Tech Paper #841656
Roach MR, Burton AC (1959) The effect of age on the elasticity of human iliac arteries. Can J Biochem Physiol 37:557–570
Robbins D, Lehman R, Nusholtz G et al (1983) Quantification of thoracic response and injury: tests using human surrogate subjects. University of Michigan, Ann Arbor, MI, UMTRI-83-26
Saul RA, Pritz HB, McFadden J et al (1998) Description and performance of the Hybrid III three-year-old, six-year-old, and small female test dummies in restraint system and out-of-position air bag environments. In: Proceedings from IRCOBI conference. Paper #98-S7-O-01
Scheuer L, Black S (2000) Developmental juvenile osteology. Academic Press Limited, London
Schmidt G (1979) The age as a factor influencing soft tissue injuries. In: Proceedings from 1979 IRCOBI conference
Schreck R, Patrick L (1975) Frontal crash evaluation tests of a five-point harness child restraint. SAE Tech Paper #751152
Seki S, Iwamoto H (1998) Disruptive forces for swine heart, liver, and spleen: their breaking stresses. J Trauma 45:1079–1083
Shaw G, Wang C, Bolton J et al (1999). Chestband performance assessment using static tests. In: Proceedings from 27th inter workshop on injury biomechanical research, NHTSA USDOT, Washington, DC
Sherwood C, Shaw C, van Rooij L et al (2002) Prediction of cervical spine injury risk for the 6-year-old child in frontal crashes. In: 46th Annual proceedings/association for the advanced automotive medicine
Sherwood C, Shaw C, van Rooij L et al (2003) Prediction of cervical spine injury risk for the 6-year-old child in frontal crashes. Traffic Inj Prev 4:206–213
Sinclair D (1978) Human growth after birth. Oxford University Press, London
Smith M, Burrington J, Woolf A (1975) Injuries in children sustained in free falls: an analysis of 66 cases. J Trauma 15
Snyder R (1963) Human tolerances to extreme impacts in free fall. J Aerosp Med 34:8

Snyder R (1969) Impact injury tolerances of infants and children in free fall. In: 13th Annual proceedings/association for the advanced automotive medicine

Stamenovic D, Yager D (1988) Elastic properties of air- and liquid-filled lung parenchyma. J Appl Physiol 65:2565–2570

Stingl J, Baca V, Cech P et al (2002) Morphology and some biomechanical properties of human liver and spleen. Surg Radio Anat 24:285–289

Stitzel J, Gayzik FS, Hoth J et al (2005) Development of a finite element-based injury metric for pulmonary contusion part I: model development and validation. Stapp Car Crash J 49: 271–290

Stolze H, Kuhtz-Buschbeck J, Mondwurf C et al (1997) Gait analysis during treadmill and overground locomotion in children and adults. Electroencephalogr Clin Neurophysiol Electromyogr Motor Contr 105:490–497

Stürtz G (1980) Biomechanical data of Children. In: Proceedings from 24th Stapp car crash conference

Tamura A, Omori K, Miki K et al (2002) Mechanical characterization of porcine abdominal organs. Stapp Car Crash J 46:55–69

Tepper R, Wiggs B, Gunst S et al (1999) Comparison of the shear modulus of mature and immature rabbit lungs. J Appl Physiol 87:711–714

Theis M (1975) Untersuchung der dynmischen und statischen Biegebelastung frischer menschlicher Rippen in Abhängigkeit zu Alter und Geschlecht. Inaugural-Dissertation. Ruprecht-Karl-Universität, Heidelberg, Germany

Tsitlik JE, Weisfeldt ML, Chandra N et al (1983) Elastic properties of the human chest during cardiopulmonary resuscitation. Crit Care Med 11:685–692

Uehara H (1995) A study on the mechanical properties of the kidney, liver and spleen, by means of tensile stress test with variable strain velocity. J Kyoto Prefect Univ Med 104:439–451

Vallis CJ, Mackenzie I, Lucas BG (1979) The force necessary for external cardiac compression. Practitioner 223:268–270

van Ratingen M, Twisk D, Schrooten M et al (1997) Biomechanically based design and performance targets for a 3-year old child crash dummy for frontal and side impact. In: Proceedings from 41st Stapp car crash conference, pp 243–260

Vawter D (1980) A finite element model for macroscopic deformation of the lung. J Biomech Eng 102:1–7

Vawter D, Fung YC, West J (1978) Elasticity of excised dog lung parenchyma. J Appl Physiol 45:261–269

Wang BC, Wang GR, Yan DH et al (1992) An experimental study on biomechanical properties of hepatic tissue using a new measuring method. Biomed Mater Eng 2:133–138

Weaver J, Chalmers JK (1966) Cancellous bone: its strength and changes with aging and an evaluation of some methods for measuring its mineral content. Part 1: age changes in cancellous bone. J Bone Joint Surg 48:299–308

Weber K, Melvin J (1982) Dynamic testing of child occupant protection systems. University of Michigan Highway Safety Research Institute USDOT, Final Report UM-HSRI-82-19

Wismans J, Maltha J, Melvin J et al (1979) Child restraint evaluation by experimental and mathematical simulation. SAE Tech Paper #791017

Yamada (1970) In: Evans FG (ed) Strength of biological materials. The Williams & Wilkins Co, Baltimore, MD

Yeh W, Li P, Jeng Y et al (2002) Elastic modulus measurements of human liver and correlation with pathology. Ultrasound Med Biol 28:467–474

Chapter 7
Pediatric Computational Models

Bharat K. Soni, Jong-Eun Kim, Yasushi Ito, Christina D. Wagner, and King-Hay Yang

Introduction and Background

A computational model is a computer program that attempts to simulate a behavior of a complex system by solving mathematical equations associated with principles and laws of physics. Computational models can be used to predict the body's response to injury-producing conditions that cannot be simulated experimentally or measured in surrogate/animal experiments. Computational modeling also provides means by which valid experimental animal and cadaveric data can be extrapolated to a living person. Widely used computational models for injury biomechanics include multibody dynamics and finite element (FE) models. Both multibody and FE methods have been used extensively to study adult impact biomechanics in the past couple of decades.

The simulations associated with multibody dynamics involve a set of rigid body segments linked by different joints in a tree structure. The governing equations of motion of these collections of rigid bodies in an acceleration field are developed using Lagrangian methods. Each rigid body defined by mass, center of gravity, and moment of inertia is geometrically represented by an ellipsoid, plane, or cylinder. The relative motion between rigid bodies is described by kinematic joints such as revolute, spherical, and universal joints. The multibody method has been an attractive technique because of its easy modeling and rapid analysis. A widely used multibody model, MAthematical DYnamic Model (MADYMO), developed at TNO, the Netherlands represents an industry standard for automotive safety analysis.

B.K. Soni, Ph.D. • J.-E. Kim, Ph.D. • Y. Ito, Ph.D.
Department of Mechanical Engineering, University of Alabama at Birmingham,
1150 10th Avenue South, BEC 257, Birmingham, AL 35294, USA

C.D. Wagner, Ph.D. • K.-H. Yang, Ph.D. (✉)
Computational Biomechanics, Biomedical Engineering, Wayne State University,
818 W. Hancock, Room 2128, Detroit, MI 48201, USA
e-mail: aa0007@wayne.edu

The software system also includes a comprehensive set of validated crash dummy models and advanced restraint systems (seat belts, airbags) (www.madymo.com). However, biofidelic geometry and material properties cannot be cooperated with the multibody models.

Meanwhile, the FE methods can provide more detailed biomechanical responses as compared to rigid body models. The methods facilitate high-fidelity simulation of linear and nonlinear sets of governing equations allowing deformable body models (as opposed to rigid body) decomposed into discrete points with connectivity information. These points form one, two or three dimensional elements on which mechanistic quantities (i.e. deformation, stress/strain, velocity, acceleration, reaction force, etc.) can be calculated. A variety of material constitutive models (i.e. elastic, plastic, viscoelastic, hyperelastic, etc.) can be cooperated with the FE methods. Hence, FE simulations provide far more accurate representation of the biomechanics system than the multibody simulations. However, the model preparation as well as computer execution are labor- and time intensive. Widely used FE systems in injury biomechanics include: LS-DYNA, originally developed at Livermore National Laboratory and distributed and maintained by Livermore Software Technology Corporation. LS-DYNA is a general-purpose transient explicit and implicit dynamic finite element program capable of simulating complex nonlinear real-world problems (www.lsdyna-portal.com). Other widely used FE systems are PAM-CRASH (www.esi-group.com) and RADIOSS (www.altair.com)—explicit, large deformation, Lagrangian dynamic finite element programs tailored for dynamic crash simulation and for nonlinear structure analysis. Detailed theoretical foundation and methodologies for computational mechanics can be found in the Encyclopedia of Computational Mechanics (www.wiley.co.uk/ecm) with a focus on computational biomechanics of soft tissues in Chap. 18.

Clearly, the development of biofidelic computational models is essential. The challenges inherent in simplifying a biomechanical system enough to model it mathematically using available computational resources leaves each and every model open to constructive criticism. Irrespective of which approach is used, the development of a computational model involves:

1. *Geometric discrete representation*: knowledge of human anatomy and anthropometry with appropriate discretization. The accurate representation of the anatomically specific geometry in the FE model enhances its function as predictive tool for bony fractures and soft tissue injuries.
2. *Material characteristics*: mechanics and material science applied to identify constitutive material properties.
3. *Numerical algorithm*: mechanical engineering and dynamics with associated mathematical equations representative of the behavior and interactions among all components of the system (such as stress–strain and/or force–deflection characteristics, and friction related to interface properties). In general, the response of the system to a loading scheme is calculated through a series of discrete time steps and is simultaneously solved for all components.
4. *Validation and verification*: experimental/measured injury/behavior biomechanical responses to verify the model and validate model predications.

Computational Models: Advantages

Computational models associated with injury biomechanics can provide quick and inexpensive prediction of how compression, tension, shear, and bending during the accident contribute to injury and an approach to conquer the complexity of injury with increased degree of confidence. Computational models also offer a cost-effective way to facilitate discovery, development, and testing of new concepts and protective equipments as well as benchmarking data of equipment codes and standards associated with design of vehicles.

Most importantly, computational models provide a solution to injury scenarios in which experimentation is difficult, limited measurement data are available, and the mechanisms to expand injury assessment and reproduce kinematics are not easily achievable via experimental techniques. Both cadavers and anthropometric test devices (ATDs, also known as crash dummies) have been used as surrogates to examine the response of humans to impact loading. However, optimized design of vehicles as well as restraint systems require greater biofidelity than provided by ATDs. ATDs provide relatively simplistic representations of the human anatomy and seldom include specific representation of the viscera and other soft tissue structures. Since dummies do not provide a method to predict localized responses, injury prediction using dummies relies on interpretation of the measured engineering response parameters relative to established injury thresholds. While cadavers have realistic anatomical structures, the mechanical properties of cadaveric tissue degrade rapidly post-mortem and, in the case of soft tissues, may compromise the injury tolerance and response. Furthermore, the muscular response of a living person involved in an impact cannot be evaluated using cadaveric tissue. Computational models provide a mechanism to address the inherent weaknesses of dummies and cadavers by explicitly representing soft and hard tissue anatomy and material properties as well as active and passive muscular attributes.

In case of pediatric studies, the limited availability of child cadaver data poses a significant critical barrier. Child ATDs incorporate few pediatric-specific material characteristics and responses under loading and have been generally designed and developed using scaling laws applied to "average size" adult data. A general lack of child ATDs biofidelity results from these scaling approximations combined with the inherent anatomical simplifications of dummy designs.

Pediatric computational models can be developed using image and scanning methods (e.g., CT, MRI, and other imaging techniques), when available, and can represent pediatric anatomy more accurately than an ATD. Once a model has been developed, it can be exercised in parametric studies to investigate the effect of each parameter on the model responses. This can be both time- and cost-effective when compared to traditional experimental research, since the sensitivity analysis using distinct parameters could be easily performed. The resulting computational simulations could provide internal stresses and strains at tissue level contributing to more detailed and accurate injury analysis and evaluations. Hence, computational modeling can be used to elucidate injury mechanisms to expand on injury assessment reference values, and to reproduce kinematics, especially for impact scenarios not easily achievable via experimental techniques thus making it a useful technique, especially for pediatric injury biomechanics.

Computational Models: Validation and Evaluation

Validation and verification of computational model is extremely critical to establish confidence level. There is a distinct relationship between experimental research and computational modeling techniques. In fact, it would be negligent to develop numerical models without experimental data or to quantify how accurate a model's predictions may be. Several types of data are needed, and at each stage the assumptions/approximations made influence the accuracy and hence must be fully validated. First, geometric information needs to be represented as a discretized model. Here, the biofidelity of the geometry and discretization process must be validated. Second, material behavior and tissue properties are necessary for implementation in the model. For human tissue, this behavior is generally strain-rate dependent and may be anisotropic or may respond differently in tension and compression, so the tissue must be tested in a variety of loading modes. Third, the loading and boundary conditions as well as body's response must be accurately documented during experiments so that the simulation results can be compared and some mathematical assumptions can be adjusted to ensure proper predictions. Lastly, since the numerical model is based on the discretized algorithm, secondary loading conditions must be considered to verify that the model is valid in more than one condition.

Real-world data from field investigations may offer a way to exercise computational models and provide possible validation in a qualitative sense. However, biomechanical detail garnered from a field investigation can be largely speculative given the number of variables and conditions that can exist in a given crash with a given occupant. In this respect, data generated from laboratory experiments using human surrogates are mainly used for model validation.

The concept of model validation is nontrivial and should be carefully considered when critiquing the results of any type of numerical or computational model. It is accepted that some simplifications and assumptions must be made when dealing with a complex system such as the human body, but the model should be robust enough to yield reasonable results in both kinetics and stress/strain response for a variety of input conditions to ensure that the equations used to define the model are valid. In summary, a close collaboration with experimentalists and computational modelers is critical for fully validating the injury biomechanics study.

Computational Models: Limitations

The use of ATD response data to attempt validation of whole body computational models leads to questions regarding the biofidelity of both physical and numerical surrogates. Age-matched human experimental results and models must be compared in order to properly quantify the degree of validation achieved and to illicit confidence in model predictions. Even in the adult realm, whole body modeling for impact purposes does not always reach the level of sophistication that component models have offered. For children, this gap seems to be even wider, with a majority of the data used to develop and validate the models coming from directly or indirectly scaled sources.

In modeling impact response of the human body, natural variation between individuals in a population has largely been neglected in favor of an "average" or 50th percentile approach. Tissue properties have been averaged from several subjects to arrive at the material properties reported in the scientific literature. However, much of these data are skewed towards the older end of the age spectrum, as post-mortem human subjects (cadavers) are the surrogate of choice and are generally elderly. For the pediatric population, this age bias is especially detrimental, as there are additional growth and development issues to contend with as the child transitions to adulthood. In children, it is much more difficult to determine what constitutes an "average" child of a certain age. The rate of a child's development is in a constant state of nonlinear flux, dependent on genetics, environment, and more. Developmental milestones relevant to impact biomechanics have not truly been established, and age-related changes in most tissue properties have never been considered for the pediatric population. Unfortunately, the breadth of research specifically related to children that could be used for modeling purposes is much narrower than for their adult counterparts. Since much of the tissue response data used for children have been scaled from adult tests or is based on an insufficient number of samples over too wide an age range, the field of pediatric computational modeling has been lagging behind the adult modeling efforts.

While pediatric computational modeling is not as advanced as that for adults, there have been many published studies highlighting the use of modeling techniques specific to pediatric population. Although these studies offer a justifiable start, it is clear that pediatric modeling will need to show advances in the future to overcome the difficulties inherent in modeling a population where so little experimental research is available.

In this chapter, pediatric computational models published in the literature will be reviewed and critiqued in the state-of-the-practice section. In addition, we will focus on the future directions of research that will help address the critical technology and knowledge barriers faced in pediatric computational modeling that are discussed in detail in the chapter.

Current State-of-the-Practice on Pediatric Computational Models

Introduction

Although there are particular challenges in pediatric computational modeling due to a lack of available material property data, quantitative age-dependent anatomical data, and pediatric impact response data, a number of pediatric computational models have been developed. A summary of specific pediatric computational models reported in the literature, including whole body, head and brain, neck/cervical spine, upper and lower extremity models, is presented in the following sections.

Pediatric Whole Body Models

Whole body models utilizing multibody and finite element methods have been created to investigate the pediatric biomechanical response to impact (Table 7.1). The use of multibody models for studying whole body response is facilitated by the ability to scale for different anthropometries. Despite inherent limitations in biofidelity for scaling from adult to children models, pediatric ellipsoid models developed by Liu and Yang (2002) and van Rooij et al. (2004) were developed based on the 50th percentile male MADYMO (www.madymo.com) human pedestrian model. The child models were developed through structural scaling of the ellipsoidal rigid bodies and joint properties based on geometry and elastic moduli and response-based scaling of the joint properties.

Unlike the approach by Liu and Yang, Van Rooij's scaling methods neglected material differences and based purely on geometry for a 6-year-old subject. In both the published works, real-world cases were used to exercise their models, due to the absence of child pedestrian biomechanical studies in the literature. While these models can potentially highlight differences in pedestrian kinematics and injury mechanism for children and adults, the scaling procedures and resulting models need to be validated against actual biomechanical data to verify the results.

A slightly more advanced model configuration than the ellipsoid multibody model is the facet model. Two studies utilizing a MADYMO human facet model have been conducted to evaluate whole body pediatric responses (Van Rooij et al. 2005; Forbes et al. 2008). These models are not multibody configurations, instead incorporating a deformable torso with a mesh representing the skin. In addition, the flexibility of the spine is more biofidelic due to additional joint models that represent the joints of the vertebral column. The child versions of these models were developed by structurally scaling the 50th percentile adult occupant model. The joint stiffnesses and damping parameters were scaled based on local geometrical changes without accounting for material or anatomical differences. Van Rooij et al. (2005) simulated 1.5- and 3-year-old human models in child restraint systems and Forbes et al. (2008) performed a validation study of a 6-year-old facet model against static neck tension and flexion/extension pediatric cadaveric tests (Ouyang et al. 2005), frontal thoracic pendulum pediatric cadaveric tests (Ouyang et al. 2006), and abdominal belt loading tests on pig surrogates (Kent et al. 2006).

Okamoto et al. (2003) reported initial efforts on an advanced 6-year-old child FE model, but the model is reportedly still under development (Ito et al. 2008). A more complex whole body child model (Mizuno et al. 2005) was developed by scaling the THUMS AM50 model, an advanced whole human body model of the 50th percentile American male developed by Iwamoto et al. (2002). The child model was structurally

Table 7.1 Pediatric whole body impact computational models

Author	Liu and Yang (2002)	Van Rooij et al. (2004)	Van Rooij et al. (2005)	Okamoto et al. (2003)	Mizuno et al. (2005)
Figure					
Software	MADYMO (Multibody)	MADYMO (Multibody)	MADYMO (Facet)	PAM-CRASH (FE)	LS-DYNA (FE)
Geometry	Geometrically scaled from 50th percentile human pedestrian model to GEBOD database anthropometry	Geometrically scaled from 50th percentile human pedestrian model to Q6 dummy dimensions	Geometrically scaled from 50th percentile adult male to Q6 dummy dimensions	Upper body unknown, lower extremities reconstructed from MRI data of a 6-year-old volunteer	Geometrically scaled from THUMS AM50
					Body region — *Scale factor*
					Head — 0.879
					Neck — 0.557
					Torso — 0.557
					Pelvis — 0.620
					Upper arm — 0.506
					Lower arm — 0.519
					Upper Leg — 0.569
					Lower leg — 0.560
					Foot — 0.560
Age	3, 6, 9, and 15 years	6 years	1.5, 3, and 6 years	6 years	3 years

(continued)

Table 7.1 (continued)

Author	Liu and Yang (2002)	Van Rooij et al. (2004)	Van Rooij et al. (2005)	Okamoto et al. (2003)	Mizuno et al. (2005)
Elements	• 15 ellipsoids • 14 kinematic joints (stiffness scaled based on response-based scaling methods of the resistive joint mechanisms)	• 64 ellipsoids • 52 kinematic joints (stiffness scaled based on dimensional scaling of local ellipsoids)		• Upper body = rigid segment linked by joint elements (dimensional scaling method from adult data) • Lower body = deformable elements – Shell: skin, medial collateral ligament – Solid: muscle mass, bones, meniscus, all other ligaments	• 65,947 nodes • 102,661 elements

scaled from the adult model to represent a 3-year-old child (Table 7.1). The biomechanical behavior of the model was compared to 3-year-old target corridors created through response-based scaling methods. The scale factors used to create the model were based mostly on height, to allow for the use of a single scale factor in all orthogonal directions. While this facilitated the preservation of the cross-sectional shape, it was noted that the proportions of a child's body in each region are not similar to the proportions of an adult. The scale factors were nevertheless adjusted so that the differences in representative lengths in each direction were within 10 % of the baseline geometry. This does not account for anatomical changes that are inherent in development and growth, which may affect biomechanics in many instances.

The large head size required a change in node connectivity from the original THUMS model. Other regions, such as the extremities, are connected with beam elements are required no modification. Scaling methods were also used to modify skin, cortical bone, and ligament thickness in shell elements. Material properties for this study were determined using both scaled (Irwin and Mertz 1997) and unscaled data, shown in Table 7.2.

For whole body multibody and finite element computational models, development of models has relied heavily on scaling methods for both geometry and material definition. To ensure that scaling produces accurate models, validation studies are needed that investigate the whole body models on a global kinematics level as well as on a component level. Currently published models overall are lacking in this area, although the results provide a first approximation of pediatric kinematics and response in an impact situation.

Head Models

In pediatric impact biomechanics, finite element models have been developed to investigate skull deformation and how it relates to skull fracture or brain injury in the pediatric population. The development of each head or skull model is summarized in Tables 7.3 and 7.4 together with references and information including model visualization, solver (software), geometry source, represented age, number and type of elements, and material properties with constitutive models and source of data. In this section, details of model development, as they pertain to conclusions drawn from the simulation results, will be discussed.

Models with simplified geometry and material properties have been used to study the effects of infant head compliance due to the material behavior of non-fused cranial sutures. Kurtz et al. (1998) created a partial infant head model representing the 3-month-old braincase as a semiellipsoid (Table 7.3). Although material properties for the bone and suture were not scaled, they were derived from a sample size of one 3-month-old subject (Runge et al. 1998) indicating elastic–plastic behavior of the bone. In this model, the sutures were represented by linear springs, a technique that has not been employed in subsequent models, given that such elements support tension but not bending. Through simulated impact scenarios, this

Table 7.2 Material properties reported for the 3-year-old FE model (modified from Mizuno et al. 2005)

Component	Constitutive model	Density (kg/m³) (kg/m)[a]	Elastic modulus (MPa) (N/strain)[b]	Yield stress (MPa)	Ultimate strain (%)
Head					
Skull	Rigid	N.R.	–	–	–
Brain	Rigid	N.R.	–	–	–
Spine					
Vertebrae	Rigid	1,000	–	–	–
Nucleus pulposus*	Elastic	1,000	0.013	–	–
Anulus fibrosus (outer)*	Elastic	1,000	13.3	–	–
Anterior longitudinal ligament	Seatbelt	0.0324[a]	101[b]	–	–
Posterior longitudinal ligament	Seatbelt	0.0052[a]	113[b]	–	–
Interspinous ligament	Seatbelt	0.0252[a]	219[b]	–	–
Intertransverse ligament	Seatbelt	0.002[a]	31[b]	–	–
Thorax					
Sternum cortical bone	Piecewise linear plasticity	2,000	5,460	98	5.3
Sternum cancellous bone	Piecewise linear plasticity	862	19	1.4	14.9
Rib cortical bone	Piecewise linear plasticity	2,000	5,460	59	2.7
Rib spongy bone	Piecewise linear plasticity	862	19	1.4	14.9
Costal cartilage*	Piecewise linear plasticity	1,000	2.3	3.8	25.5
Abdomen					
Abdominal muscle and fat*	Viscoelastic	1,210	–	–	–
Lower and upper abdomen*	Crushable foam	1,000	38	–	–
Pelvis					
Pelvis cortical bone	Piecewise linear plasticity	2,000	8,080	17.9	–
Pelvis cancellous bone	Elastic	1,000	70	–	–
Lower extremity					
Femur cortical bone	Piecewise linear plasticity	2,030	8,220	52	2.4
Femur cancellous bone	Isotropic elastic-plastic	1,030	14	1.5	–

N.R. not reported, *asterisk* without scaling

Table 7.3 Pediatric head impact FE models with simplified geometry

Author	Kurtz et al. (1998)	Margulies and Thibault (2000)	McPherson and Kriewall (1980)
Figure			
Software	LS-DYNA3D	ANSYS/LS-DYNA3D	
Geometry	Idealized	Idealized	Orthographic projection from three orthogonal radiographs
Age	3 months	1 month	Newborn
Elements	• 25,279 eight-node hexagonal solids for the indentor, brain, and CSF • 5,514 four-node quadrilateral shells for the bone, foramen magnum, and dura • 137 two-node, 1D springs	12,772 total elements	
Materials	Elasto-plastic bone: $E = 880$ MPa $\sigma_y = 12$ MPa $\sigma_t = 47$ MPa $v = 0.28$ Suture (springs): $k = 189$ N/mm Linear viscoelastic brain (from porcine): $G_0 = 5.99\text{e-}3$ MPa $G = 2.32\text{e-}3$ MPa $\beta = 9.43\text{e-}2$ s $K = 2.110$ MPa Dura and foramen magnum: $E = 100$ MPa	Bone: $E = 1.300$ MPa $\rho = 2.150$ kg/m^3 $v = 0.28$ Suture: $E = 200$ MPa $\rho = 1.130$ kg/m^3 $v = 0.28$ Linear viscoelastic brain (from porcine): $G_0 = 5.99\text{e-}3$ MPa $G = 2.32\text{e-}3$ MPa $\beta = 9.43\text{e-}2$ s $K = 2.110$ MPa Dura and foramen magnum: $E = 100$ MPa	Orthotropic elastic bone: $E_{\text{radial}} = 3.86$ GPa $E_{\text{tangential}} = 965$ Mpa

Table 7.4 Pediatric head impact FE models with complex geometry

Author	Lapeer and Prager (2001)	Klinich et al. (2002)	Roth et al. (2007a, b; 2008)	Coats et al. (2007)	Zhang et al. (2007)
Figure					
Software	–	LS-DYNA	–	ABAQUS/Explicit	MSC Patran
Geometry	Laser scanning of an infant skull replica	CT of 27 week subject; facial geometry from Zygote model	CT of 6 month subject	CT and MRI of 5-week-old subject; suture geometry idealized	CT of 7-year-old subject, craniofacial sutures reconstructed
Age	Newborn	6 months	6 months	1.5 months	7 years
Elements	Triangular shells	• Shell: scalp • Solid: CSF, dura, and brain • Thick shell: skull and suture	• 69,324 bricks (brain, CSF, scalp) • 9,187 shells (falx, tentorium, fontanels, sutures, skull)	• 11,066 ten-node tetrahedral solids for the brain • 18,706 eight node hexagonal continuum shells for the skull • 2,485 four-node membrane elements for the sutures • 624 eight-node hexagonal solids for the scalp	• 57,481 nodes • 52,901 elements

Materials

In-plane orthotropic skull: Elasticity matrix (GPa) = 3.901 0.215 0 0.215 0.977 0 0 0 1.582 Hyperelastic Mooney-Rivlin, sutures and fontanels: $C_1 = 1.18$ MPa $C_2 = 0.295$ MPa	Skull (from porcine): $\rho = 2,150$ kg/m^3 $E = 3.0$ GPa $v = 0.22$ Skull (from porcine): $\rho = 2,150$ kg/m^3 $E = 1.95$ GPa $v = 0.22$ Brain (from porcine): $G_0 = 5.99$ kPa $G = 2.32$ kPa $\beta = 9.43e\text{-}2$ s $K = 2,110$ MPa CSF (from adult): $\rho = 1,040$ kg/m^3 $E = 70$ kPa $v = 0.499$ Dura (from adult): $\rho = 1,133$ kg/m^3 $E = 31.5$ MPa $v = 0.45$ Scalp (from adult): $\rho = 1,200$ kg/m^3 $E = 17$ MPa $v = 0.42$ Face (estimated): $\rho = 9,000$ kg/m^3 $E = 30$ kPa $v = 0.22$	Skull: $\rho = 2,150$ kg/m^3 $E = 2.5$ GPa $v = 0.22$ Skull: $\rho = 2,150$ kg/m^3 $E = 1.5$ GPa $v = 0.22$ Brain (from porcine): $G_0 = 5.99$ kPa $G = 2.32$ kPa $\beta = 9.43e\text{-}2$ s $K = 2,110$ MPa CSF (from adult): $\rho = 1,040$ kg/m^3 $E = 12$ kPa $v = 0.49$ Dura (from adult): $\rho = 1,040$ kg/m^3 $E = 31.5$ GPa $v = 0.45$ Scalp (from adult): $\rho = 1,200$ kg/m^3 $E = 16.7$ MPa $v = 0.42$	Nonlinear isotropic viscoelastic hyperelastic (Odgen) brain: (μ and α scaled from adult, using porcine ratios) $\mu = 559$ Pa $\alpha = 0.00845$ $\rho = 1,040$ kg/m^3 $v = 0.499$ Orthotropic linear elastic bone: $\rho = 2,090$ kg/m^3 $v_{12} = 0.19$ Parietal $E1 = 453$ MPa Parietal $E2 = 1,810$ MPa Parietal $G = 662$ MPa Occipital $E1 = 300$ MPa Occipital $E2 = 1,200$ MPa Occipital $G = 503$ MPa Linear elastic suture (ρ and v from adult primate): $\rho = 1,130$ kg/m^3 $v = 0.49$ $E = 8.1$ MPa Linear elastic scalp (from adult primate): $\rho = 1,200$ kg/m^3 $v = 0.42$ $E = 16.7$ MPa	Cortical bone: $E = 13.7$ GPa $v = 0.3$ Cancellous bone: $E = 7.9$ GPa $v = 0.3$ Teeth: $E = 20.7$ GPa $v = 0.3$ Suture: $E = 7.1$ GPa $v = 0.45$

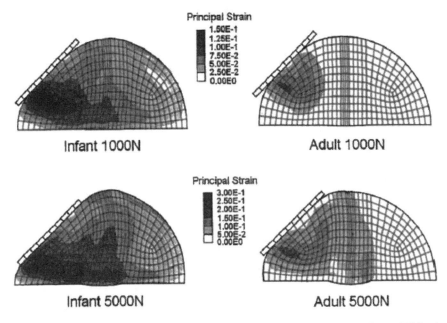

Fig. 7.1 Strain distributions in adult and infant braincase models (Image: Margulies and Thibault 2000)

study suggests that occipital fracture is more likely than parietal fracture given similar oblique loading to the parietal area.

Another anatomically simplified model was created by Margulies and Thibault (2000) to compare biomechanical response between adults and infants (Table 7.3). For the skull, the bone was considered purely elastic, with the adult model's elastic modulus of 10 GPa compared to the infant model's moduli of 1.3 GPa for bone and 200 MPa for suture. Unlike the Kurtz model, the sutures were modeled as a soft elastic material. A homogenous brain mass was included with the same material properties for infant and adult, directly connected to the skull with a no-slip interface, similar to the Kurtz model. A half-sine load at two amplitudes was applied to the models, as shown in Fig. 7.1. Skull deformations and the accompanying intracranial strains affecting the brain were noticeably affected by the change in elastic modulus from adult to infant. Peak intrusion was more than double for the infant as compared to the adult, and the response of the brain showed diffuse bilateral strain distribution in the infant and focal unilateral distribution in the adult. Model predictions therefore indicate that impacts causing focal brain injuries in adults may yield diffuse injuries in children, due to the more compliant braincase.

More anatomically detailed models have been developed as well. To combat the difficulties of validating a pediatric model, Klinich et al. (2002) developed a FE model of a 6-month-old infant head model to reconstruct real-world cases of automotive impact. They compared the results from sled reconstructions of the same cases using a 6-month-old crash dummy, and drew conclusions about infant head impact response and skull fracture tolerance. The geometry for this model was based on CT images, with the average skull thickness measured from each CT slice and used to map the inner skull surface with nodal connectivity to the dural layer and homogeneous brain. The face is modeled as rigid, using geometry from the Zygote infant model (Zygote Media Group, Inc., Lindon, UT). Material properties included a combination of young porcine and adult human values from the literature (Table 7.4). A series of parametric studies utilizing loading conditions similar to a frontal impact (based on head CG accelerations measured in the CRABI dummy) was performed to quantify the effects of using adult and estimated properties for the brain, CSF, dura, and scalp. Although some degree of model assessment was obtained by combining real-world injury with dummy kinematics, no internally measured brain-specific biomechanical data was available, which greatly limited the validation process. Although the real-world cases were occipital impacts, parietal fractures occurred as a result of outward bending; this behavior could not be predicted by the model based on the von Mises stress criterion employed.

Other researchers have foregone the validation of their models and simply used the models to make qualitative comparisons between child and adult biomechanics. Roth et al. (2007a) developed a detailed 3-year-old child head model and compared it with a scaled adult head model. The numerical results showed that scaling an adult head to obtain a child head is not accurate. Roth et al. (2007b) evaluated traumatic injuries from shaking and impact loading based on angular and linear velocity, acceleration, and relative motion between brain and skull using a FE model of a 6-month-old child. The FE model was created from computed tomography data including main anatomical features of the skull, tentorium, fontanels, falx, cerebrospinal fluid (CSF), scalp, cerebrum, and cerebellum. The brain, CSF, and scalp were modeled as three-dimensional brick elements, while the remaining features were represented by a layer of shell elements and the bridging veins with spring elements. Material properties used were similar to the Klinich model, though the skull and sutures were slightly less stiff, and the additional anatomic features utilizing adult properties when pediatric data were not available. By keeping the level of detail similar between the child and adult models, some degree of uncertainty was removed, but it should be noted that the mesh densities of the two models were visibly disparate, which will affect model predictions and possibly discount comparisons.

Roth et al. (2008) further investigated injury mechanisms under frontal, lateral, and occipital impacts using the FE model of a 6-month-old child. Frontal, lateral, and occipital impacts against a rigid wall at 1 m/s were simulated. Comparisons between the scaled adult model and the infant model showed differing pressure and stress distributions and time histories for all cases. "Fracture prediction" from stress criteria showed differing locations in each simulation, with the infant model showing a more reasonable prediction compared to the scaled model in terms of fracture orientation, based on one real-world case (Fig. 7.2). It is unclear whether any concrete conclusions can be drawn from a single case, especially when considering the results of the Klinich model.

Another detailed infant head model was developed by Coats et al. (2007), and validation of the model was attempted. The geometry for this model was based on

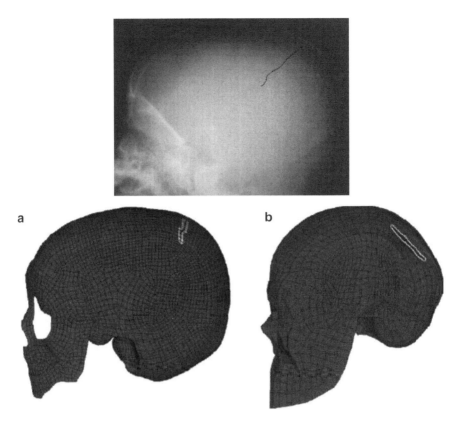

Fig. 7.2 Fracture prediction from a real-world case (*upper*), showing better agreement with the infant model (*lower*, **a**) than with the scaled adult model (*lower*, **b**) (image: Roth et al. 2008)

MRI and CT images of a 5-week-old, and the MRI data were used to create closed boundary contours for the outer skull surface using a closed-boundary edge-detecting algorithm. The thickness of each skull plate was measured from CT, and the outer surface projected inward accordingly to create internal cranial structures, such as a homogeneous brain. Unlike previously published models, tetrahedral elements and an Ogden constitutive law instead of hexahedral elements and a linear viscoelastic material law were used for the brain. Additionally, the bone was modeled as orthotropic, stiffer in the radial direction than tangential with different values for the parietal and occipital bones (parietal being stronger). Stress-based fracture predictions from the model were compared to experimentally produced skull fractures in infant cadavers from the German literature (Weber 1984, 1985) to match general location. While this was relatively subjective, it represented a way to use the limited pediatric cadaver data available. Additionally, a series of parametric studies was performed to investigate the effects of changing five parameters (brain stiffness, brain compressibility, suture thickness, suture width, and scalp inclusion). The parametric study results showed that increasing brain stiffness or altering brain incompressibility had an effect on peak stress and force parameters, as well as an effect on contact

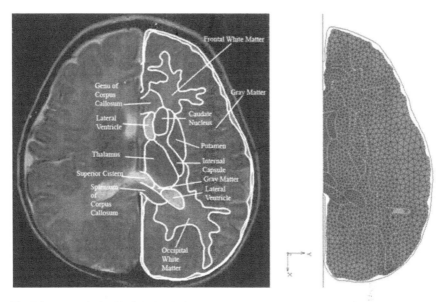

Fig. 7.3 Anatomic detail of the 2D SBS model (image: Couper and Albermani 2008)

duration when incompressibility was changed. The suture parameters were shown to affect overall biomechanical response as well.

In addition to the braincase models described, a model specifically focusing on craniofacial impact has been developed (Zhang et al. 2007). CT scan images of a 7-year-old female were used to model frontal impact to the zygoma and to elucidate the role of the craniofacial sutures in pediatric facial injury biomechanics, although few details on model development were given. The reported information is shown in Table 7.1. The model predicted that the immature sutures acted to reduce stress transmission into the deep layer of the skull. However, the large stress gradient observed in the craniofacial sutures also made them vulnerable for injury due to high shear and tensile stresses. It is unclear how to interpret this result, as facial fractures are biomechanically related to sinus pneumatization in developing children (Chan et al. 2004), which is not taken into account by this model, nor is a brain included.

Computational models developed in other areas of biomechanics may be relevant to research interests in impact biomechanics. For example, studies on Shaken Baby Syndrome (SBS) and fetal head molding during birth (Fig. 7.3). Couper and Albermani (2008) developed 2D FE models of unilateral transverse cross-sections of the infant brain to study a mechanism of the SBS. The geometry for the model is based on MRI data from a 3-month-old infant, yielding identification of the dura mater, falx, white and gray matter, corpus callosum, lateral ventricle, superior cistern, caudate nucleus, internal capsule, thalamus, and putamen. Each mesh yielded a model of approximately 4,000 triangular plane strain elements, used due to the incompressibility of the brain material and reasonable in a two-dimensional model. Material properties were complex and considered assumptions of hyperelastic, viscous, incompressible brain behavior showing evidence of strain conditioning.

White matter and grey matter were defined separately. Although quasilinear viscoelastic theory was discussed as a possible brain tissue constitutive model, the authors elected to use a large strain Ogden formulation based on other reports in literature and its ability to model tension–compression behavior with different moduli. For small strains, a Maxwell model was utilized. The properties used were based on porcine data from various researchers, not necessarily age-matched. Lubrication theory was applied to model the CSF motion and interface as this was considered most realistic for this component, which is critical to SBS research. Tests were conducted to investigate interface types, thickness of CSF layers, dissipation variants, brain stiffness variants, nonlinearity effects, effects of myelination, and inclusion of the pia mater. The results indicated that it was imperative to represent CSF as a fluid when modeling oscillatory loads. The morphology of the gyri and inclusion of the pia mater were shown to have a large effect on biomechanics as well. Further work was suggested to determine load path. The authors of this paper also cautioned against using skull–brain interface techniques common for blunt impact modeling for oscillatory loading.

Another paper utilizing modeling to investigate SBS did not consider different skull–brain interfaces for different loading modes, and the applicability cannot be proven without the availability of intracranial validation data. Roth et al. (2007b) used the FE model of a 6-month-old child described previously, but used a different approach to investigate the issue of SBS. Both shaking and an impact to the occiput were simulated. The results showed that brain pressure and shear stress were lower during shaking than during impact, but bridging vein strain was equal for both cases. However, the maximum strain was reached late in the shaking event compared with immediately upon impact. Although further investigation would be necessary for these results to be conclusive, this study highlights the ability of computational models to simulate a variety of input conditions and play a role in elucidating injury mechanisms.

Although dynamic models are generally highlighted in this chapter, there are other applications specific to infant head biomechanics that are more static in nature and may give insight into skull deformations at higher loading rates as well. McPherson and Kriewall (1980) used rudimentary FE techniques to study fetal head molding during birth. In this process, the parietal bone undergoes most of the deformation, and a quantitative understanding of biomechanics of fetal molding was desired. Results showed bone deformation, as opposed to rigid body motion, with strains ranging from −0.0133 to 0.01. Validation was subjective, using kinematic trends, and the boundary conditions were not chosen properly even according to the authors, so unrealistic parietal overlapping occurred during simulation. Much later, Lapeer and Prager (2001) used a much more anatomically complex static model to investigate pressure distribution on the fetal skull during the first stage of labor (see Table 7.4). These birth models showing deformation of the skull can aid in the understanding of the role of cranial sutures in impact biomechanics, which is not completely understood at present. Quasi-static loading during birth compared with the dynamic loading in impact events may hint at the rate-dependent nature of pediatric suture tissue during growth.

In summary, pediatric head models in impact biomechanics can offer useful information to researchers, but it is currently difficult to consider quantitative results from the models due to lack of rigorous validation. Many of the models have been used to predict the risk of skull fracture, using stress or strain as a failure criterion, but these criteria are not well established. Although these models seek to provide insight into fracture tolerance and pediatric skull biomechanics, crack propagation and accurate fracture pattern prediction are currently beyond the scope of these models. Moreover, given the nature and purpose of these models, little attention has been paid to detailed modeling of the brain, which lags behind the complexity of currently published adult brain models. None of the published pediatric head models have utilized different properties to represent gray and white matter, although this distinction has been reported to influence injury prediction in the adult (e.g. Zhou et al. 1996; Zhang et al. 2001b). The interface conditions between the brain (pia) and skull (dura and arachnoid) also need to be examined carefully and tuned to the purpose of the model in order to achieve realistic predictions. The choice between frictionless and friction contact may affect brain response significantly.

These simplified models may offer insight into pediatric head biomechanics, but there are several limitations that should be noted. Firstly, the material properties of all the tissues represented in these models have not been completely characterized. Directional, regional, and strain rate-dependence were disregarded, and brain tissue was considered homogenous. The simplified geometry itself may induce an unacceptable degree of error, especially in the sutures. It is difficult to model the cranial sutures in an anatomically precise manner, due to thickness variations, interdigitation, and non-discrete boundary inherent in skull growth. Whether the suture representation techniques used in FE modeling of the pediatric head are appropriate remains to be confirmed. Smaller scale models may be appropriate to investigate this further and help validate the biomechanical results of these models.

Cervical Spine Models

Finite element models of the neck have been created to highlight the importance of anatomical accuracy in pediatric modeling, but often these models are created using the adult models as a baseline, as opposed to creating models directly from medical images. The complexity of the cervical spine and surrounding structures makes modeling difficult, and a lack of material behavior data for the ancillary ligaments and other important features currently requires the use of scaling methodologies. The efforts on development of pediatric cervical spine models are summarized in Table 7.5 together with references and information including model visualization, solver (software), geometry source, represented age, number and type of elements, and material properties with constitutive models and source of data.

Kumaresan et al. (1997, 2000) published two studies that investigate the effects of anatomical development on spine biomechanics. These models were created by modifying validated adult models by incorporating the local geometrical and material

Table 7.5 Pediatric neck FE models

Author	Dupuis et al. (2006)	Meyer et al. (2007)	Kumaresan et al. (1997, 2000)
Figure			(baseline adult model pictured)
Software	N.A.	Radioss	ABAQUS
Geometry	CT of 3-year-old, modified from adult model	CT of 3-year-old	Manually modified from adult model
Age	3 years	3 years	1, 3, and 6 years
Elements	• 7,340 rigid shell elements for vertebral bodies and coarse head mass • 1,068 eight-node solids for the intervertebral discs • Nonlinear shock-absorbing springs for 11 principle ligaments	• 24,758 shell elements • 712 spring elements • 2,826 volume elements	• 8,878 solids for articular cartilage, costals, and bone • 976 incompressible fluid for disc nuclei and synovial fluid • 694 composites for annulus • 517 cables for ligaments • 98 membranes for synovium

Materials	Bone: rigid Discs (scaled from adult): $E = 100$ MPa	Bone: rigid Discs (scaled from adult): $E = 100$ MPa Ligaments: similar to Dupuis model, assumed viscoelastic behavior same as adult, but failure properties lower than adult	Vertebral centrum: $E = 75$ MPa, $v = 0.29$ Growth plate and costals, posterior synchondrosis, and neurocentral cartilage: $E = 25$ MPa, $v = 0.40$ Neural arches: $E = 200$ MPa, $v = 0.25$ Ground substance: $E = 4.2$ MPa, $v = 0.45$ Disc annulus fibers: $v = 0.30$, $E =$ 1YO: 400 MPa (10 %) 3YO: 425 MPa (15 %) 6YO: 450 MPa (20 %) Articular cartilage: $E = 10.4$ MPa, $v = 0.40$ Synovial membrane: $E = 12$ MPa, $v = 0.40$ Synovial fluide and uncovertebral joint: $E = 1{,}666.7$ MPa Ligaments: Proportionate to adult values: 1YO = 80 % 3YO = 85 % 6YO = 90 %

Fig. 7.4 Developmental features included in Kumaresan's cervical spine models (Kumaresan et al. 2000)

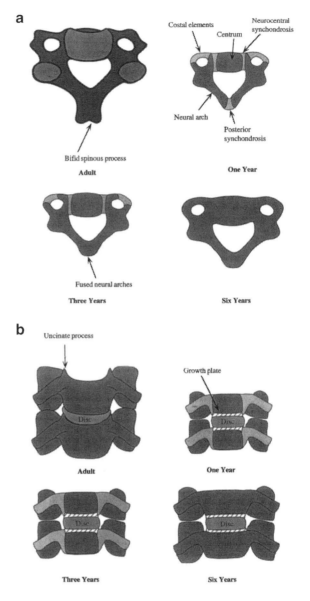

characteristics of the child anatomy (see Fig. 7.4). Using a combination of structural scaling and representative anatomical features, models of the C4–C6 motion segment for children of 1, 3, and 6 years of age were created from a baseline adult model (Table 7.4). The developmental milestones involve fusion of the primary and secondary ossification centers, as well as changes in facet angle (1 year = 60°, 3 years = 53°, 6 years = 48°). The material properties for cartilage in the pediatric spine were taken from literature (Yamada 1970; Melvin 1995), while the ligament properties were obtained by scaling down from ones of the adult model according to the developmental characteristics of the pediatric spine.

The authors made a comparison between three methods of pediatric model development based on an adult model: pure structural scaling, anatomical and material changes without scaling, and a combination of these approaches. Only some of the material properties used in this study were directly measured from pediatric subjects, and the sample sizes for these experiments were quite small. While the accuracy of these values may be questionable, trends due to growth and development can still be examined. Compressive loads were applied, and changes in normalized axial flexibility of the spine were observed. Structural scaling produced linearly related flexibility results that were within 119 % of the adult response, demonstrating that structural scaling alone has little effect on response. The anatomical and material changes produced a more nonlinear response within 465 % of the adult values. The addition of scaling to this approach yielded an increase in axial flexibility with 534 % of the adult response. Angular flexibility under flexion and extension loading was also compared.

Using the combined approach, several observations could be made that were not apparent when using structural scaling alone. Under flexion and extension, the 1-year-old was most flexible, followed by the 3-year-old and the 6-year-old. In contrast, during high compressive loading, the 3-year-old was most flexible, followed by the 6-year-old and the 1-year-old. The more horizontal facet angle in the 1-year-old model was offered as an explanation for this result. In compression and extension, flexibility decreases with increasing load, but no such trend was seen in flexion. This model creation method offers a way to investigate the biomechanical differences in different age groups even in the absence of solid material property data, but none of these conclusions can be validated due to the lack of experimental data.

Dupuis et al. (2006) developed a 3-year-old child FE model. The model included the seven cervical vertebrae (C1–C7), first thoracic spine (T1), intervertebral disc, and the principal ligaments. The cervical column of 3-year-old child was reconstructed from CT images and was used as a reference to morph and remesh an adult model. No comparisons between the geometry of the child and adult models were reported. The mass and inertia of the vertebral bodies was scaled from adult data using the response-based scaling method proposed by Irwin and Mertz (1997). The vertebral bodies were modeled as rigid structures, while the intervertebral discs were given an elastic modulus from adult data scaled in terms of force response. The behavior of the ligamentous structures represented by spring elements was defined using adult values as well. To validate this model, results were compared with experimental data from a Q3 dummy (a crash dummy) sled test (frontal and rear) because no pediatric data was available. The authors stated that no set of material parameters could be found to yield good agreement in all three orthogonal directions. This may be related to dummy biofidelity issues, especially since dummies are not designed to be omnidirectional, and underscores the problems with validating models of humans against dummy data or models of dummies.

Meyer et al. (2007) used the same pediatric image data as that in Dupuis et al. (2006) to generate a 3-year-old child neck model meshed directly from the reconstructed geometry as opposed to remeshing and morphing an adult model. The material properties were similar to the Dupuis's model. This model was validated against acceleration and displacement response corridors for lateral, oblique, rear, and frontal automotive impact tests generated by scaling adult responses without changing the magnitude of mechanical inputs. The model time histories fell reasonably within these corridors, but whether these corridors have any basis remains unknown.

Thorax and Abdomen Models

Although impact biomechanics studies have been performed on pediatric cadavers and animal surrogates related to thorax and abdomen response, to our knowledge, no computational models specific to children have been developed for these anatomical components outside of the whole body models.

Thoracic and Lumbar Spine Models

In pediatric impact biomechanics, the thoracic and lumbar spines are generally not considered as separate anatomic entities. Dynamic loading to these structures is most relevant in context with the whole body. Orthopedic applications, with quasi-static loads, have been modeled, usually on a subject-specific basis without validation of results.

Upper Extremity Models

To our knowledge, no child upper extremity FE models have been reported in literature specifically related to impact biomechanics. However, a geometrically simplified model of the infant clavicle was published by Meghdari et al. (1992) to investigate shoulder dystocia during birth for the interested reader.

Lower Extremity and Pelvis Models

A pediatric lower limb finite element model was developed using MR images of a 6-year-old (Okamoto et al. 2003). This model included anatomical features unique to pediatric anatomy and biomechanics, such as growth plates, epiphyseal cartilage, and ossification centers. No material properties or validation, however, were included in the publication. In 2008, the model was modified to include a segmented, rigid upper body for pedestrian impact simulation (Ito et al. 2008). Joint characteristics were scaled from an adult pedestrian model based on a dimensional scaling rule (Langhaar 1980). The ranges for material and failure properties for the bone were scaled from adult and canine data, limiting the confidence in this unvalidated model.

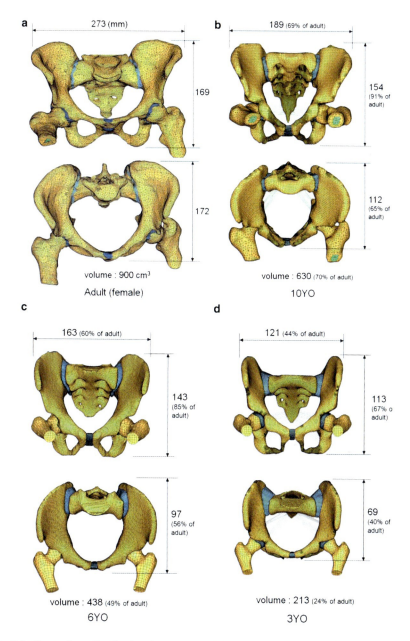

Fig. 7.5 Comparison of pediatric pelves with an adult model: (**a**) adult (Li et al. 2007); (**b**) 10YO; (**c**) 6YO; (**d**) 3YO

Additionally, 3- and 6-, and 10-year-old pelvis models have been developed from pediatric geometry (Kim et al. 2009). A geometric comparison of the pediatric pelves with an adult model is shown in Fig. 7.5, including articular cartilage and ligaments. In order to avoid intrinsic scaling error determining material properties from sparse or nonexistent datasets, an advanced technique for parameter identification was utilized based on matching model response to available cadaveric data. The details of model development process are described in Sect. "Optimization-Based Material Identification." While this technique offers a unique solution to the difficulties posed by a lack of pediatric material properties available, comparison with data from a different loading scheme would be necessary for model validation.

Conclusion

Compared with the adult, relatively few studies have been reported for modeling pediatric impact biomechanics, and the degree of anatomical detail lags behind that of the adult models. This does not, however, negate the value of such models to investigations into pediatric impact biomechanics. Any computational model inherently involves some degree of estimation, an acceptable practice. Where the pediatric models need improvement is in systematic component and global validation against the data which are available. Whether utilizing scaling or optimization methods to bridge the gap in material behavior and constitutive properties, the model predictions must be validated against well-characterized, age-matched data.

Future Directions

Introduction

Novel techniques are needed to bridge the gap between data available and data needed for pediatric modeling, and researchers are beginning to explore such ideas. High-fidelity and fully validated finite element (FE) algorithms are very critical for future development.

In this respect the following critical barriers must be addressed:

- Geometry: Anatomical geometry with topologically water-tight discrete definition of the surface is of fundamental importance. Here the term topologically water-tight geometry implies consistent orientation of all discretized decomposed subsurface normal and tangent characteristics such that resulting sum of all areas covered by subsurfaces is equivalent to the area associated with the surface of the geometry. Since many of the pediatric geometry information is obtained from medical image data, additional effort is involved in generating associated discretized water-tight geometry in three dimensions. This process is complex and time-consuming. Further research and tools are needed to simplify and improve the efficiency of this process.
- Meshing: Triangular and tetrahedral meshes are widely used for computational simulations because of their flexibility for complex geometries (Ito et al. 2006).

However, triangles and tetrahedra are usually considered to be constant strain elements in FE methods, i.e., only one value of strain is assigned for each element without any gradient. Many simplicial elements are needed to achieve the same accuracy as bilinear elements. Hence, quadrilateral and hexahedral meshes are preferred in computational structural mechanics (CSM) simulations. Novel algorithms for generating high-quality hexahedral meshes on these geometries, including both soft and hard tissues, are required for high-fidelity computational simulations.

- Material properties: Accurate representation of nonlinear and nonhomogeneous tissue properties and development of associated constitutive equations are required for appropriate material identification. As described before, this is a challenging task especially for children and needs further development with right strategies.
- Computational algorithms: Computational modeling and model optimization algorithms need considerable improvements for efficient and accurate simulations especially when multiscale analysis is involved. This is true for pediatric injuries requiring simulations of dynamic tissue behavior and musculoskeletal/ neuromusculoskeletal aspects.

These critical barriers have been addressed in the "Digital Child Project" sponsored by the Southern Consortium for Injury Biomechanics (SCIB) at the University of Alabama at Birmingham (UAB). The objective of this project is to develop age-dependent pediatric FE models aiming to a more comprehensive understanding of injury mechanisms experienced by children with a focus on the development of pediatric models. Results from model simulations will contribute to the long-term goal of improving child safety and subsequent establishment of tolerances and injury criteria among children by revising or improving current standards and regulations that lead to more efficient designs of child restraint systems. The progress realized in the development of discretized geometry from medical image data and/or cadaver data, hexahedral high-fidelity meshing algorithm, and optimization-based material identification utilizing limited experimental data under the SCIB sponsored project is described in the following sections.

Anatomical Geometry and Repair of Geometric Defects

The ongoing research efforts with the focus on defining geometry and anatomy of children under SCIB include collaboration between Wayne State University and UAB. The pediatric geometrical models have been developed based on medical image datasets in either computer tomography (CT) or magnetic resonance imaging (MRI). In particular, the datasets for 3-, 6-, and 10-year old subjects have been collected. However, since these datasets are mainly for clinical use with poor image

Fig. 7.6 Geometric defects in a set of vertebrae

resolution, the preparation of water-tight surface models was challenging. Moreover, the data represented only certain portions of body of clinical interest.

There are three tedious and labor-intensive steps in the geometry preparation: (1) the registration of medical image datasets, (2) geometry transformation after geometry components are extracted from the image datasets, and (3) the repair of geometric defects to make geometric models water-tight. The mixed-element grid generator in three dimensions (MEGG3D) (Ito et al. 2006, 2009a, b) developed at the UAB was used to manipulate discrete surface models and to create missing joints, which could not be extracted from low-resolution image data, such as sacroiliac, pubic symphisis and hip joints. The MEGG3D was also used to generate associated computational meshes.

Figure 7.6 shows a triangulated surface model for a set of vertebrae as an example of the geometric defects. Some of the triangular faces overlap each other, which are highlighted in red. Obviously, those should be modified because the vertebrae should not intersect each other. Figure 7.7 shows unreasonable sharp corners in a skull model, which is colored based on surface curvature. The red color indicates very sharp corners. The MEGG3D enables to identify and highlights these kinds of defects and provides tools for repairing some of the problems easily, but tedious manual operations sometimes have to be done to modify the geometry individually to fix these problems.

Fig. 7.7 Undesirable and inaccurate sharp corners in a skull model

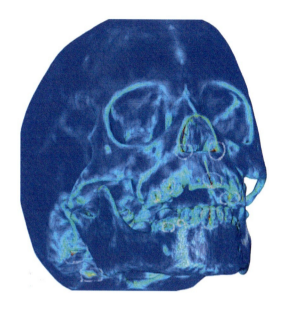

High-Quality Mesh Generation

After surface models are created, the next step is mesh generation. In FE simulations, hexahedral meshes are often preferred over tetrahedral meshes for better accuracy of numerical simulations. The meshing research community is well focusing on developing algorithms and associated software for generation of high-quality hexahedral meshes. In this respect, under SCIB sponsorship, the UAB team has developed an octree-based hexahedral mesh generation algorithm tailored for biomedical components and has been cast as a software module into the MEGG3D system (Ito et al. 2009a, b). The mesh generation algorithm is combined with a new, easy-to-understand, easy-to-implement set of refinement templates to automatically create hexahedral meshes for complex biomedical geometries with reasonable mesh quality. The addition of a buffer layer on the boundary surface and the node smoothing methods with certain restrictions improve the resulting mesh quality significantly. The starting point of the mesh generation can be either original medical image data or triangulated surface models extracted from the medical image data. Since the latter approach is more flexible to control noise in the medical image data, we use the triangulated surface models as input.

To create a hexahedral mesh automatically from a triangulated surface model using our approach, a user can specify the following parameters:

- The size of elements to create a baseline uniform mesh, L.
- The number of local refinement, l (≥ 0; no local refinement if $l=0$), based on one or the combination of the following sensors:
  - Angle α as a threshold for the normals of any two of the triangles that are contained or intersected by each leaf cube to detect geometrical features.

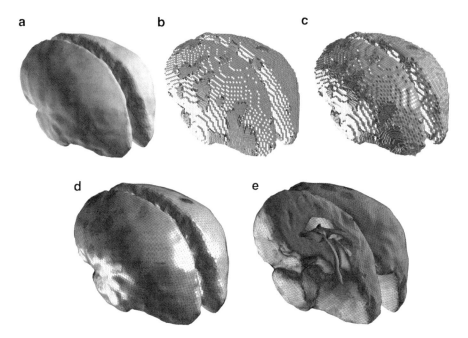

Fig. 7.8 Hexahedral mesh generation process for a 10YO brain model: (**a**) original triangulated surface model; (**b**) hexahedral bumpy core mesh; (**c**) core mesh after applied smoothing and added a buffer layer; (**d**) front and (**e**) back sides of final mesh after conforming to the boundary of the geometry

- Surface curvature.
- Thickness of local volume: thin sections will be locally inflated to create coarse hexahedral meshes (temporal local inflation).
- The position and the size of the octree (the center and the length of the surface model as default).
- Flag to specify whether or not the octree is aligned with the oriented bounding box of the surface model (temporal rotation).

Figures 7.8 and 7.9 show the mesh generation process for a 10-year-old brain model and a 6-year-old cervical vertebrae model, respectively. There are four major steps (Ito et al. 2009a): (1) the generation of a hexahedral core mesh (Figs. 7.8b and 7.9b) inside an input triangulated surface model (Figs. 7.8a and 7.9a) using six refinement templates; (2) the removal of special hexahedra that become poor-quality elements in the later steps; (3) the addition of a buffer layer (Figs. 7.8c and 7.9c) on the boundary surface to improve the quality of the final mesh; and (4) the application of node smoothing and boundary projection methods to create a boundary conforming mesh (Figs. 7.8d,e and 7.9d). The geometry preparation and mesh generation were performed for 3-, 6-, and 10-year-old child geometric models. Figs. 7.10, 7.11, and 7.12 show examples of meshes generated with the MEGG3D.

7 Pediatric Computational Models

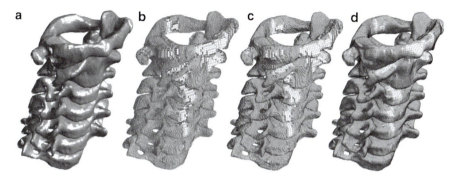

Fig. 7.9 Hexahedral mesh generation for 6YO cervical vertebrae: (**a**) triangulated surface model; (**b**) hexahedral bumpy core mesh; (**c**) core mesh after applied smoothing and added a buffer layer; (**d**) hexahedral mesh after conforming to the boundary of the geometry

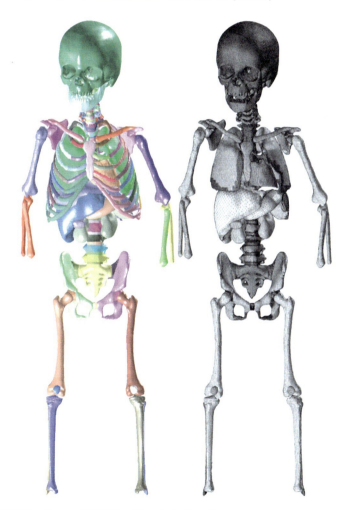

Fig. 7.10 3YO surface model (*left*) and hexahedral meshes (*right*)

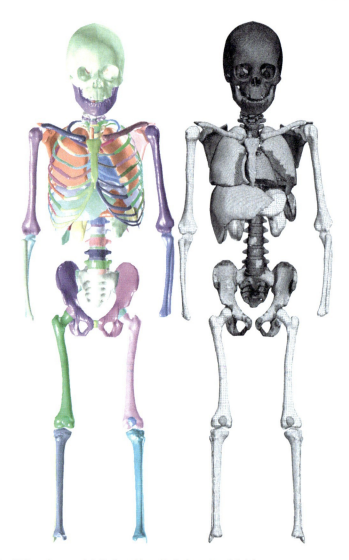

Fig. 7.11 6YO surface model (*left*) and hexahedral meshes (*right*)

This hexahedral mesh generation method can be applied to any other surface models without sharp features. A pictorial view of hex-mesh generated on human brain is presented in Fig. 7.13. Human brain is constructed from a set of CT-scan images and 2.2 M hexes are generated. A view of the mesh sliced through the brain is presented in Fig. 7.14. As displayed good-quality hex meshes could be generated with high resolution essential for accurate simulations.

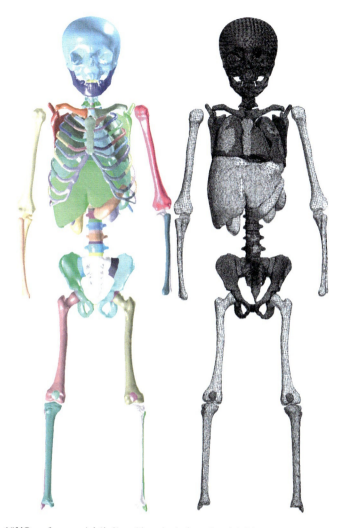

Fig. 7.12 10YO surface model (*left*) and hexahedral meshes (*right*)

Model Practice

An FE model of a 10-year-old child head was developed based on the mesh shown in Fig. 7.8. Cadaveric impact tests using adult heads by Nahum et al. (1977) were replicated to simulate a head impact scenario as a preliminary study. Figure 7.15 shows the head and impactor models. For model validation, a generic cylindrical impactor (with a Young's modulus of 76 GPa) and padding material with a shear modulus of 1.7 MPa was attached to the impactor. The total mass of the impactor is 5.59 kg, which is the same as the mass used in the test. For the preliminary test, the brain (white and grey matters), cerebellum, and brain stem were assumed as a linear

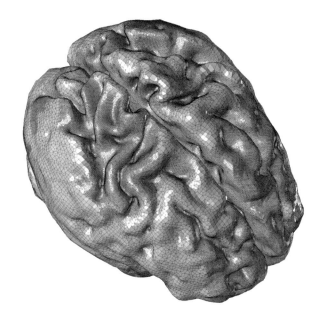

Fig. 7.13 Hex-mesh generated on human brain

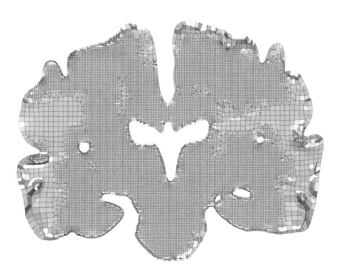

Fig. 7.14 Mesh sliced through the brain

viscoelastic material. The gray matter properties from Zhang et al. 2001a, were used in this test; short-term shear modulus = 10 kPa, long-term shear modulus = 2 kPa, decay constant = 80/s, and bulk modulus = 2.19 GPa for all of the brain components. The CSF was assumed as a fluid material. The density and bulk modulus of the CSF were set to 1,004 kg/m^3 and 2.19 GPa, respectively (Zhang et al. 2001b). The skull was assumed as a linear elastic material with a Young's modulus of 15 GPa and Poisson's ratio of 0.22. Contact interfaces were defined between the skull and CSF

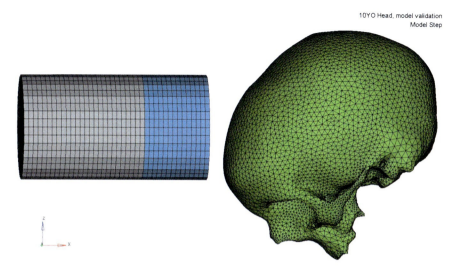

Fig. 7.15 FE models of the head and impactor for preliminary study and model validation

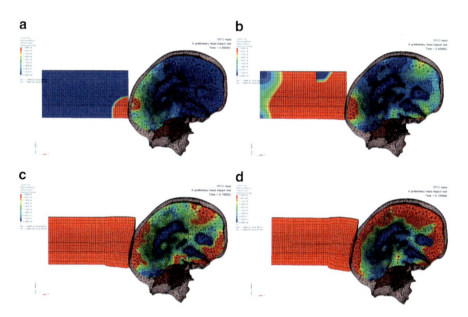

Fig. 7.16 Von-Mises stress distribution during a 10YO head impact (sectional view): (**a**) time at 4.95 ms; (**b**) 5.4 ms; (**c**) 6.75 ms; (**d**) 9.15 ms

and between CSF and the brain components. Figure 7.16 shows Von-Mises stress distribution within the brain during the impact test. However, this study has a potential limitation to validate the model due to no experimental test data regarding age-specific pediatric heads.

Optimization-Based Material Identification

For the maximum use of very limited experimental data on children, an efficient methodology is warranted. A systematic method, *inverse FE and optimization-based material identification method*, is derived to determine material properties that should otherwise be estimated by ineffective heuristic means. The objective is to find an optimal combination of the material property set of the pediatric bone and its associated soft tissues, with which FE model responses were well agreed with available experimental data.

An application of this method to the model development of a child pelvis is demonstrated here. Pelvic bone fractures and injury mechanisms in children may differ substantially from those for adults due to anatomical differences between pediatric and adult populations (Silber and Flynn 2002; Vitale et al. 2005). Many experimental side impact studies have been conducted to investigate the biomechanical properties, structural responses, and injury tolerances of the adult pelvis using whole cadavers or isolated pelves under different test scenarios such as drop tower (Beason et al. 2003), pendulum (Viano et al. 1989; Zhu et al. 1993), and sled tests (Cavanaugh et al. 1990; Yoganandan and Pintar 2005). FE models of the adult pelvis were successfully used to evaluate the pelvic structural responses under quasi-static or impact load scenarios that complement experimental studies (Dawson et al. 1999; García et al. 2000; Majumder et al. 2004; Anderson et al. 2005; Song et al. 2006; Li et al. 2007).

None of these studies, however, involved pediatric pelves, and relatively little information is available on the material properties, structural responses, and injury tolerance of the pediatric pelvis. One experimental study on lateral impact responses of pediatric pelves was reported using 12 whole child cadavers with ages ranging from 2- to 12-year old (Ouyang et al. 2003). The data from the experimental study were used to find an optimal combination of the material properties of a pediatric pelvic bone and its associated soft tissues.

Summary of the Experimental Study

The data from four cadaveric specimens (6-, 7.5-,and two 12-year-olds) in the experimental study (Ouyang et al. 2003) were used in the optimization process. All pelves of the child cadavers were laterally impacted by a free plate with diameter of 18 cm, an average mass of 3,28 kg, and an average impact velocity of 7.5 m/s. To reduce stress concentrations which may result in local pelvic bone fractures at the impact areas, a layer of 3/8-in. thick sorbothane foam was used to cover the impactor surface, and a layer of 3/8-in. thick neoprene foam was placed behind the impacted cadavers. The contralateral side was firmly constrained while the femurs of the cadavers were perpendicular to the impact direction and free to move. The response of the iliac wing to the lateral impact, impact force, pelvic deformation, and the pelvic viscous criteria (compression times velocity of deformation) were recorded to investigate the possibility of pelvic injury.

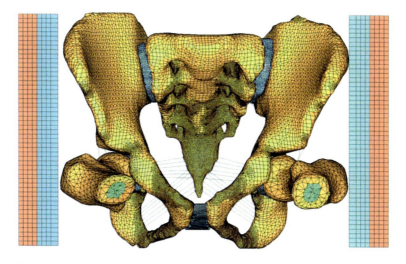

Fig. 7.17 Lateral impact FE model of 10YO-pelvis created based on experimental settings

FE Model Development

An in-house FE model of a 10-year-old (10YO) child pelvis was developed based on an axial computed tomography (CT) scan of a 10YO female abdomen and pelvis. The automatic octree-based hexahedral mesh generator (Ito et al. 2009a, b) was used to create eight-node hexahedra for the bony structure. The hexahedral FE model of the pelvis represented the trabecular bone, and a uniform thin layer of quadrilateral surface shell elements was used to represent the outer cortical bone. The pelvic joints, including the pubic symphysis and sacroiliac and hip joints, could not be reconstructed from the CT data as mentioned before. Thus, the joints were manually reconstructed by filling gaps between the bony structures and meshed by hexahedral elements. In accordance with the experimental boundary and loading conditions (Ouyang et al. 2003), a side impact FE model of the 10YO pelvis was developed as shown in Fig. 7.17. Tension-only spring or beam elements were used for the ligaments. Hypermesh software (Altair Engineering, Inc., Troy, MI) was used for model development. Computational analyses were performed under dynamic impact loading conditions in accordance with the experimental settings using a nonlinear explicit dynamic FE code, LS-DYNA3D. The details of modeling process can be found in the published paper (Kim et al. 2009).

Sensitivity Analysis

The FE model of the pediatric pelvis was simulated using trial material properties scaled by a factor of 0.7 from the adult pelvic data. The material properties of the adult pelvis are listed in Table 7.6. Poisson's ratios were not scaled or varied subsequently.

Table 7.6 Material properties of the adult pelvis from literature

Components	Constitutive model	Properties	References
Cortical bone	Isotropic elastic	Young's modulus: 17 GPa Poisson's ratio: 0.3 Average thickness: 2.0 mm	Dalstra et al. (1995) Majumder et al. (2004)
Trabecular bone	Isotropic elastic	Young's modulus: 70 MPa Poisson's ratio: 0.2	Dalstra et al. (1995)
Interpubic joint cartilage	Hyperelastic (Monney-Rivlin)	$C_{10}=0.1$ MPa, $C_{01}=0.45$ MPa, $C_{11}=0.6$ MPa	Li et al. (2006)
Sacroiliac joint cartilage	Hyperelastic (Monney-Rivlin)	Two parametric $C_1=4.1$ MPa, $C_2=0.41$ MPa	Anderson et al. (2005)
Hip joint cartilage	Hyperelastic (Monney-Rivlin)	Two parametric $C_1=4.1$ MPa, $C_2=0.41$ MPa	Anderson et al. (2005)
Interpubic ligament	Four spring elements	Spring constant: 0.543 kN/mm	Dakin et al. (2001)
Sacroiliac ligament	16 Discrete truss elements	Young's modulus: 250 MPa Poisson's ratio: 0.4 Area: 320 mm^2	Yamada (1970) Bechtel (2001)
Hip ligament	18 Discrete truss elements	Young's modulus: 181 MPa Poisson's ratio: 0.4 Area: 300 mm^2	Hewitt et al. (2001)
Sacrospinous ligament	12 Spring elements	Spring constant: 1.5 kN/mm	Phillips et al. (2007)
Sacrotuberous ligament	12 Spring elements	Spring constant: 1.5 kN/mm	Phillips et al. (2007)
Surrounding soft tissue	Hyperelastic (Monney-Rivlin)	Two parametric $C_1=85.5$ kPa, $C_2=21.38$ kPa	Majumder et al. (2007)

Sensitivity studies were conducted to quantify the effects of varying properties by a range of scaling factors from 0.4 to 1.0 (1.0 = adult data), when other properties were not varied. Linear regression analysis was then performed to evaluate the association between the variation of material properties and model outcomes.

Three outcomes were calculated: peak impact force (F_{max}), peak pelvic compression (C_{max}), and peak pelvic viscous criteria (VC_{max}). Sensitivity was determined when significant probabilities ($p \leq 0.05$) occurred between the variation of material properties of the components and the model outcomes. The sensitivity results indicated that F_{max} was sensitive to the modulus (E_C) and thickness (T_C) of the cortical bone. C_{max} and VC_{max} were sensitive to the moduli of trabecular bone (E_T) and the sacroiliac ligament (E_{SIL}) as well as E_C and T_C. Therefore, these four properties were selected for further optimization.

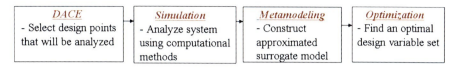

Fig. 7.18 Procedure of surrogate based optimization

Optimization-Based Material Identification

Through the sensitivity analysis, key material parameters (E_C, T_C, E_T, and E_{SIL}) that affect model responses significantly were selected as independent variables in an optimization process. The surrogate-based optimization process, which is an integration of the metamodeling (surrogate model) and optimization techniques, was used to find an optimal combination of material properties. Figure 7.18 illustrates the process of the surrogate-based optimization.

The DACE, which stands for design and analysis of computer experiments, is a process to choose the number and distribution of the design points to be analyzed. Based on the simulation results, metamodeling strategies can be employed such as polynomial-based response surface methodology (RSM) (Myers 1971; Roux et al. 1998), radial basis functions (Powell 1987; Gutmann 1999), Kriging (Sacks et al. 1989; Koehler and Owen 1996), multivariate adaptive regression splines (Friedman 1991), and artificial neural network (Zhang and Gupta 2000).

For this study, 20 analysis points according to scaling factors (0.4–1.0) of E_C, T_C, E_T, and E_{SIL} determined by Latin hypercube sampling (a DACE) (Iman et al. 1981) were used as listed in Table 7.7 along with the impact simulation results. Based on the simulation results, second-order quadratic polynomials (a RSM) were used for the meta-modeling. The multi-start-based global optimization technique (Haftka and Gürdal 1992) was used to find the best solution among many local optimums, and the global criterion formulation (Rao 1984) was used to minimize multiple objective functions simultaneously. The objective functions to be minimized were the percent differences between F_{max}, C_{max}, and VC_{max} comparing simulation and experimental data, where the averaged data from the four subjects (mean age 9.4 years), $\overline{F}_{max}$ (3,327.5 N), $\overline{C}_{max}$ (29.25 %) and $\overline{VC}_{max}$ (0.80 m/s), were used (Ouyang et al. 2003). Table 7.8 lists the resulting optimized material properties for the 10YO child pelvis through the optimization process. Table 7.9 shows the comparison of model responses using the optimized properties with the experimental results, indicating good agreement (within 5 %) for the three output parameters. Figure 7.19 shows a lateral impact simulation with the optimized material properties.

Discussion

A limitation of this study was that the other properties (not listed in Table 7.7) were not estimated since they did not substantially affect the model responses under the lateral impact ($p \geq 0.05$). Therefore, including these parameters in the optimization

Table 7.7 Analysis points and percent differences between simulations and averaged experimental data

Analysis number	E_C	T_C	E_T	E_{SIL}	$F_{max}\|_{\%diff}$	$C_{max}\|_{\%diff}$	$VC_{max}\|_{\%diff}$
1	0.829	0.402	0.733	0.814	−6.236	2.701	14.625
2	0.627	0.478	0.464	0.537	−11.357	9.915	21.250
3	0.991	0.548	0.880	0.988	−0.210	−8.308	1.750
4	0.641	0.919	0.985	0.657	4.748	−8.855	−0.500
5	0.940	0.551	0.836	0.718	−0.457	−5.778	4.625
6	0.544	0.896	0.432	0.746	0.905	−7.111	2.625
7	0.874	0.492	0.580	0.961	−4.754	−1.538	9.375
8	0.924	0.721	0.711	0.881	3.585	−10.598	−0.750
9	0.464	0.652	0.565	0.594	−5.202	4.718	13.875
10	0.413	0.622	0.507	0.781	−8.427	6.735	18.125
11	0.757	0.850	0.778	0.404	5.758	−7.966	1.250
12	0.432	0.827	0.880	0.617	−1.566	−4.957	5.500
13	0.581	0.582	0.943	0.863	−3.159	−1.744	10.000
14	0.708	0.777	0.408	0.511	1.025	−3.487	6.250
15	0.772	0.944	0.522	0.453	4.838	−9.470	−0.625
16	0.894	0.696	0.916	0.842	1.247	−12.513	−2.625
17	0.819	0.457	0.691	0.482	−3.892	2.222	14.500
18	0.675	0.811	0.665	0.575	4.111	−6.530	3.375
19	0.507	0.732	0.636	0.676	−2.476	−0.615	11.125
20	0.551	0.979	0.790	0.928	5.256	8.615	1.375

Table 7.8 Estimated material properties of the 10YO pelvis

	Properties of 10YO pelvis	Scaling factor
E_C	12.24 (GPa)	0.72
T_C	1.6 (mm)	0.80
E_T	44.8 (MPa)	0.64
E_{SIL}	140.0 (MPa)	0.56

Table 7.9 Model validation

	Experiment	Model responses	% Difference
F_{max} (N)	3,327.5	3,387.2	1.79
C_{max} (%)	29.25	27.84	4.82
VC_{max} (m/s)	0.80	0.816	2.00

process is useless. Instead, they were assumed as 70 % of the adult based on a geometrical scaling factor. The total volume of the 10YO pelvis is approximately 70 % of that of an in-house adult female pelvis. Another limitation of this study is that the experimental data used for validating the FE model were from the whole pelvic impact response. Responses of individual joints/ligaments/tissues are essential to validate an FE model joint by joint and tissue by tissue; however, they

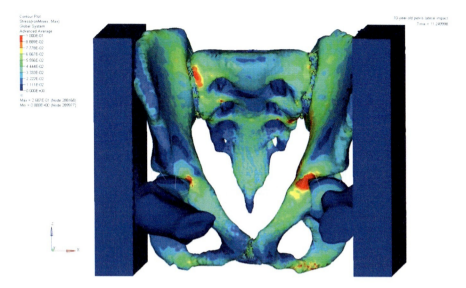

Fig. 7.19 Model simulation with optimized properties, color contour: Von-Mises stress distribution

were not measured in the experimental study nor anywhere else. The material properties estimated in this study were a result of the material identification process, with which the overall model responses to a lateral impact were well matched with the pelvic responses from the experimental study (Ouyang et al. 2003). More improved quantification of the pediatric properties, especially interpubic and sacroiliac joint properties, would be performed when joint- and tissue-level experimental data for children are available. These limitations may yield imperfection in the material property estimation. Nevertheless, this study presents an important and necessary step towards age-dependent computational models for children, which may ultimately serve to evaluate injury mechanisms and means of protection for the pediatric population.

Multiscale Computational Simulations

Computational models associated with dynamic tissue behavior, blood vessels, membranes, and musculoskeletal/neuromusculoskeletal as well as pathology of organs involve fluid, structure, and molecular simulations. The modeling and simulation of such complex behavior involves disparate time- and length scales associated with fluid and structural body dynamics, tissue/membrane/muscle biomechanics, and pathology of organs (Imielinska et al. 2006). Musculoskeletal computational models with simplified joints (e.g., hip joints as spherical joint), passive structure (e.g. modeling of ligaments as nonlinear springs), muscle, and motor control strategies are widely used in injury biomechanics studies (Tawhai et al. 2009). However, understanding of stress–strain profile at the joints (Tawhai et al. 2009) or scenarios in which models of muscle coordination coupled with the detailed representation of joints and tissues are needed, warrants multiscale modeling. In injury biomechanics, the coupling of multiscale models is relatively novel (Cloots et al. 2008) especially in pediatrics modeling where understanding of injury and loading on hard and soft tissues as well as muscles and interactions with entire body and skeletal framework is critical.

While it is established that forces at higher scales influence behavior at lower scales and that lower-scale properties influence higher-scale response, these relationships and interactions are rarely included explicitly in computational models (Tawhai et al. 2009). As a result, robust multiscale computational models ranging from nanoscale to macroscale and seamless bridging methods between the different scale domains are still lacking. Two different type of multiscale modeling approaches have been explored in literature: sequential and concurrent approaches (Belytschko et al. 2000; Xiao and Belytschko 2004). The sequential approach attempts to build several different-scale computational models and then define the passing-parameters across the scales. The concurrent approach attempts to build a hybrid model by linking several-scale computational models using multigrid and domain decomposition. A handshake region is defined to combine and communicate between the different-scale models. However, many critical issues such as efficient computational model for molecular dynamics (MD) concerning atomistic level of tissue and cellular behavior, automatic synthesis of reliable multiscale simulation, propagation of physical, statistical and model uncertainty across scales, and detailed measurement of cortical and trabecular structures needed for high-fidelity simulation models are still challenging. Near-term progress in improving the accuracy and distribution of stresses within the tissues will likely involve multiscale coupling of models at the tissue and micro-structural levels. The invention of micro-computed tomography (μCT) will enable more detailed material geometry (Crandall 2009) essential for these simulations.

As described, the development of computational algorithms and software associated with multiscale modeling require strong or weak/loose coupling of numerical models, data sharing, and very efficient and accurate simulation schemes along with very high performance computing. If simulation studies within injury biomechanics will entail simultaneous interactions across spatial scales including cellular, tissue, organ, and multiorgan systems, a computational framework in which to link and to interpret these multiscale results must be generated (Crandall 2009). In case of pediatric injury biomechanics these requirements are compounded by validation and availability of detailed material geometry and characteristics and constitutive models associated with different components (e.g. different tissues/membranes). In this respect, multiscale computational modeling is still in research arena, especially for pediatric injury biomechanics, requiring further development of algorithms and strategies.

Conclusion

Computational models are used to study many different aspects of the fields of biomechanics and medicine. Research utilizing computational modeling can be greatly beneficial in the study of human impact response, injury mechanism, and tolerance. The information presented in this chapter makes it clear that modeling the pediatric population is both a worthwhile and necessary pursuit. However, following critical barriers must be addressed for robust validation and hence routine application of pediatric models for hypotheses-driven research, development, and analysis:

1. *Increased availability of experimental data*: This requires more systematic cadaveric testing to occur, however, because conducting enough experiments on pediatric cadavers may not be a near-term solution, exploration of alternate means to obtain experimental data may be necessary. Examples include the reconstruction of case studies of pediatric accidents and associated injury case study reports, the use of chest deformation data while performing life-saving CPR, etc. Parametric studies made possible by computational models can guide experimental studies and provide new directions for research.
2. *Quantification of the effects of anatomical development and changing material properties as child ages*: The ability to account for developmental changes in a way that allows for accurate injury assessment has not been demonstrated. The geometry and material properties incorporated in a model must be specific to the age and developmental stage of the modeled child for accurate results.

    One such area of concern is in skeletal growth, especially the growth plate and ossification centers. For adult models, these are not of concern, but for children, they may play a significant role in the biomechanical response. The application of quantitative computed tomography (QCT) for both geometry and material definitions should be considered for future modeling efforts to best represent these structures within the continuum. Understanding these issues not only throughout childhood but also within inter-subject variation is necessary to substantiate any claim of a model representing the average child within a subset and to define the parameters in a probabilistic setting.
3. *Improved biofidelity*: In addition to the anatomical and material property issues addressed above, incorporating muscle activation into models for improved biofidelity has been discussed for applications in a greater variety of scenarios. This would be especially important for children where there is a high head to body mass ratio and neck strength is a known concern. Currently, few muscle strength studies have been performed on children, and even fewer on activation timing and activation level. In order to pursue this direction further, it seems that perfecting the techniques among the adult population would be advantageous so that there would be a clear path for future research in children.

    Accurate modeling of joints is also essential to perform high-fidelity simulations. Unfortunately, the joints, especially small ones such as facet joints, cannot be clearly seen in clinically available, low-resolution medical image data. CT or MRI scanning of cadavers is desired to obtain high-resolution image data.

4. *Population representation*: One of the most pertinent issues related to pediatric modeling is the idea of how to actually represent a population. Current pediatric modeling efforts are purely deterministic meaning that for each model and input condition, only one answer can be achieved rather than results relevant for an average child of a certain age.

   In the future, probabilistic approaches can be used to account for individual anatomical and physiologic development at a specific chronological age. Future advances may also allow subject-specific models to be developed quickly based on these probabilistic datasets and individual geometry to aid in treatment of pediatric patients. Although probabilistic simulation is feasible from a technical standpoint, data limitations in both the adult and pediatric populations currently limit its usefulness. In order to be able to utilize a probabilistic approach in the pediatric population, issues related to uncertainty in pediatric material properties must be addressed.

5. *Interpretation of results*: Arguably the largest hurdle for pediatric modeling at this time is how to interpret the results. In all cases, true *quantitative validation* is difficult, given the little post-mortem human subject biomechanical data obtainable for this purpose. Even if a model is "fully" validated, some analysis is necessary to decipher what the model predictions mean in the real world.

   Generally, injury assessment reference values and other measures of human tolerance are hypothesized in laboratory, and model predictions of tissue or kinematic response are compared. These tolerances have been studied in adults for decades to provide a link between measureable biomechanical parameters and injury. However, to analyze the results of a pediatric model, these tolerance values may not exist. Tissue-level injury thresholds for the pediatric population are sorely lacking and do hamper the usefulness of current model predictions.

   Some of the models are validated at component levels against data from material property experiments (which do not account for geometric influence) or against subjective data from real-world cases of accidental injury such as automobile crashes. Although this may be an acceptable first step, increased focus on component-level validation utilizing the limited cadaveric data that is available should be explored as well as other means of procuring validation data.

In summary, pediatric modeling is disadvantaged due to a widespread lack of impact biomechanics data throughout the population and an inability to define where the limited data available stand statistically. This is true on both the whole body and at tissue level and for kinematic or injury response. Difficulties specific to pediatric modeling are related to growth and development and how they can be defined throughout an age continuum. Current models do not address this sufficiently. That is not to say that the models presented here have no value. On the contrary, even an unvalidated model, if interpreted correctly, can offer insight into better understanding of impact response in pediatric populations. However, in terms of absolute, concrete predictions based on model results, there is a long road ahead. In order for pediatric modeling to realize its full potential, extensive cadaveric or tissue testing must be performed on a large number of samples. Because this is unlikely to occur, new techniques, such as the ones described in this text must be developed to bridge the gaps and provide better validation for current models.

References

Anderson AE, Peters CL, Tuttle BD, Weiss JA (2005) Subject-specific finite element model of the pelvis: development, validation and sensitivity studies. J Biomech Eng 127:364–373

Beason DP, Dakin GJ, Lopez RR, Alonso JE, Bandak FA, Eberhardt AW (2003) Bone mineral density correlates with fracture load in experimental side impacts of the pelvis. J Biomech 36(2):219–227

Bechtel R (2001) Physical characteristics of the axial interosseous ligament of the human sacroiliac joint. Spine J 1(4):255–259

Belytschko T, Guo Y, Liu WK, Xiao SP (2000) A unified stability analysis of meshless particle methods. Int J Num Meth Eng 48(9):1359–1400. doi:10.1002/1097-0207(20000730) 48:9 1359::AID-NME829>3.0.CO;2-U

Cavanaugh J, Walilko T, Malhotra A, Zhu Y, King A (1990) Biomechanical response and injury tolerance of the pelvis in twelve sled side impacts. Stapp Car Crash Conference, Orlando, FL, pp 1–12

Chan J, Putnam M, Feustel P, Koltai P (2004) The age dependent relationship between facial fractures and skull fractures. Int J Pediatr Otorhinolaryngol 68:877–881

Cloots RJH, Gervaise HMT, van Dommelen JAW, Geers MGD (2008) Biomechanics of traumatic brain injury: influences of the morphologic heterogeneities of the cerebral cortex. Ann Biomed Eng 36(7):1203–1215. doi:10.1007/s10439-008-9510-3

Coats B, Margulies SS, Ji S (2007) Parametric study of head impact in the infant. Stapp Car Crash J 51:1–15

Couper Z, Albermani F (2008) Infant brain subjected to oscillatory loading: material differentiation, properties, and interface conditions. Biomech Model Mechanobiol 7(2):105–125

Crandall J (2009) Simulating the road forward: the role of computational modeling in realizing future opportunities in traffic safety. In: Proceedings of 2009 international IRCOBI conference on the biomechanics of injury, York, UK, Sept 9–11, pp 3–30

Dakin GJ, Arbelaez RA, Molz FJ, Alonso JE, Mann KA, Eberhardt AW (2001) Elastic and viscoelastic properties of the human pubic symphysis joint: effects of lateral impact loading. J Biomech Eng 123(3):218–226

Dalstra M, Huiskes R, van Erning L (1995) Development and validation of a three-dimensional finite element model of the pelvic bone. J Biomech Eng 117(3):272–278

Dawson JM, Khmelniker BV, McAndrew MP (1999) Analysis of the structural behavior of the pelvis during lateral side impact using the finite element method. Accid Anal Prev 31:109–119

Dupuis R, Meyer F, DecK C, Willinger R (2006) Three-year-old child neck finite element modelization. Eur J Orthop Surg Traumatol 16(3):193–202

Forbes PA, Van Rooij L, Rodarius C, Crandall J (2008) Child human model development: a hybrid validation approach. ICRASH, 22–25 July, Kyoto, Japan

Friedman (1991) Multivariate adaptive regression splines. Ann Stat 19(1):1–141

García JM, Doblaré M, Seral B, Seral F, Palanca D, Gracia L (2000) Three-dimensional finite element analysis of several internal and external pelvis fixations. J Biomech Eng 122(5):516–522

Gutmann H-M (1999) A radial basis function method for global optimization. J Glob Opt 19(3):201–227

Haftka RT, Gürdal Z (1992) Elements of structural optimization. Kluwer Academic Publisher, Dordrecht/Boston/London

Hewitt J, Guilak F, Glisson R, Vail TP (2001) Regional material properties of the human hip joint capsule ligaments. J Orthop Res 19:359–364

Iman RL, Helton JC, Campbell JE (1981) An approach to sensitivity analysis of computer models, part 1. Introduction, input variable selection and preliminary variable assessment. J Qual Technol 13(3):174–183

Imielinska C, Przekwas A, Tan XG (2006) Multi-scale modeling of trauma injury. Lecture Notes in Computer Science, Volume 3994/2006, pp 822–830. doi:10.1007/11758549_110

Irwin A, Mertz HJ (1997) Biomechanical basis for the CRABI and hybrid III child dummies. In: Proceedings of the 26th Stapp car crash conference, pp 261–272

Ito Y, Shum PC, Shih AM, Soni BK, Nakahashi K (2006) Robust generation of high-quality unstructured meshes on realistic biomedical geometry. Int J Num Meth Eng 65(6):943–973. doi:10.1002/nme.1482

Ito O, Okamoto M, Takahashi Y, Mori F (2008) Validation of an FE lower limb model for a child pedestrian by means of accident reconstruction. Pedestrian safety, vehicle aggressivity and compatibility in automotive crashes, SAE SP 2165, pp 51–65

Ito Y, Shih AM, Soni BK (2009a) Octree-based reasonable-quality hexahedral mesh generation using a new set of refinement templates. Int J Num Meth Eng 77(13):1809–1833. doi:10.1002/nme.2470

Ito Y, Shih AM, Soni BK (2009b) Efficient hexahedral mesh generation for complex geometries using an improved set of refinement templates. In: Proceedings of the 18th international meshing roundtable, Salt Lake City, UT, pp 103–115. doi:10.1007/978-3-642-04319-2_7

Iwamoto M, Kisanuki Y, Watanabe I, Furusu K, Miki K, Hasegawa J (2002) Development of a finite element model of the total human body model for safety (THUMS) application to injury reconstruction. In: 2002 International IRCOBI conference, pp 31–42

Kent R, Stacey S, Kindig M, Forman J, Woods W, Rouhana SW, Higuchi K, Tanji H, Lawrence SS, Arbogast KB (2006) Biomechanical response of the pediatric abdomen, part 1: development of an experimental model and quantification of structural response to dynamic belt loading. Stapp Car Crash J 50:1–26

Kim JE, Li Z, Ito Y, Huber CD, Shih AM, Eberhardt AW, Yang KH, King AI, Soni BK (2009) Finite element model development of a child pelvis with optimization-based material identification. J Biomech 42(13):2191–2195

Klinich KD, Hulbert GM, Schneider LW (2002) Estimating infant head injury criteria and impact response using crash reconstruction and finite element modeling. Stapp Car Crash J 46:165–194

Koehler JR, Owen AB (1996) Computer experiments. In: Ghost S, Rao CR (eds) Handbook of statistics, vol 13. Elsevier Science, Amsterdam, pp 261–308

Kumaresan S, Yoganandan N, Pintar FA (1997) Age-specific pediatric cervical spine biomechanical responses: three-dimensional nonlinear finite element models. SAE Paper 973319, pp 31–61

Kumaresan S, Yoganandan N, Pintar FA, Maiman DJ, Kuppa S (2000) Biomechanical study of pediatric human cervical spine: a finite element approach. J Biomech Eng 122:60–71

Kurtz SM, Thibault KL, Giddings VL, Runge CF, Thibault LE et al (1998) Finite element analysis of the deformation of the human infant head under impact conditions. In: CDC 8th injury prevention through biomechanics symposium, Detroit, MI

Langhaar HL (1980) Dimensional analysis and theory models. R. E. Krieger Pub. Co.

Lapeer RJ, Prager RW (2001) Fetal head moulding: finite element analysis of a fetal skull subjected to uterine pressures during the first stage of labor. J Biomech 34:1125–1135

Li Z, Alonso JE, Kim JE, Davidson JS, Etheridge BS, Eberhardt AW (2006) Three-dimensional finite element models of the human pubic symphysis with viscohyperelastic soft tissues. Ann Biomed Eng 34(9):1452–1462

Li Z, Kim JE, Davidson JS, Etheridge BS, Alonso JE, Eberhardt AW (2007) Biomechanical response of the pubic symphysis in lateral pelvic impacts: a finite element study. J Biomech 40(12):2758–2766

Liu XJ, Yang JK (2002) Development of child pedestrian mathematical models and evaluation with accident reconstruction. Traffic Injury Prevention 3(4):321–329

Majumder S, Roychowdhury A, Pal S (2004) Dynamic response of the pelvis under side impact load – a three-dimensional finite element approach. Int J Crashworthiness 9(1):89–103

Majumder S, Roychowdhury A, Pal S (2007) Simulation of hip fracture in sideways fall using a 3D finite element model of pelvis-femur-soft tissue complex with simplified representation of whole body. Med Eng Phys 29:1167–1178

Margulies SS, Thibault KL (2000) Infant skull and suture properties: measurements and implications for mechanisms of pediatric brain injury. J Biomech Eng 122(4):364–371

McPherson GK, Kriewall TJ (1980) The elastic modulus of fetal cranial bone: a first step towards an understanding of the biomechanics of fetal head molding. J Biomech 13(1):9–16

Meghdari A, Davoodi R, Mesbah F (1992) Engineering analysis of shoulder dystocia in the human birth process by the finite element method. J Eng Med 206(4):243–250

Melvin JW (1995) Injury assessment reference values for the CRABI6-month infant dummy in a rear-facing infant restraint with airbag deployment. SAE Paper 95082

Meyer F, Bourdet N, Roth S, Willinger R (2007) Three years old child neck FE modelling under automotive accident conditions. In: IRCOBI conference, Maasricht, The Netherlands

Mizuno K, Iwata K, Deguchi T, Ikami T, Kubota M (2005) Development of three-year old child human FE model. Traffic Inj Prev 6(4):361–371

Myers RH (1971) Response surface methodology. Allyn and Bacon, Inc., Boston

Nahum AM, Smith R, Ward CC (1977) Intracranial pressure dynamics during head impact. In: Proceedins of the of 21st Stapp car crash conference, pp 339–366

Okamoto M, Takahashi Y, Mori F, Hitosugi M, Madeley J, Ivarsson J, Crandall JR (2003) Development of finite element model for child pedestrian protection. In: 18th ESV conference, paper number 151

Ouyang J, Zhu Q, Zhao W, Xu Y, Chen W, Zhong S (2003) Experimental cadaveric study of lateral impact of the pelvis in children. Acad J First Med Coll PLA 23(5):397–408

Ouyang J, Zhu Q, Zhao W, Xu Y, Chen W, Zhong S (2005) Biomechanical assessment of the pediatric cervical spine under bending and tensile loading. Spine 30(24):E716–E723

Ouyang J, Zhao W, Xu Y, Chen W, Zhong S (2006) Thoracic impact testing of pediatric cadaveric subjects. J Trauma 61(6):1492–1500

Phillips ATM, Pankaj P, Howie CR, Usmani AS, Simpson AHRW (2007) Finite element modeling of the pelvis: inclusion of muscular and ligamentous boundary conditions. Med Eng Phys 29:739–748

Powell MJD (1987) Radial basis functions for multivariable interpolation: a review. In: Mason JC, Cox MG (eds) Algorithms for approximation. Oxford University Press, London, pp 143–167

Rao SS (1984) Multiobjective optimization in structural design with uncertain parameters and stochastic processes. AIAA J 22(11):1670–1678

Roth S, Raul JS, Ruan J, Willinger R (2007a) Limitation of scaling methods in child head finite element modeling. Int J Veh Saf 2(4):404–421

Roth S, Raul JS, Ludes B, Willinger R (2007b) Finite element analysis of impact and shaking inflicted to a child. Int J Legal Med 121:223–228

Roth S, Raul JS, Willinger R (2008) Biofidelic child head FE model to simulate real world trauma. Comput Methods Programs Biomed 90:262–274

Roux WJ, Stander N, Haftka RT (1998) Response surface approximations for structural optimization. Int J Num Meth Eng 42:517–534

Runge CF, Youssef A, Thibault KL, Kurtz SM, Magram G, Thibault LE (1998) Material properties of human infant skull and suture: experiment and numerical analysis. In: CDC 8th injury prevention through biomechanics symposium, Detroit, MI

Sacks J, Welch WJ, Mitchell TJ, Wynn HP (1989) Design and analysis of computer experiments. Stat Sci 4:409–435

Silber JS, Flynn JM (2002) Changing patterns of pediatric pelvis fractures with skeletal maturation: implications for classification and management. J Pediatr Orthop 22(1):22–26

Song E, Xavier T, Hervé G (2006) Side impact: influence of impact conditions and bone mechanical properties on pelvic response using a fracturable pelvis Model. Stapp Car Crash J 50:75–95

Tawhai M, Bischoff J, Einstein D, Erdemir A, Guess T, Reinbolt J (2009) Multiscale modeling in computational biomechanics. Eng Med Biol Mag IEEE 28(3):41–49. doi:10.1109/MEMB.2009.932489

Van Rooij L, Meissner M, Bhalla K, Crandall JR, Longhitano D, Takahashi Y, Dokko Y, Kikuchi Y (2004) A comparative evaluation of pedestrian kinematics and injury prediction for adults and children upon impact with a passenger car. SAE Paper 2004-01-1606

Van Rooij L, Harkema C, de Lange R, de Jager K, Bosch-Rekveldt M, Mooi H (2005) Child poses in child restraint systems related to injury potential: investigations by virtual testing. In: 19th International technical conference on the enhanced safety of vehicles (ESV 2005)

Viano DC, Lau IV, Asbury C, King AI, Begeman P (1989) Biomechanics of the human chest, abdomen and pelvis in lateral impact. Accid Anal Prev 21(6):553–574

Vitale MG, Kessler MW, Choe JC, Hwang MW, Tolo VT, Skaggs DL (2005) Pelvic fractures in children: an exploration of practice patterns and patient outcomes. J Pediatr Orthop 25(5):581–587

Weber W (1984) Experimental studies of skull fractures in infants. Z Rechtsmed 92(2):87–94

Weber W (1985) Biomechanical fragility of the infant skull. Z Rechtsmed 94(2):93–101

Xiao SP, Belytschko T (2004) A bridging domain method for coupling continua with molecular dynamics. Comput Meth Appl Mech Eng 193(17–20):1645–1669. doi:10.1016/j.cma.2003.12.053

Yamada H (1970) Strength of biological materials. Williams & Wilkins, Baltimore, MD, pp 99–104

Yoganandan N, Pintar FA (2005) Deflection, acceleration, and force corridors for small females in side impacts. Traffic Inj Prev 6:1–8

Zhang QJ, Gupta KC (2000) Neural networks for RF and microwave design. Artech House Publishers, Norwood, MA

Zhang L, Yang KH, Dwarampudi R, Omori K, Li T, Chang K, Hardy WN, Khalil TB, King AI (2001a) Recent advances in brain injury research: a new human head model development and validation. Stapp Car Crash J 45:369–394

Zhang L, Yang KH, King AI (2001b) Comparison of brain responses between frontal and lateral impacts by finite element modeling. J Neurotrauma 18(1):21–30

Zhang HZ, Hou M, Bai SP, Ma CS, Liu CM, Bu RF (2007) The finite element study on zygomatic injury by impact in child. Zhonghua Yi Xue Za Zhi 87(20):1420–1422

Zhou C, Kahlil TB, Dragovic LJ (1996) Head injury assessment of a real world crash by finite element modelling. In: Proceedings of the AGARD conference, New Mexico, pp 81–87

Zhu JY, Cavanaugh JM, King AI (1993) Pelvic biomechanical response and padding benefits in side impact based on a cadaveric test series. In: Proceedings of 37th Stapp car crash conference, SAE933128, Washington, DC, pp 223–233

Index

A
AAP. *See* American Academy of Pediatrics (AAP)
Abbreviated Injury Score (AIS), 43, 209, 257, 258, 260–262, 265, 272, 276, 279
Abdomen, 6, 15, 28, 49–53, 56, 57, 62, 64–66, 68, 221–281, 296, 310, 323
Abdominal, 50, 52–54, 131, 222, 223, 239, 249–253, 256–258, 260, 279, 280, 292, 296
Abdominal aorta, 53, 238
Abdominal artery, 239
Abrasion, 58, 262, 265, 267
Acceleration, 45, 172, 182, 205, 208, 209, 242, 247, 257–260, 267, 272, 273, 276, 278–280, 287, 288, 301, 309
Acetabulum, 131
Adrenal glands, 222
Airbag, 9, 40, 57, 192, 212, 258, 271, 272, 274–280, 288
AIS. *See* Abbreviated Injury Score (AIS)
Algorithm, 166, 288, 290, 302, 312, 313, 315, 328
American Academy of Pediatrics (AAP), 42
Anatomy, 28, 43, 158–160, 174–177, 183–184, 192–197, 211, 215, 223, 288, 289, 308, 310, 313
Animal, 88, 99–100, 105–108, 114–116, 118–126, 151, 157, 163, 166, 169, 172–174, 184, 185, 212–215, 223, 224, 228, 232, 235, 258, 270–274, 276–279, 287, 310
Anisotropy, 166–167, 177, 180, 185
Ankle, 5, 127, 135, 198
Annulus, 195, 306, 307

Anterior tubercle, 194
Anthropometric data, 1, 2, 4, 7, 27
Anthropometry, 1–28, 137, 138, 140, 141, 147, 199, 202, 288, 292, 293
Anthropomorphic test devices (ATDs), 6, 16, 17, 26–28, 45, 48, 87, 88, 151, 208, 210, 213, 216, 221, 223, 289, 290
Apparent density, 105, 106
Arachnoid, 159, 168–170, 305
Arm, 5, 8, 18, 54, 135, 151, 293
Artery, 171, 238, 239, 272
Articular cartilage, 88, 127–131, 151, 306, 307, 311
Ash density, 105
ATDs. *See* Anthropomorphic test devices (ATDs)
Atlanto-axial, 192, 203
Atlanto-occipital, 57, 192
Atlas, 192, 194, 195, 211
Axis, 18, 89, 90, 92–95, 100, 105, 116, 120, 122, 128, 177, 180, 181, 192, 194, 195, 203, 211, 243
Axons, 159, 165–167, 173

B
Basilar skull, 46
Belt fit, 24, 26, 28, 48
Belt geometry, 53
Belt-positioning boosters, 26, 40, 41, 48–50, 56, 64
Belt tension, 251, 253, 255, 260, 267
Belt webbing, 47
Bending, 21–25, 95–99, 111, 114, 136, 137, 139, 141, 144, 147, 148, 150, 151, 182, 192, 198, 199, 213, 225, 238, 240, 241, 249, 266, 280, 289, 295, 301

Biofidelic, 87, 221, 266, 280, 288, 292
Biofidelity, 87, 221, 242, 247, 270, 289, 290, 292, 309, 329
Body mass index (BMI), 7, 8, 18
Body segment, 1–3, 9, 11, 19–21, 27, 287
Booster seats, 24, 26, 33, 40–43, 48–50, 53, 54, 56, 57, 61, 62, 64, 66, 260, 268, 271
Boundary conditions, 233, 266, 267, 290, 304
BPB. *See* High back belt-positioning booster seats (BPB)
B-pillar, 65
Brain, 11, 45, 57, 61, 121, 133, 141, 157–185, 202, 236, 242, 292, 295–305, 316, 318–321
Braincase, 295, 300, 303
Brain stem, 57, 164, 166, 167, 173, 174, 192, 319
Brain volume, 158, 173
Brick element, 301
Bridging vein, 161, 170, 171, 185, 301, 304
Bulk moduli, 231

C
C1, 192, 194, 195, 210, 261, 309
C2, 192, 194, 195, 197, 201, 203–207, 210, 214, 215
C3, 23, 193, 194, 196–198, 204, 206, 209, 210
C4, 194, 196–198, 200, 201, 203–206, 208, 212, 308
C5, 194, 196–198, 200, 201, 203–206, 209, 210, 212
C6, 194, 196–198, 200, 201, 203–206, 210, 212, 267, 308
C7, 23, 193, 194, 196–198, 200, 201, 203–206, 210, 212, 267, 309
Cadaver, 20, 170, 171, 179, 182, 183, 198, 223, 226, 229, 232, 236, 238, 241, 243, 244, 246–251, 253, 256–258, 260–271, 279, 289, 291, 302, 310, 313, 322, 329
CAESAR study, 28
Cartilage, 66, 88, 108–116, 127–131, 151, 175, 222, 226–228, 267, 296, 306–308, 310, 311, 324
CCIS. *See* Cooperative Crash Injury Study (CCIS)
Center of gravity (CG), 1, 3–5, 17, 18, 20, 21, 198, 264, 287, 301
Center of mass, 18–20
Center of volume, 20
Center rear, 48, 52, 62
Central nervous system, 171
Cerebral, 113, 114, 133, 141, 167, 171, 173, 174, 202, 242
Cerebral spinal fluid (CSF), 157–174, 185, 297–299, 301, 304, 320, 321
Cerebral vasculature, 161, 170–171
Cerebrum, 158, 159, 162, 165, 167, 174, 301
Cervical spine, 21–23, 46, 54, 57, 65, 70, 72, 191–201, 203, 205, 207–216, 224, 267, 292, 305–310
CG. *See* Center of gravity (CG)
Chance fracture, 54
Chestbands, 243, 244
Child Led Injury Design (CHILD) projects, 37, 39, 77
Children's Hospital of Philadelphia (CHOP), 33, 34, 41, 42, 48, 61, 76, 247
Child restraint, 1, 6–8, 40–47, 50, 51, 53, 57, 61, 62, 64, 65, 71, 72, 75, 261, 263
Child restraint system (CRS), 27, 28, 33, 39–42, 44, 45, 47, 48, 62, 65, 67, 68, 208, 209, 274, 292, 313
Child Restraint System in Cars (CREST), 37, 39, 43, 50, 65, 66, 77
Child seat, 45, 260, 263, 264, 267, 272–274, 276
CHOP. *See* Children's Hospital of Philadelphia (CHOP)
CIREN. *See* Crash Injury Research and Engineering Network (CIREN)
Clavicle, 5, 16, 17, 310
Collagen, 105, 116, 119, 120, 127, 128, 195
Compression, 52–54, 95, 98, 100, 102, 104–106, 109, 127, 132, 133, 135, 162, 164, 182, 184, 209, 210, 213, 226, 232, 234, 235, 237, 241, 247, 248, 251, 289, 290, 304, 309, 322, 324
Compression fractures, 52, 54
Computational model, 1, 87, 88, 151, 157, 165, 167, 168, 171, 185, 208, 221, 223, 266, 287–330
Computed tomography (CT), 13, 16, 17, 28, 289, 298, 301–303, 306, 309, 313, 318, 323, 329
Constitutive properties, 215, 216, 312
Consumer Product Safety Commission, 2
Contact injuries, 45
Convex, 195
Cooperative Crash Injury Study (CCIS), 37, 39, 77
Coronal suture, 175, 176, 181, 183
Corona radiate, 162, 164, 167, 168
Coronary artery, 238
Corpus callosum, 159, 162, 164, 167, 168, 303

Index

Cortex, 135, 160–164, 168
Cortical bone, 89–100, 103, 108, 151, 176, 224–226, 240, 279, 295, 296, 299, 323, 324
Costal cartilage, 222, 226–228, 267, 296
CRABI, 27, 212, 301
Cranial bone, 175–181, 185
Craniofacial, 298, 303
Crash, 1, 33, 174, 191, 221, 288
Crash direction, 49, 54, 58, 63–68, 72
Crash Injury Research and Engineering Network (CIREN), 37, 38, 46, 61, 65, 66, 75
Crash test dummies, 1, 280
CREST. *See* Child Restraint System in Cars (CREST)
Cross section, 98, 100, 110, 117, 122, 147, 303
CRS. *See* Child restraint system (CRS)
CSF. *See* Cerebral spinal fluid (CSF)
CT. *See* Computed tomography (CT)
Cyclic, 119, 122, 124, 126, 168, 174, 234
Cylinder, 19, 105, 182, 287

D

DAI. *See* Diffuse axonal injury (DAI)
Datasets, 6, 7, 13, 16, 38, 44, 48, 58, 61, 63, 64, 66, 67, 73–75, 240, 249, 258, 312–314, 330
Deceleration, 53, 191, 257, 273, 274, 276
Deformable body, 288
Deformation, 2, 67, 98–100, 122, 135, 136, 160, 162, 165–168, 171, 175, 182, 222, 232–234, 238, 242, 243, 250, 260, 288, 295, 300, 304, 322, 329
Density, 3, 18–20, 74, 90, 91, 95–98, 103, 119, 182, 296, 301, 320
Deploy, 40, 57, 58
Deployment, 33, 57, 58, 65, 75, 192, 258, 274, 276
Developmental, 43, 88, 103, 108, 116, 157, 158, 172, 174, 176, 183, 193–197, 210–212, 214–216, 224, 240, 291, 308, 329
Diaphragm, 222, 276
Diffuse axonal injury (DAI), 57, 172, 174
Digital Child, 313
Disability, 33, 34, 71, 75, 157
Disc, 195, 196, 306, 307, 309
Discretization, 288, 290
Discretized, 290, 312, 313
Dislocation, 113, 114, 192, 203

Displacement, 18, 46, 109, 119, 120, 122, 160, 173, 182, 196, 199, 201, 203, 204, 207, 210, 234–238, 241, 243, 247, 248, 266, 267, 269, 270, 279, 280, 309
Drop height, 137, 140
Drop test, 137, 140
Drop tower, 322
Dura, 159, 161, 168–170, 297–299, 301, 303, 305
Dynamic, 63, 99, 137, 151, 163, 169, 170, 179, 182, 184, 216, 224, 225, 227, 234, 238, 241, 249, 251, 253, 274–280, 287, 288, 304, 310, 313, 323, 327
Dynamic impact testing, 98

E

Ejection, 50, 66–67
Elastic, 89, 90, 100, 106, 107, 115, 118, 119, 164, 165, 168–170, 177, 179–181, 221, 226, 227, 229, 231, 233, 234, 236, 238, 240, 247–249, 288, 292, 295–297, 299, 300, 309, 320, 324
Elbow, 5, 6, 135
Ellipsoid, 27, 287, 292, 294
Elongation, 117–121, 123, 125, 171, 226, 232, 237–239, 253, 262
Energy absorption, 98, 103, 105–108, 122, 226
Energy management, 45
Ephemoid bone, 175
Epidemiology, 33–77, 193, 199
Epidural hematoma, 45
Epiphyseal, 110–112, 198, 310
Epiphyseal plate, 88, 108–113
Epiphysis, 108–110, 114, 115, 122, 124, 135
Esophagus, 222
Extremities, 4, 18, 20, 43, 45, 46, 50, 51, 54, 56–58, 62, 63, 65, 67–70, 87–151, 225, 292, 293, 295, 296, 310–312

F

Facet, 196, 292, 293, 329
Facet angle, 192, 196, 308, 309
Falx, 159, 168, 298, 301, 303
Far side, 63, 64, 66, 68
Fatality, 33–35, 37–39, 44–46, 51, 61, 63, 65, 67, 69, 71, 73, 77
Fatality risk, 50, 63, 64, 66, 71
Federal Motor Vehicle Safety Standard (FMVSS), 40, 58, 213, 214

Femur, 70, 88, 98, 101, 102, 110, 115, 135, 137–140, 143, 144, 147, 148, 183, 296, 322
FFCRS. *See* Forward facing CRS (FFCRS)
Fiber-oriented, 177
Fibula, 88, 135, 143, 144, 147, 148
Finite element model (FEM), 28, 166, 168, 215, 216, 287, 288, 292, 295–298, 300, 301, 303–306, 309, 310, 313, 319, 321–326
Flexibility, 75, 223, 249, 292, 309, 312
Flexion, 21–25, 196, 198, 199, 210, 213, 266, 270, 292, 309
Fontanelles, 175
Foot, 5, 18–20, 135, 270, 293
Force, 40, 88, 164, 197, 232, 288
Forearm, 5, 18, 135, 151
Forward facing CRS (FFCRS), 41, 43–48, 50, 54–56, 64, 65
Fracture, 43, 90, 182, 192, 242, 288
Frontal bone, 176, 179, 180, 182
Frontal impact, 43, 55, 58, 63, 66, 208, 258, 268, 270–272, 276, 301, 303
Frontal lobe, 45, 173

G
Galea, 183
Geometry, 3, 11, 16, 17, 28, 53, 195, 288, 290, 292, 293, 295, 297, 298, 301, 303, 305, 306, 309, 311–317, 328–330
German In Depth Accident Study (GIDAS), 37, 39, 77
GIDAS. *See* German In Depth Accident Study (GIDAS)
Gray matter, 158, 160, 162, 164, 167, 168, 173, 303, 320
Greater trochanter, 132
Greenstick fracture, 135
Growth charts, 7, 240
Growth patterns, 1–4, 173
Growth plate, 88, 108–116, 135, 136, 151, 307, 310, 329

H
Hand, 4–6, 18, 135, 247
Hard tissue, 289, 313
Head, 3, 42, 111, 157–185, 192, 249, 292
Head contours, 13, 27
Head excursion, 45, 50, 258, 266, 274, 275
Head Injury Criterion (HIC), 182, 209, 258, 274, 278
Heart, 35, 133, 141, 202, 222, 242, 247, 258, 274, 276, 278–280
Height, 3–7, 15, 19, 24, 26, 28, 42, 66, 95, 99, 105, 132, 137, 140, 182, 183, 196, 202, 205, 234, 235, 237, 257, 270, 272–274, 295
HIC. *See* Head Injury Criterion (HIC)
High back belt-positioning booster seats (BPB), 41, 50, 55
Humerus, 88, 109, 135, 148–150
Hybrid III, 16, 17, 26–28, 212, 213, 244, 266–268, 270, 271
Hyperestic, 160, 164, 235, 288, 299, 303, 324
Hyperextension, 57
Hypermesh, 323

I
Iliac crest, 15, 26, 131
Iliac spine, 15
Iliac wing, 16, 54, 132, 322
Ilium, 131
Imaging, 3, 28, 289, 313
Inertia, 9, 18–21, 135, 172, 287, 309
Inertial injuries, 45
Inertial properties, 1, 3, 27, 216
Injury causation, 37–39, 63, 64, 67, 72, 76
Injury criteria, 213, 245, 247, 281, 313
Injury risk, 43, 44, 49–51, 56–68, 213, 240, 256, 260, 266
Installation, 47
International Road Traffic And Accident Database (IRTAD), 37, 39, 77
Intervertebral disc, 195, 210, 215, 306, 309
Intestines, 222, 251
Intra-abdominal injuries, 52, 53
Intrusion, 45, 46, 65–67, 131, 300
IRTAD. *See* International Road Traffic And Accident Database (IRTAD)
Ischial rami, 131
Ischium, 131

J
Joint, 20, 66, 109, 116, 118, 121, 122, 127, 130–132, 192, 196, 198, 287, 292, 294, 307, 310, 314, 323, 324, 326–329

K
Kidneys, 222, 232–237, 251, 258
Kinematics, 34, 48, 50, 53, 69, 208, 209, 223, 225, 266–268, 271, 273–275, 289, 290, 292, 295, 301

Index 339

Kinetic, 290
Knee, 5, 6, 46, 51, 70, 118, 127, 128, 135, 137, 270

L

Lagrangian method, 287
Lambdoid suture, 175, 176
Lamellar structure, 176
Lap belt, 15, 26, 48, 52, 53, 251, 253, 260, 261, 263, 267, 274
Lap belt angle, 53
Lap/shoulder belt, 52, 53
LATCH. *See* Lower anchor attachment (LATCH)
Lateral impact, 132, 135, 137, 151, 271, 272, 322, 323, 325, 327
Lateral mass, 193, 194
Leg, 19, 21, 70, 132, 135, 151, 198, 293
Ligament, 88, 114, 121–127, 192, 210, 215, 241, 257, 265, 279, 294–296, 305–309, 311, 323, 324, 326, 327
Link, 39, 126, 127, 328, 330
Liver, 222, 232–237, 251, 261, 265, 272, 276, 279
Lordosis, 23, 196
Lower anchor attachment (LATCH), 47
LS-DYNA, 288, 293, 297, 298, 323
Lumbar spine, 24, 49, 52, 54, 210, 241, 267, 310
Lumbosacral spinal, 53
Lung, 65, 222, 228–231, 260, 272, 276
Lymph nodes, 222

M

MADYMO. *See* MAthematical DYnamic Model
Magnetic resonance imaging (MRI), 13, 15, 28, 158, 289, 293, 298, 302, 303, 313, 329
Manubrium, 222
Mass, 1–3, 9, 26–28, 132, 137, 173, 192–194, 198, 199, 202, 209, 237, 242, 244, 251, 258, 261, 262, 273, 274, 276, 287, 294, 300, 306, 309, 319, 322, 329
Mass distribution, 1, 7, 9, 17–21, 27, 28
Material behavior, 290, 295, 305, 312
Material properties, 87–132, 136, 157, 160–172, 177–181, 183, 185, 215–216, 225, 229, 232, 240, 288, 289, 291, 295, 296, 300, 301, 303, 305, 308–313, 322–327, 329, 330
MAthematical DYnamic Model (MADYMO), 287, 292, 293

Mechanical properties, 88, 106, 107, 117, 119, 122, 171, 184, 228, 230, 231, 239, 289
Mechanism of injury, 53, 192, 313
Medical imaging, 3, 28, 305, 312–315, 329
MEGG3D. *See* Mixed-element grid generator in three dimensions (MEGG3D)
Meninges, 157–174
Mesenchyme, 175
Mesenchymel cells, 175
Mesenteric tears, 52, 54
Mesh(ing), 16, 127, 292, 301, 303, 312–320, 323
Metamodeling, 325
Metaphysis, 108, 109, 111, 112, 114, 115, 135
Mineral, 98, 99, 105, 106
Mineralization, 98
Misuse, 44–48, 52, 54
Mixed-element grid generator in three dimensions (MEGG3D), 314–316
Moments of inertia, 9, 18, 19
Morphing, 309
Morphology, 193, 280, 304
Motor vehicle, 33–77, 174, 191, 192, 221
MRI. *See* Magnetic resonance imaging (MRI)
Multibody, 216, 287, 288, 292, 293, 295
Multiscale, 313, 327–328
Muscle, 20, 114, 116, 118, 177, 183, 211, 222, 251, 267, 294, 296, 327–329
Musculature, 198, 265
Myelin, 159, 166
Myelination, 159, 165, 166, 304

N

National Automotive Sampling System-Crashworthiness Data System (NASS-CDS), 37, 38, 43, 46, 49, 58, 63, 65–67, 73, 74
National Automotive Sampling System-General Estimates System (NASS-GES), 37, 38, 59, 62, 67, 73, 74
National Center for Health Statistics, 7
National Highway Traffic Safety Administration (NHTSA), 40, 42, 49–51, 57, 58, 69, 72–76, 192, 273
National Survey of the Use of Booster Seats (NSUBS), 42
Near side, 63, 64, 66–68
Neck, 5, 6, 11, 13–15, 19, 23, 42, 43, 45–47, 50, 56, 57, 62, 65, 68, 70, 109, 111, 131, 174, 191–216, 253, 266, 267, 276, 280, 292, 293, 305, 306, 309, 329
Neonate, 17, 176, 180, 182, 193, 197, 198, 223

Nerve, 121, 183, 222
Neural arch, 192, 194, 307
NHTSA. *See* The National Highway Traffic Safety Administration (NHTSA)
Nucleus pulposus (NP), 195, 296

O

Oblique impact, 46
Occipital bone, 176, 179, 182, 302
Odontoid, 46
Omentum, 260
Omni-directional, 309
OOP. *See* Out-of-position (OOP)
Optimization, 28, 45, 54, 67, 166, 221, 312, 313, 322, 324, 325
Orthogonal, 106, 177, 295, 297, 309
Ossification, 43, 66, 108, 109, 131, 174–176, 194, 308, 310, 329
Outboard, 52, 58, 59, 62, 63, 67
Out-of-position (OOP), 192, 212, 258, 274–279
Ovaries, 222

P

PAM-CRASH, 288, 293
Pancreas, 222
Parameter, 18, 50, 58, 62, 74, 103, 115, 166, 231, 235, 253, 260, 274, 289, 292, 302, 303, 309, 312, 315, 325, 328–330
Parametric studies, 168, 289, 301, 302, 329
Parenchyma, 232, 236, 279
Parietal bone, 176–183, 304
Partners for Child Passenger Safety (PCPS), 37, 39, 41, 42, 48, 56, 58–62, 65, 67, 68, 76
Peak, 26, 69, 124, 132, 158, 173, 182, 201, 209, 214, 234, 236, 238, 241, 244, 251, 257–260, 267, 270, 273, 274, 276, 278, 279, 300, 302, 324
Peak force, 124, 132, 137, 140, 251
Pedestrian, 37, 69–73, 75, 292, 293, 310
Pelvic ring, 66, 131
Pelvis, 15–16, 26, 28, 42, 53, 66, 70, 87–151, 222, 223, 267, 293, 296, 310–312, 322–326
Pendulum test, 292
Percentile, 7, 8, 13, 27, 133, 139–142, 145, 146, 213, 270, 291–293
Perforation, 52, 54, 251
Perinatal, 202, 203, 205, 210, 211
Periosteum, 114, 115, 183
Physical properties, 88, 105, 106

Pia, 168–170, 305
Plane, 69, 109, 114, 115, 127, 141, 147, 198, 205, 223, 287, 303
Plastic, 57, 98–102, 135, 136, 288, 295, 297
Plastic deformation, 98, 99, 136
PMHS. *See* Post-mortem human subjects (PMHS)
Pons, 164, 166
Porosity, 88, 98, 99
Post mortem, 170, 198, 208, 211, 221, 251, 289, 291, 330
Post-mortem human subjects (PMHS), 87, 89–91, 95, 98, 99, 104, 105, 109, 117, 127, 132–137, 140, 141, 145–147, 149, 151, 198, 201–206, 208–211, 214, 215, 221, 242, 244, 245, 247, 249, 253, 268, 270, 291
Posture, 1–3, 9, 13, 16, 17, 23–26, 28, 279
Pre-impact, 57, 253
Pressure, 2, 9, 54, 131, 168, 228–232, 236, 239, 258–260, 272, 274, 276, 278–280, 301, 304
Prevention, 33, 37, 57, 71, 75, 133
Pubic rami, 66, 131, 132
Pubis, 52, 66, 131
Pulmonary, 202, 245, 272, 279

Q

Q-series child dummy, 27
Quasi-static, 15, 89–97, 99, 101, 102, 104, 105, 109, 113, 115, 117, 119, 136, 138, 151, 169, 177, 179, 182, 184, 210, 238, 251, 252, 304, 310, 322

R

RADIOSS, 288
Radius, 88, 135, 148–150
Radius of gyration, 18, 19, 21
Range of motion, 1, 3, 21–25, 210
Rate, 18, 46, 89, 157, 191, 225, 291
Rear-facing child restraint (RFCRS), 43, 47, 55–57
Rear impact, 67
Restrained, 34, 41–43, 45–50, 52–55, 57–60, 62–67, 71, 209, 238, 253, 263, 266, 267, 280
Restraint, 1, 33, 208, 221, 288
Revolute joint, 287
RFCRS. *See* Rear-facing child restraint (RFCRS)
Rib-cage, 16, 17, 222, 240, 249, 251, 265
Ribs, 222, 223, 226, 238–241, 249, 267, 279

Index

Rigid barrier test, 40
Rigid body, 287, 288, 304
Rollover, 51, 63, 66, 67
Rotation, 21–25, 65, 67, 141, 147, 157, 168, 172–174, 199, 240, 316

S

Sacroiliac, 66, 132, 314, 323, 324, 327
Sacrum, 131
Sagittal suture, 175
SBS. *See* Seat belt syndrome (SBS)
Scaling, 87, 88, 166, 173, 185, 198, 199, 201, 208, 210, 213, 214, 216, 221, 222, 244, 247, 289, 292, 294–296, 301, 305, 308–312, 324–326
Scalp, 157, 159, 183–184, 298, 299, 301, 302
Scan, 3, 11, 13, 16, 196, 303, 318, 323
SCI. *See* Special Crash Investigations (SCI)
Seat belt sign, 52
Seat belt syndrome (SBS), 50, 52–54, 57, 64, 303, 304
Seated posture, 3, 9, 16, 26, 28
Second-generation air bags, 40, 58
SED. *See* Strain energy density (SED)
Segmental inertial properties, 3
Segment(al) masses, 2, 18, 20, 27
Segmental moment of inertia, 7
Segments, 1–3, 7, 9–11, 18–21, 27, 54, 117, 175, 181, 200, 201, 203, 205, 206, 208, 210–213, 240, 241, 287, 294, 308, 310
Shear, 108, 160, 229, 303
Shear moduli, 160, 162, 165, 168, 231
Shell element, 295, 301, 306, 323
Shield booster, 50, 264
Shoulder, 1, 5, 6, 11, 13, 14, 16–17, 28, 44, 48, 50, 52, 54, 56, 261, 262, 265, 267, 270, 310
Shoulder belt, 48, 50–54
Side impact, 51, 62–66, 71, 72, 132, 280, 322, 323
Simulation, 28, 87, 168, 185, 233, 287–290, 295, 301, 304, 310, 312, 313, 315, 318, 325–330
Sitting height, 3, 28, 270
Skinfold, 8, 19
Skull, 11, 12, 45, 46, 57, 157, 159, 174–184, 201, 202, 210, 221, 225, 295, 296, 298–305, 314, 315, 320
Skull fracture, 45, 46, 182, 183, 295, 301, 302, 305
Sled, 253, 258, 259, 264, 265, 271, 273, 274, 276, 300

Sled test, 40, 44, 45, 48, 132, 205, 209, 238, 253, 258, 260–274, 309, 322
Small occupant, 57
Soft tissue, 2, 11, 28, 88, 114, 121, 126, 147, 183, 198, 211, 222, 240, 245, 274, 280, 281, 288, 289, 322, 324, 328
Southern Consortium for Biomechanics (SCIB), 313, 315
Special Crash Investigations (SCI), 37, 38, 57, 74–76, 192
Specimen, 17, 89, 160, 198, 229, 322
Sphenoid bone, 175
Spherical joint, 327
Spinal column, 46, 54, 131
Spinal cord injury without radiographic abnormalities (SCIWORA), 46
Spleen, 222, 232–237
Spring element, 301, 306, 309, 324
Static, 124, 126, 274, 292, 304
Stature, 3–9, 13, 15, 27, 69, 70, 133, 135, 138–142, 146, 147, 199, 261, 262
Sternebrae, 222, 249
Sternum, 50, 222, 223, 243, 247, 249, 267, 276, 279, 296
Stomach, 222, 237, 251
Strain, 89, 160, 224, 288
Strain energy density (SED), 119, 224, 226, 227, 232, 233, 235, 237
Stress, 88, 160, 226, 288
Structural properties, 88, 122–125, 131–136, 151, 157, 184, 197–212, 215
Subarachnoid, 159, 171, 173, 174
Subcutaneous tissue, 4
Subdural, 170, 171, 173, 174
Subluxation, 52, 267
Surveillance, 34, 37, 39, 71, 72, 76
Sutures, 175, 176, 181–183, 295, 297–305
Symphysis pubis, 66
Synchondroses, 193–195, 203, 215, 216

T

Tangent modulus, 114–116, 121
TBI. *See* Traumatic brain injury (TBI)
Temporal bone, 174, 175
Temporal cortex, 160, 161
Tendon, 88, 116–121, 177, 213
Tensile loading, 89, 99–100, 198
Tensile strength, 90, 114–122, 124, 129, 198, 203, 210, 226, 232, 233, 236, 237, 239
Tensile stress, 93, 94, 98, 114, 116, 127–129, 238, 303

Tension, 57, 92–95, 98, 99, 113–117, 127, 136, 161, 168–171, 181, 184, 192, 199, 201, 209–211, 213, 228, 241, 251, 253, 255, 260, 267, 289, 290, 292, 295, 304, 323
Tentorium, 159, 168, 298, 301
Tether, 43, 47, 48
Thigh, 5, 6, 15, 18, 20, 26, 48, 50, 135, 137, 138, 140, 142, 147, 151
Thoracic, 65, 66, 69, 70, 210, 221, 222, 225, 231, 238–245, 247, 249, 260, 270, 276, 292
Thoracic artery, 238, 239
Thoracic duct, 222
Thoracic spine, 223–225, 261, 265–267, 279, 280, 309, 310
Thoracic vertebrae, 222, 226
Thoracoabdominal, 225, 238, 257–279
Thorax, 16, 28, 46, 49, 221–281, 296, 310
Three-point belt, 26
Three-point bending, 95–99, 136–138, 141–144, 147–150, 177, 179, 238
Thymus gland, 222, 242
Tibia, 88, 101, 102, 104, 105, 114, 135, 136, 143, 144, 147, 148, 183
Tissue density, 88, 92, 105, 106
Tissue properties, 88, 90, 100, 157, 167, 225–237, 290, 291, 313
Torsion, 109, 136
Trabeculae, 103, 175–177, 180, 181
Trabecular bone, 88, 89, 103–108, 226, 228, 323, 324
Trachea, 222
Trajectory, 158, 263, 264, 266
Transverse moment of inertia, 18
Traumatic brain injury (TBI), 157, 168, 170, 171, 174
Triadiate cartilage, 131

U
UAB. *See* University of Alabama at Birmingham (UAB)
Ulna, 89, 136, 148–150
Ultimate strain, 89, 91, 92, 94, 104, 105, 107, 113, 115, 119, 120, 122, 124, 127, 180, 181, 224, 226, 235, 237, 296
Ultimate stress, 88, 89, 91, 92, 98, 104, 105, 107, 113, 115, 117, 119, 161, 170, 179–181, 224, 226, 235, 237
Unintentional injury, 33–35, 71
Universal joint, 287
University of Alabama at Birmingham (UAB), 313–315
University of Michigan Transportation Research Institute, 2

Unrestrained, 43, 44, 49, 55, 57–59, 66
Upper extremities, 18, 51, 54, 56–58, 62, 65, 68, 70, 87, 135–151, 310
Urinary bladder, 222
Uterus, 222

V
Validation, 88, 151, 208, 221, 288, 290–292, 295, 301, 304, 305, 310, 312, 319, 321, 326, 328–330
Vasculature, 161, 170–171, 238, 239, 279
Vein, 161, 170, 171, 185, 260, 301, 304
Velocity, 133, 137, 140, 173, 174, 179, 209, 232, 233, 236, 247, 249, 253, 274, 280, 288, 301, 322
Ventricle, 158, 171, 303
Verification, 129, 288, 290
Vertebra, 23, 24, 104, 193, 194, 201, 226, 243, 268
Viscoelastic, 106, 107, 124, 126, 162–166, 199, 201, 227, 232, 234–236, 288, 296, 297, 299, 302, 304, 307, 320
Volunteer, 15, 24, 26, 205, 223, 251, 252, 293

W
Wayne State University, 313
Weight, 5–9, 13, 15, 19, 27, 28, 42, 43, 48, 89, 90, 92, 103, 105, 115, 119, 133, 135, 138–142, 145, 172, 173, 198, 209, 212, 257, 272, 273
White matter, 158, 159, 162–165, 167, 168, 173, 174, 304, 305
Whole-body, 3, 21, 28, 202, 223, 225, 251, 258–279, 290–291, 310, 330
Whole body model, 291–295, 310
Wrist, 5, 135

X
Xiphoid, 222

Y
Yield point, 100
Yield stress, 100–102, 104, 226, 227, 240, 296
Young's modulus, 87–97, 100–107, 117, 119–121, 123, 125–127, 239, 319, 320, 324

Z
Zygoma, 303

Printed by Publishers' Graphics LLC